ACE Inhibitors: Current Use and Future Prospects

Cover: En face specimen of a hamster aortic arch containing macrophage foam cells stained with oil red O.

ACE Inhibitors: Current Use and Future Prospects

Edited by
Michael Schachter MRCP
Senior Research Fellow
Department of Clinical Pharmacology and Therapeutics
St Mary's Hospital Medical School
London, UK

MARTIN DUNITZ

First published in the United Kingdom in 1995 by
Martin Dunitz Ltd
The Livery House
7–9 Pratt Street
London NW1 0AE

A CIP catalogue record for this book is available from the British Library

ISBN 1-85317-121-2

Composition by Scribe Design, Gillingham, Kent
Printed and bound in Great Britain by
Biddles Ltd, Guildford and King's Lynn

CONTENTS

List of Contributors

Gunnar Aberg PhD
Docent, Department of Pharmacology, Linköping University, Linköping, Sweden; and Vice President, Research and Development, Sepracor Inc, Marlborough MA, USA

Dr Stephen C Bain MA MD MRCP
Senior Lecturer in Medicine, The University of Birmingham, and Honorary Consultant Physician, Department of Medicine, Birmingham Heartlands Hospital, Birmingham, UK

Professor Anthony H Barnett BSc MD FRCP
Professor of Diabetic Medicine, The University of Birmingham, and Honorary Consultant Physician, Department of Medicine, Birmingham Heartlands Hospital, Birmingham, UK

Professeur A Berdeaux MD PhD
Professor, Faculté de Médicine Paris–Sud, Département de Pharmacologie, Le Kremlin–Bicêtre, France

Julie Blasioli BSc(Hons)
PhD Student, St Vincent's Institute of Medical Research, Fitzroy, Victoria, Australia

Dr Duncan J Campbell MB BS BMedSci PhD FRACP
Senior Research Fellow, St Vincent's Institute of Medical Research, Fitzroy, Victoria, Australia

Alexander W Clowes MD
Professor of Surgery, Department of Surgery, School of Medicine, University of Washington, Seattle WA, USA

Denis deBlois PhD
Assistant Professor, Department of Pharmacology, Université de Montrèal, Centre de Recherche de l'Hôtel–Dieu de Montrèal, Montrèal PQ, Canada

P Demolis MD
Maître de Conférence, Faculté de Médicine, Département de Pharmacologie, Le Kremlin–Bicêtre, France

Oliver Edling Dipl. chem.
PhD Student, Pharmakologisches Institut, Ruprecht–Karls–Universität Heidelberg, Heidelberg, Germany

Peter Gohlke PhD
Research Fellow, Institut für Pharmakologie, Christian–Albrechts–Universität zu Kiel, Kiel, Germany

Professeur J-F Giudicelli MD PhD
Professor, Faculté de Médecine Paris–Sud, Département de Pharmacologie, Le Kremlin–Bicêtre, France

W Dallas Hall MD
Department of Medicine, Division of Hypertension, Emory University School of Medicine and Grady Memorial Hospital, Atlanta GA, USA

Zafar H Israili PhD
Department of Medicine, Division of Hypertension, Emory University School of Medicine and Grady Memorial Hospital, Atlanta GA, USA

Mark C Kowala PhD
Senior Research Investigator II, Bristol–Myers Squibb Pharmaceutical Research Institute, Princeton NJ, USA

Prof. Dr.med. Eckhard P Kromer
Professor, Klinik und Poliklinik für Innere Medizin (Kardiologie, Nephrologe und Pneumologie), Universität Regensburg, Regensburg, Germany

Dr Graham W Lipkin BSc MB ChB MRCP
Senior Registrar in Nephrology, Department of Nephrology, Queen Elizabeth Hospital, Birmingham, UK

Dr John J V McMurray BSc MD MRCP
Consultant Cardiologist, Western General Hospital, and Honorary Senior Lecturer, Department of Medicine, University of Edinburgh, Edinburgh, UK

Christopher J Molloy PhD
Senior Research Investigator II, Bristol–Myers Squibb Pharmaceutical Research Institute, Princeton NJ, USA

Dr Alyn H Morice MA MD FRCP
Senior Lecturer in Medicine (Chest Diseases), University of Sheffield Department of Medicine & Pharmacology; Head of Section of Respiratory Medicine and Honorary Consultant Physician, Royal Hallamshire Hospital, Sheffield, UK

Dr H J Marjorie G Nelissen–Vrancken PhD
Assistant Professor, Faculty of Medicine, Department of Pharmacology, University of Limburg, Maastricht, The Netherlands

Martin Paul MD
Privatdozent, Max–Delbrück–Centrum für Mol. Forschung, Berlin, Germany

E Pussard PhD
Maître de Conférence, Faculté de Médicine Paris–Sud, Département de Pharmacologie, Le Kremlin–Bicêtre, France

Professor Anthony E G Raine D Phil FRCP
Professor of Renal Medicine, Department of Nephrology, The Royal Hospital of St Bartholomew, London, UK

Dr Simon G Ray BSc MD MRCP
Senior Registrar, Department of Cardiology, Cardiothoracic Centre, Broadgreen, Liverpool, UK

Professor Rainer Rettig MD
Professor, Institut für Physiologie, Ernst–Moritz–Arndt–Universität Greifswald, Greifswald, Germany

C Richer PhD
Maître de Conférence, Faculté de Médicine Paris–Sud, Département de Pharmacologie, Le Kremlin–Bicêtre, France

Prof. Dr.med. Günther A J Riegger
Director and Professor, Klinik und Poliklinik für Innere Medizin II (Kardiologie, Nephrologe und Pneumologie), Universität Regensburg, Regensburg, Germany

Stephen M Schwartz MD PhD
Professor of Pathology, School of Medicine, University of Washington, Seattle WA, USA

Prof.dr Jos F M Smits
Professor of Pharmacology, Faculty of Medicine, Department of Pharmacology, University of Limburg, Maastricht, The Netherlands

Harald M Stauss MD
Research Fellow, Humboldt–Universität Berlin, Berlin, Germany

Ulrike M Steckelings MD
Research Fellow, Virchow–Krankenhaus, Berlin, Germany

Prof.dr Harry A J Struijker Boudier
Professor of Pharmacology, Faculty of Medicine, Department of Pharmacology, University of Limburg, Maastricht, The Netherlands

Prof. Dr.med. Thomas Unger
Professor, Institut für Pharmakologie, Christian–Albrechts–Universität zu Kiel, Kiel, Germany

Dr Wilfred W Yeo MB ChB MRCP
Lecturer, University of Sheffield Department of Medicine & Pharmacology, Section of Pharmacology & Therapeutics, Royal Hallamshire Hospital, Sheffield, UK

Preface

The success of the angiotensin converting enzyme (ACE) inhibitors in clinical medicine must have surprised researchers and manufacturers alike. These drugs have rapidly become established as first-line therapy in both hypertension and heart failure and it is clear that their therapeutic potential is greater than this. They have also contributed to a great resurgence of interest in the renin-angiotensin-aldosterone system, in every imaginable context. At the recent meeting of the International Society of Hypertension, it was estimated that at least a third of the presentations were in some way related to this system. As a part of this picture, there is also intense interest in other ways of interfering with the system, notably with selective angiotensin II receptor antagonists. However, as always in medicine, enthusiasm for new treatments must not obscure the fact that they are not panaceas and are not suitable for everyone. It is as important to identify the patient who should avoid a drug as the one who must have it.

The aim of this book is to place the ACE inhibitors in their current context, not only with respect to the well-recognized roles already mentioned but also in areas where their use is more speculative and experimental: for instance, cardioprotection, the modification of the response to vascular injury and in preventing the progression of atherosclerosis. There is also emphasis on specific patient groups, such as black hypertensives, diabetics and patients with renovascular disease. Some overlap and repetition is inevitable, since the topics cannot be assigned to watertight compartments. Overall, the aim has been to provide a balance between fundamental research and clinical perspectives; the exact proportions differ considerably from topic to topic, but generally there is clinical emphasis for the non-specialist. The authors are extremely well placed to make these judgements and have made significant contributions to our knowledge of these valuable therapeutic agents. Most continue to work actively on their further development and I am very grateful to them for finding the time to write their contributions in the midst of already hectic schedules. I would also like to thank Alison Campbell, Alan Burgess and Martin Dunitz of Martin Dunitz Ltd for their help, encouragement and perseverance at every stage of this project.

Michael Schachter
May 1994

SECTION I

BACKGROUND

1

The renin–angiotensin system: physiology and pathophysiology

Physiology of the renin-angiotensin system

Oliver Edling, Peter Gohlke, Martin Paul, Rainer Rettig, Harald M Stauss, Ulrike M Steckelings and Thomas Unger

History

The renin-angiotensin system (RAS) has long been considered a classical endocrine system with angiotensin II (ANG II), its effector peptide, as the circulating hormone. Renin was discovered in 1898, when Tigerstedt and Bergmann found that injection of rabbit kidney extract into rabbits caused an increase in blood pressure.[1] In the following decades the existence of the vasopressor substance was discussed controversely, and in 1934 Goldblatt rediscovered the importance of renin for blood pressure regulation.[2] His group was able to show a relationship between renin secretion and renal ischaemia. Only a few years later, in 1937, Blalock and Levy discussed the link between renin secretion and renal perfusion pressure,[3] and renal baroreceptors were postulated.[4] The physiological and biochemical characterization of renin was accomplished in the following years. The enzymatic properties of renin were described by Page and Helmer[5] and Braun-Menendez et al.[6] At the same time Page et al.[7] identified the renin substrate angiotensinogen as a plasma protein and interpreted the RAS as a functional unit. In those days the opinion prevailed that the cleavage product of angiotensinogen, the decapeptide angiotensin I (ANG I), was causing vasoconstriction. This view was revised in the mid-fifties by Skeggs et al., who identified angiotensin converting enzyme (ACE) in lungs.[8] ACE catalyses the conversion of ANG I to ANG II, which mediates the actions of the RAS.

In 1958, Gross postulated a physiological link between the RAS and the aldosterone secretion in the zona glomerulosa of the adrenal gland, where ANG II stimulates the secretion of aldosterone and, therefore, enhances sodium retention.[9] More recently, local RASs in various organs, including those of cardiovascular regulation — the heart, vascular wall, kidney, adrenal gland and brain — have been postulated.[10–12] This assumption is based on a substantial body of experimental findings obtained with biochemical, immunocytochemical, and pharmacological techniques and most recently, with the help of molecular biology. The proteins of the RAS — angiotensinogen (the high molecular weight precursor of ANG I), the enzymes renin and ACE, and angiotensin peptides and angiotensin

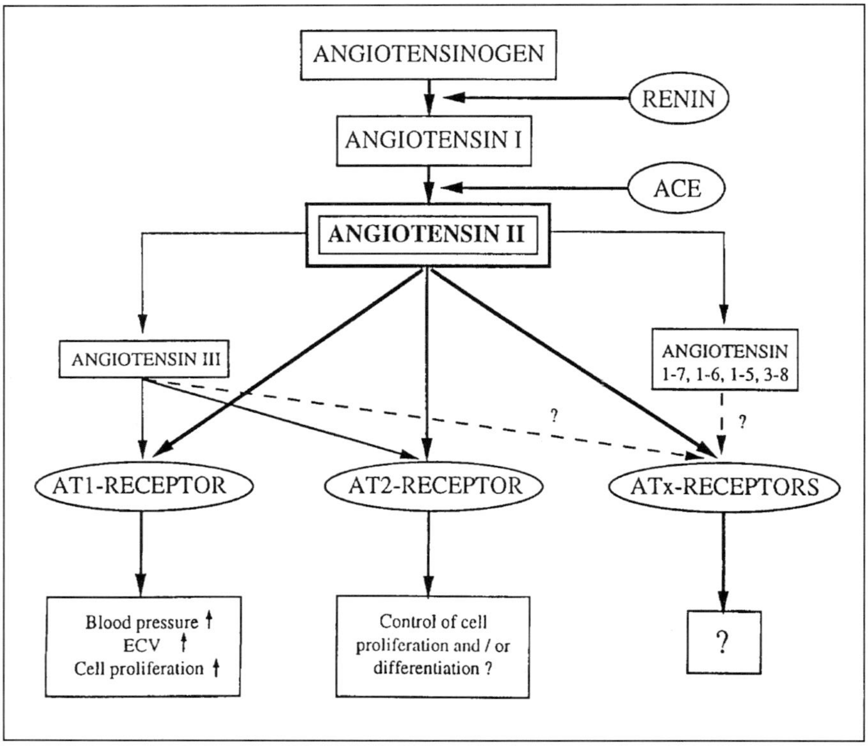

Figure 1.1
Schematic representation of the interactions between the different components of the RAS. The question marks indicate uncertain or controversial interactions or effects

receptors — have been localized in most of these tissues. In addition, some authors have reported that the regulation of tissue RASs was different from and independent of the regulation of the circulating RAS.

Biochemistry of the renin-angiotensin system

The biosynthesis of ANG II is initiated with the enzymatic cleavage of the decapeptide ANG I from angiotensinogen by renin. ANG I is converted to ANG II by ACE. ANG II then acts on its specific receptors (Fig. 1.1).

Renin

Protein structure

Renin belongs to the family of aspartyl proteases, sharing structural homologies such as gene sequence, amino acid composition and three-dimensional structure with other members of the family.[13–15] Its molecular weight is, differing from species to species, between 37 000 and 42 000 daltons. The active form of renin consists of two polypeptide chains linked by a disulphide bridge to form a bilobal structure with one active site.[16] In contrast to other aspartyl proteases, renin is most active at neutral pH and shows an extremely high substrate specificity for angiotensinogen, the only known endogenous substrate for renin. This specificity may be related to the presence of a so-called 'flap region' covering the cleft. The 'pro' part of the renin molecule forms a plug in the cleft between the two lobes of the molecule and hinders the admission of renin substrate to the catalytic site.[17]

Renin genes

Human,[18–20] rat[21,22] and mouse[23–25] renin genes have been cloned and sequenced. In humans and rats only one renin gene exists; in mice, two genes, Ren-*1* and Ren-*2*, have been identified. Ren-*1* and Ren-*2* are products of a gene duplication,[26–28] and the coding sequences of these genes show high similarities (up to 99%). Significant dissimilarities are found in the 5′-regulatory regions, and they are due to insertional elements. Both genes conserve a cis-acting core sequence mediating cyclic 3′, 5′-adenosine monophosphate (cAMP) responsiveness of renin gene expression. A major difference between the proteins renin-1 and renin-2 is the presence of glycosylation sites on renin-1.

Biosynthesis and processing of renin

Renin is produced and stored in granules in the juxtaglomerular apparatus of the kidney (Fig. 1.2). Translation of the renin gene yields a preprorenin consisting of 340 amino acids which is processed to enzymatically inactive prorenin (MW 57 000). Prorenin can be activated to renin by acidification[29] or prolonged cooling.[30] Renal ischaemia enhances prorenin transformation to renin.[6] Prorenin can also be converted in vitro to active renin by neutral serine proteases such as trypsin, glandular kallikrein, plasma kallikrein and plasmin, or by acid proteases such as pepsin and cathepsin D. More than 80% of the total circulating renin is prorenin,[31] and the kidneys secrete 10 times more prorenin than renin, whereas extrarenal tissues exclusively secrete prorenin at very low rates.[32]

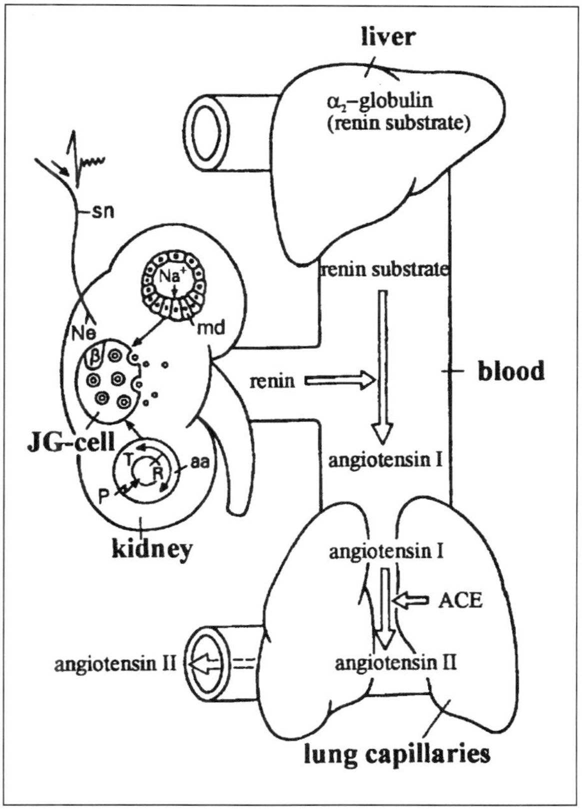

Figure 1.2
General biochemical metabolism of the endocrine RAS. aa = afferent arteriole, sn = sympathetic nerve, md = macula densa, Ne = noradrenaline, P = wall pressure, R = radius, T = tonicity

High concentrations of renin occur in female reproductive organs, and the ovarian prorenin secretion changes dynamically during the menstrual cycle and during gestation. Secretion occurs during the LH surge, and when hCG is present in the blood during pregnancy. Only low levels of active renin are found in association with very high concentrations of prorenin in ovarian follicular fluid after gonadotropin stimulation.[32]

Mechanisms controlling renin release

An overall view of the mechanisms controlling renin release allows the following statements (for detailed review, see Hackenthal et al:[33]

- renin is released in defence of extracellular fluid volume or local blood flow,
- renin secretion is inhibited by vasoconstrictors and by expansion of extracellular fluid volume.

The mechanisms involved in the regulation of renin release can be classified as follows:

Intrarenal mechanisms A major mechanism of renin release involves sensing blood pressure in the afferent glomerular arteriole. Vascular wall stretching, transmural pressure gradients, circumferential stress or volume strain at this site translate into stimulation of renin release. Prostaglandins produced by vascular cells of the afferent arteriole may mediate this renin release, because cyclo-oxygenase inhibitors interfere with this mechanism at least within the autoregulatory range of perfusion pressures. Another mechanism of renin release is related to the macula densa. This structure, formed by a specialized group of cells at the apex of the thick ascending loop of Henle, reacts to active transport of luminal Cl^-. Increased Cl^- delivery to the macula densa, such as in the case of sodium loading, inhibits renin release. Reduced Cl^- concentrations, such as in sodium chloride depletion or after administration of loop diuretics, stimulate renin release.

Sympathetic mechanisms The sympathetic nervous system and catecholamines interfere with renin release. Stimulation of renal nerves or the direct infusion of β-adrenergic agonists brings about the release of renin via increments of intracellular cAMP in juxtaglomerular cells. Alpha-adrenergic stimulation directly inhibits renin release. A decrease in blood pressure at the carotid sinus also leads to renin secretion.

Humoral mechanisms ANG II can exert a negative feedback action on renin release, affecting the juxtaglomerular cells directly, even at non-pressor doses. Arginine vasopressin (AVP) also inhibits renin release, both by direct effects on juxtaglomerular cells and, indirectly, by changes in plasma volume. The inhibitory effect of ANG II and other vasoconstrictors is mediated predominantly by receptor-operated increments of cytosolic calcium in juxtaglomerular cells. Atrial natriuretic peptide (ANP) and potassium in high concentrations inhibit renin release, the latter even in the state of sodium depletion.

Prostaglandins, histamine, dopamine, β-adrenoceptor agonists and forskolin, a specific stimulator of adenylate cyclase, stimulate renin release. The effect is mediated by the increase of the intracellular second messenger, cAMP. Factors affecting plasma renin levels are shown in Table 1.1.

Adult non-juxtaglomerular vascular smooth muscle cells in the kidney do not express renin mRNA under basal conditions, but chronic administration of ACE inhibitors can restore gene transcription at these sites.[34]

Angiotensinogen

Protein structure

Angiotensinogen is glycosylated α_2-globulin with a molecular weight of 54–60 kD; it is the only known substrate for renin.[35]

Increased plasma renin levels:
upright posture
sodium deficiency, dehydration, hypovolaemia, haemorrhage
adrenal insufficiency
renal artery stenosis
malignant hypertension
renin-secreting tumour
therapy with ACE inhibitors, ANG II antagonists, renin inhibitors, diuretics, vasodilators
pregnancy
food intake
high protein diet
Decreased plasma renin levels:
supine posture
increased sodium intake
mineralocorticoid excess
glucocorticoid excess
hyporeninaemic hypoaldosteronism
liquorice, carbenoxolone ingestion
therapy with β-blockers, α-methyldopa

Table 1.1
Factors affecting plasma renin levels

Increased plasma angiotensinogen levels:	
pregnancy	
oestrogen therapy, oral contraceptives	
glucocorticoid therapy	
increased adrenal secretion of glucocorticoids (e.g. surgical stress)	
malignant hypertension	
essential hypertension	
anephric state	
Decreased plasma angiotensinogen levels:	
high renin states:	adrenal insufficiency
	ACE inhibition
	congestive cardiac failure
cirrhosis of the liver	
nephrotic syndrome	

Table 1.2
Factors affecting plasma angiotensinogen levels

Angiotensinogen genes

The genes for angiotensinogen have been isolated and characterized in rats[36] and other species.[37,38] The rat angiotensinogen gene is approximately 11.8 kilobases in length and consists of five exons and four introns. The primary structure derived from the analysis of the angiotensinogen cDNA and gene indicates the presence of an angiotensinogen precursor consisting of 453 amino acids and a signal peptide of 24 amino acids.[39] Up to four different angiotensinogen mRNA species have been detected which were presumably generated by alternative splicing.[40] In contrast to the mRNA, only one species of angiotensinogen protein has been isolated. A recent report by Jeunemaitre et al.[41] points to a linkage of the human angiotensinogen gene to the development of high blood pressure in hypertensive families.

Biosynthesis and regulation of angiotensinogen

Angiotensinogen is mainly synthesized and stored in the liver, although its mRNA and protein is also found in other organs, such as the brain, heart, kidney, adrenals, ovaries and testes[42–45] (see Fig. 1.2). In rat plasma two forms of angiotensinogen, differing in the degree of glycosylation, have been found.[46] Angiotensinogen is secreted constitutively, and its formation and release is stimulated by bilateral nephrectomy, glucocorticoids, oestrogens, thyroid hormones, and ANG II.[35,43]

Adrenalectomy and ACE inhibition are accompanied by an increased peripheral consumption of plasma angiotensinogen. In isolated hepatocytes, the secretory response was shown to be preceded by an increase in angiotensinogen mRNA.[47] ANG II is supposed to have a trans-acting effect by stabilizing the angiotensinogen mRNA. The level of circulating angiotensinogen can be decreased in liver disease,[48] adrenal insufficiency,[49] nephrotic syndrome and during treatment with ACE inhibitors.[50–53] The angiotensinogen mRNA level in kidney, but not in liver, is sensitive to the sodium status: angiotensinogen mRNA is decreased by high sodium intake and increased by low sodium intake.[54] Factors affecting plasma angiotensinogen levels are shown in Table 1.2.

Angiotensin converting enzyme (ACE)

Protein structure

ACE is a dipeptidyl peptidase (MW 150 kD), generating ANG II from ANG I by cleaving the dipeptide His-Leu from ANG I. ACE is a rather unspecific enzyme, since it also degrades bradykinin, enkephalins, substance P and LHRH, and the β-chain of insulin.

ACE genes

The ACE gene has been cloned from lung and testes,[55–57] and the nucleotide sequence has been determined.[58] According to its DNA structure, ACE has similarities with thermolysin, neutral endopeptidase and collagenase. Genetic studies demonstrated the linkage of a region on chromosome 10 containing the ACE gene to hypertension in spontaneously hypertensive rats.[59] A similar study carried out in human hypertensives failed to demonstrate the linkage of the region containing the ACE gene to hypertension, however, an association between a specific mutation of the ACE gene within a population and an increased risk for myocardial infarction could be demonstrated.[60]

Location in the organism, function and regulation of ACE

ACE is predominantly localized in vascular endothelium (see Fig. 1.4).[61–65] The presence of ACE has also been shown in epithelial cells of the kidney, on the brush border of proximal tubules,[66] in the mitral and tricuspid valves of rat heart,[67] in the adrenal cortex and medulla, and in various other sites including specific brain areas and the reproductive organs.[68–72] For further review see Unger et al.[10] and Erdös.[73]

ACE is an ectoenzyme which contains two active catalytic sites, one towards the carboxyl terminus and the second near the amino terminus.[74,75] Testicular ACE contains only one catalytic site equivalent to the carboxyl site.[76,77] There is evidence from enzymatic and chloride activation studies that the two catalytic sites may be different. Furthermore, with Ro 31-8472, a small radioligand for ACE, one can detect twice the number of binding sites in somatic ACE compared with iodinated 351A, a derivative of the ACE inhibitor lisinopril, which has, due to its size and conformational requirements, access to the carboxyl terminal site only.[78] The two catalytic sites on a single protein enzyme may turn out to be of functional and clinical significance, in that each site may have specific substrates. The major endogenous substrates, ANG I and bradykinin, seem to be equally metabolized at both catalytic sites.[78] However, there are many other peptides that can act as substrates in vitro for ACE-like substance P, enkephalins, neurotensin and some of the reproductive peptide hormones.[79,80]

Testicular ACE may be involved in the control of spermatogenesis and sperm function since it occurs in high concentrations in both ovarian follicles and in spermatozoa and germinal cells of the testis.[81] Plasma ACE is increased in patients with sarcoidosis,[82] and ACE can be induced by glucocorticoids[83] and ACE inhibitor therapy.[84,85]

Generation and metabolism of angiotensin peptides

The major ANG II generating pathway is through the action of ACE. However, ANG II may also be generated by pathways other than the conversion from ANG I and by enzymes other than ACE. Neutral and acidic non-renin proteases capable of generating ANG I from angiotensinogen have been described in lysosomes.[86] Recently, a highly specific chymase has been identified and characterized in the human heart.[87] The chymase is inhibited by soybean trypsin inhibitor, PMSF and chymostatin, but not by aprotinin. Other mammalian chymases display a much broader substrate specificity.[88] A second class of non-renin angiotensinogenases generates ANG II directly from the renin substrate. Cathepsin G,[89] tonin and tissue plasminogen activating factor (TPA)[90] have been shown to possess this activity in vitro. The generation of ANG II in human plasma by tissue kallikrein derived from porcine pancreas has been reported.[91] Okunishi and colleagues[92] described two types of ANG II-generating enzymes, ACE and chymostatin-sensitive ANG II-generating enzyme (CAGE), in the aortic wall. ACE was found mainly in the endothelium, while CAGE was distributed predominantly in adventitia. Prolylendopeptidase (EC 3.4.21.26), an acid serine protease, generates ANG (1–7) from either ANG I or ANG II. This enzyme can be blocked by Z-proprolinal, a specific inhibitor of prolylendopeptidase,[93] but not by ACE inhibitors. In the presence of ACE inhibitors, the amount of ANG (1–7) in blood increases threefold.[94] Thus, numerous aminopeptidases, carboxypeptidases and endopeptidases located on endothelial and in vascular smooth muscle cells are involved in angiotensin peptide metabolism.[64,95–97]

The plasma half-life of ANG II in rat was previously estimated to be as short as 10 seconds[98] but recently, Admiraal et al[99] determined the half-lives of ^{125}I-labelled ANG I and ANG II, as well as the half-lives of unlabelled ANG I and ANG II in human plasma in the presence and in the absence of the ACE inhibitor, captopril, by adding a known amount of peptide to fresh plasma. In the absence of captopril, the half-life of ^{125}I-ANG I ranged from 1.7 to 2.6 minutes, and the half-life of ANG II was between 2.8 and 3.2 minutes. In the presence of captopril, the half-life for ^{125} I-ANG I ranged from 11 to 16 minutes and from 13 to 18 minutes for ANG I added to a plasma sample. The half-lives in plasma of ^{125}I-ANG II and ANG II ranged from 11 to 15 minutes.

Angiotensin receptors

Subtypes of angiotensin receptors

Two different receptor subtypes have been characterized by means of newly developed peptide and non-peptide antagonists. According to the present nomenclature, they are termed AT_1 and AT_2. Receptors with the highest affinity for losartan (K_i = 10–50 nM) and the lowest affinity for CGP42112A (K_i > 0.5 μM) and PD123177 (K_i > 10 μM) are referred to as AT_1. Binding sites displaying highest affinity for CGP42112A (K_i < 1 nM) and PD123177 (K_i = 10–100 nM) and lowest affinity for losartan (K_i > 1 μM), are referred to as AT_2.[100] Several authors have reported that CGP42112A has partial agonistic effects, especially at higher concentrations, and that it should not be considered as a pure AT_2-receptor-antagonist.[101] The AT_1 receptor has been cloned from cDNA derived from a variety of species.[102,103] The receptor proved to be a membrane protein with seven transmembrane domains. The structure of the AT_2 receptor is still unknown.

Signal transduction mechanism

The signal transduction pathway of the AT_1 receptor comprises the interaction with *G proteins*, inhibition of adenylyl cyclase activity and stimulation of the phospholipases C, A_2 and D, leading to a decrease in cellular cAMP, the formation of diacylglycerol and inositol trisphosphate (IP_3, as well as prostaglandin synthesis and increasing intracellular calcium levels.[104,105] AT_2 receptors do not interact with *G proteins*[106] but may induce the stimulation of a phosphotyrosine phosphatase and subsequent inhibition of a particulate guanylate cyclase, since reduced cGMP levels have been observed after stimulation of AT_2 receptors.[107] In the brain, it is not clear whether ANG II receptors are coupled to the same signal transduction mechanisms as in the periphery. For example, in primary glial cell culture, ANG II, as in peripheral tissues, leads to stimulation of phosphatidyl inositol hydrolysis.[108] However, in in vitro experiments on cerebral cortex and a combined preparation of hypothalamus, thalamus, septum and midbrain of Sprague-Dawley rats, ANG II inhibited IP_3 generation,[109] and in neuronal cell cultures it had no significant effect on phosphatidyl inositol hydrolysis.[110] It is now clear that ANG II also stimulates tyrosine phosphorylation.[111,112]

Tissue distribution of angiotensin receptors

Angiotensin receptors have been identified in a great number of tissues, including lung, adrenal glands, kidney and brain.[113–119] AT_1 and AT_2 receptors differ in their tissue distribution. Liver, lung, kidney, placenta, urinary bladder, the gastrointestinal tract and aortic smooth muscle cells express mostly the AT_1 receptor, whereas in pancreas and ovarian granulosa the AT_2 receptor predominates and in the adrenals, heart, renal arteries and uterus both receptor subtypes are present.[115,120–135]

Differentiation between AT_{1A}, AT_{1B} and AT_2 receptor subtypes

Two subtypes of the AT_1 receptor have been identified by cloning and binding studies in mice[136] and rats,[137] termed AT_{1A} and AT_{1B}. Both subtypes show differences in tissue distribution and regulation.[138] The AT_{1A} receptor seems to be identical with the 'classical' AT_1 receptor, whereas the AT_{1B} receptor, although it is also G-protein-coupled, shows differences in the binding properties of various specific ligands: PD123319 > losartan > CGP42112A.[139] Recent data also report coupling of a subgroup of central AT_2 receptors to G proteins.[140] These receptors were localized in the ventral thalamic and geniculate nuclei and in the locus coeruleus, while AT_2 receptors in the inferior olive were not coupled to G proteins.[140]

Although the physiological role of the AT_2 receptor remains unclear at present, new data from binding studies indicate a possible involvement in processes of fetal cell growth and cell differentiation. Fetal rat tissues contain considerably more AT_2 than AT_1 receptors. Within a few weeks after birth, dramatic changes occur in the proportion of AT_1 to AT_2 receptors within most tissues, and in the adult animal the AT_1 receptor dominates.[141,142]

Other angiotensin receptors

More data point to the existence of additional ANG II receptor subtypes. The affinities of losartan, CGP42112A and PD123177 for ANG II binding sites on amphibian (*Xenopus laevis*) cardiac membranes differ markedly from their affinities for binding sites in

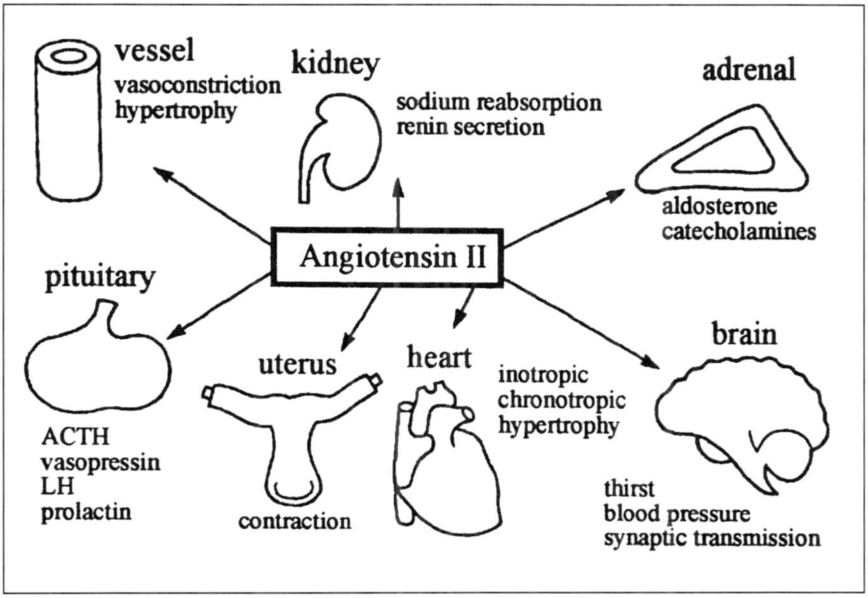

Figure 1.3
Different effects of ANG II on various organs

mammalian tissues.[143] Furthermore, in most areas of the rat brain, ANG II and (Sar[1], Ile[8])ANG II have a tenfold lower affinity for AT_1 receptors than in peripheral tissues.[144] Recently, a third receptor, AT_3, has been postulated in neuroblastoma cells.[145] The authors suggest that the new receptor subtype mediates cGMP formation through activation of cytosolic guanylate cyclase. Furthermore, a novel binding site for angiotensin (3–8) has been claimed to exist in the hippocampus of rats. The authors termed it the AT_4 receptor;[146] additionally, a novel ANG II receptor subtype from the pituitary has been cloned.[147]

Physiology of the renin-angiotensin system

The circulating endocrine renin-angiotensin system with ANG II as the effector peptide is involved in the maintenance and regulation of salt, volume, and cardiovascular homeostasis.

Since this regulation is a complex phenomenon and since many organs such as the heart, brain, kidneys, adrenals and blood vessels contribute to it, local actions of ANG II in these organs with paracrine, autocrine and intracrine functions for ANG II have been postulated.[12] A critical discussion of these 'tissue RASs' has been written by Unger et al.[148]

Dzau[12] hypothesized two compartments of the RAS: the circulating RAS compartment and the tissue RAS compartment, the former mediating short-term cardiorenal homeostasis, and the latter controlling vascular tone and tissue function. In this view, short-term effects of ANG II, such as vasoconstriction, elevation of blood pressure, and release of aldosterone from the zona glomerulosa of the adrenals can be distinguished from long-term trophic effects, such as induction of vascular and cardiac hypertrophy and proliferation of vascular smooth muscle cells[149–151] and the induction of immediate early genes (IEGs) and transcription factors.[152–154]

Effects of ANG II

An overview of the known and hypothesized effects of ANG II is shown in Fig. 1.3.

Vasoconstriction

ANG II causes direct arterial vasoconstriction, resulting in increased peripheral vascular resistance. Sustained high doses of ANG II result in tachyphylaxis.[155]

Administration of low non-pressor doses of ANG II over several days leads to a gradual increase in blood pressure,[156] possibly mediated by a decreased vascular compliance, sympathetic activation, baroreceptor resetting and enhanced aldosterone secretion and sodium retention. ANG II may facilitate reflex neurogenic venoconstriction by a sympathetic presynaptic mechanism.[157] Direct venoconstriction undergoes tachyphylaxis, not so the reflex. ANG II can stimulate release of catecholamines from the adrenal medulla.[158]

Sodium retention

Circulating ANG II leads to sodium retention, either directly by an action on the proximal tubules, or indirectly by stimulating aldosterone release from the zona glomerulosa of the adrenal gland. Aldosterone release is more pronounced in the state of sodium depletion due to possible upregulation of receptors.[159]

Growth

ANG II can stimulate vascular growth. It has been hypothesized that it participates in the hypertrophic response of the vessel wall in hypertension, and even in the hyperplastic response after balloon angioplasty.

In cultured smooth muscle cells derived from spontaneously hypertensive rats (SHR) but not from the normotensive Wistar-Kyoto (WKY) strain, the peptide was shown to produce hyperplasia.[160] ANG II is capable of stimulating DNA, RNA and protein synthesis in isolated cardiac tissue[161] and in cultured smooth muscle cells,[162] and has been reported to bind to specific chromatin receptors where it may act as a transcription factor.[163]

ANG II has a mitogenic effect on fibroblast 3T3 mouse cells[164] and on human vascular smooth muscle cells in vitro,[149] as well as a hypertrophic effect on quiescent cultured rat aortic smooth muscle cells.[150] In vascular smooth muscle cells the peptide stimulates the expression of PDGF A-chain, which is associated with mitogenesis, but paradoxically induces hypertrophy and not hyperplasia.[150] This may be due to the fact that the growth-promoting effects of ANG II may be counteracted by parallel activation of an antiproliferative pathway, that is the stimulation of transforming growth factor β1 (TGF-β1).[165] Endogenous PDGF has been described as an autocrine mediator of ANG II-mediated vascular hypertrophy.[165]

It has recently been demonstrated that ANG II and ANG III can also inhibit the proliferation of serum-stimulated microvascular endothelial cells from rat hearts.[167] This effect could not be blocked by the AT_1-receptor antagonist, losartan. ANG II was also shown to inhibit the proliferation of macrovascular serum-stimulated bovine aortic endothelial cells, an effect that is mediated by the AT_2 receptor, since it could be blocked with the AT_2-receptor antagonist, PD123319, but not with losartan.[168]

Interference with neurotransmission

ANG II can enhance sympathetic nerve activity in the periphery by pre- and postsynaptic mechanisms.[169,170] The effects of ANG II on the sympathetic outflow from the brain are complex.[171,172]

Auto- and paracrine actions of ANG II in different organs

Brain

The existence of a brain RAS is now established and has been extensively reviewed.[19,119,173] More than 30 years ago, it was first demonstrated that ANG II was able to act not only on peripheral vascular structures, but also in the central nervous system to increase blood pressure.[174] In the meantime, all components of the RAS in the brain have been identified,[175] but they occur in different locations. Angiotensinogen, for example, is mainly localized in astrocytes and ependymal cells,[176] however, there is evidence pointing to a neuronal localization of angiotensinogen in ANG II-positive brain nuclei.[177] Renin has been identified mainly in neurons and nerve terminals[71] but the renin mRNA level in the brain is low. High concentrations of the enzyme are reported for the pituitary and pineal glands, the choroid plexus,[68] the hypothalamus, the cerebellum and the amygdaloid nucleus.[173] ACE has been localized, biochemically and by means of autoradiography, especially in the choroid plexus, the circumventricular organs and in synaptosomes, where it is co-localized with renin in neurosecretory vesicles.[71] ANG II is located in synaptic vesicles in nerve terminals in the circumventricular organs, where it possibly acts as a neurotransmitter or neuromodulator, since it excites neurons at very low doses with high specificity.[178]

Although ANG II is thought not to cross the blood–brain barrier, it has been supposed that ANG II enhances the permeability of this barrier by stimulating fluid-phase endocytosis in endothelial cells lining the microvessels of the brain, an effect that appears to be mediated by prostaglandins.[179] Nonetheless, certain regions of the brain, such as the area postrema and the circumventricular organs, lack this barrier, and circulating ANG II may exert an effect at these sites. ANG II is present in the CSF and its concentration appears to be independent of the plasma concentration. Schelling et al. reported 17.6 $\pm$ 1.5 fmol/ml ANG I and 29.5 $\pm$ 5.9 fmol/ml ANG II in rat CSF.[180] Five days after bilateral nephrectomy in rats, ANG II in the brain has been reported to be elevated.[181] According to the authors, this effect depends on plasma potassium levels.

The distribution of angiotensin receptors in humans and rats is similar, as verified in several studies with autoradiography or competitive radioligand binding techniques.[116,141,182–184] In rat brain, structures related to the known central actions of ANG II mainly or exclusively contain AT_1 receptors, suggesting that this receptor subtype is responsible for mediating the cardiovascular effects of intracerebroventricular (ICV) ANG II. There are different subtypes of AT_2 receptors in the brain. AT_2 receptors have been localized in areas such as the cerebellum, locus coeruleus, ventral geniculate nucleus and inferior olive, which are not directly involved in the processing of central cardiovascular effects.[116,140]

Injection of ANG II into the ventricles of the brain elicits several distinct physiological responses including thirst, salt appetite, an increase in blood pressure, release of vasopressin and other pituitary hormones such as oxytocin,[185] TSH, FSH, LH, STH and ACTH.[10,119] Stimulation of brain ANG II receptors also elicits a blood pressure-independent natriuresis, which is in contrast to the salt-retaining effects of circulating ANG II.[186] The mechanisms mediating the central pressor response to ANG II include the release of vasopressin, modulation of central sympathetic outflow and inhibition of the baroreceptor reflex.[10] The pressor response to ANG II administered by the ICV route is mediated by an AT_1 receptor, since it can be blocked by pretreatment with losartan.[187]

Recently, a pathway of ANG II-induced AVP release in which catecholamines play an important role has been described.[188] Stimulation of periventricular AT_1 receptors results in the release of ANG II which, in turn, evokes a dose-dependent release of noradrenaline (norepinephrine), an effect which is AT_1-receptor-mediated. The released noradrenaline activates α_1- and α_2-adrenoceptors, leading to the secretion of AVP via the pituitary gland into the circulation.[189,190]

The drinking response to ICV ANG II is abolished by ICV treatment with losartan, again indicating the involvement of AT_1 receptors. The physiological drinking response after 24 hours of thirst is attenuated by losartan and atropine, suggesting that both endogenous angiotensinergic and cholinergic mechanisms are involved.[191]

These findings suggest that ANG II most probably participates in central osmoregulation. Administration of hyperosmotic saline ICV induces a dose-dependent natriuresis and, at higher concentrations, an increase in blood pressure and urine volume. The natriuretic effect of ICV-injected saline at concentrations below 0.225 mol/l sodium chloride (pressor threshold) can be completely abolished by pretreatment with the ANG II AT_1-receptor antagonist, losartan, and the natriuretic effect of saline at higher concentrations can be diminished. Injection of losartan into the subfornical organ completely inhibits the natriuresis induced by ICV administration of saline.[192] In addition, local administration of saline into the paraventricular nucleus using the microdialysis technique elicits a concentration-dependent release of ANG II and ANG III in the perfusate, which conforms with the idea that ANG peptides play a transmitter role in central osmoregulation.[193]

We have recently shown that ICV administered ANG II also induces a dose-dependent, AT_1-receptor-mediated expression of the immediate-early genes, (IEGs) *c-fos*, *c-jun* and *krox 24*, in the hypothalamus of the rat.[194]

Heart

All components of the RAS have been found in the heart, although the presence of renin mRNA is still controversial. Renin and angiotensinogen mRNA have been detected in cardiac tissue utilizing Northern hybridization analysis, S1 nuclease protection assay and in situ hybridization. In neonatal rat hearts, in situ hybridization has revealed that renin and angiotensinogen mRNA are evenly distributed between the four cardiac chambers,[195] whereas in adult rats they are detectable in the four chambers of the heart, but with the highest levels in the right atrium.[196] Both renin and angiotensinogen mRNA were demonstrated in atrial and ventricular myocytes.[197] According to Dzau et al.,[198] the renin mRNA and angiotensinogen mRNA content of adult rat heart homogenates represents approximately 2% of kidney renin mRNA and approximately 5% of liver angiotensinogen mRNA content. Whereas several investigators have described specific renin mRNA expression by Northern blot and solution hybridization analysis in mouse and rat hearts,[199,200] other authors were unable to detect it.[201,202] Sodium depletion causes a parallel increase in cardiac renin activity and renin mRNA. Angiotensinogen mRNA is increased in atria and ventricles after glucocorticoids, oestrogen and thyroid hormones.[203]

ACE occurs in cardiac endothelium of rats and is particularly concentrated around the mitral and tricuspidal valves.[204] Angiotensins have been detected in rhesus monkey heart.[205]

Concentrations were highest in the right atrium, followed by the right ventricle, left atrium, septum and left ventricle. Similar findings were reported for the rat heart by Speakman et al.[206]

Angiotensin receptors have been found in cardiac tissue of many mammalian species including man,[207] and the concentration averages 50 fmol of receptor per milligram of myocardium.[208] These binding sites fulfill the criteria for hormone receptors since binding of ^{125}I-ANG II is specific, saturable, reversible, modulated by guanine nucleotides and correlated to physiological actions in bioassays. In addition, a soluble, cytosolic, high-affinity binding protein for ANG II with a molecular weight of 78 kDa was described in neonatal rat cardiomyocytes.[209] The function of this protein is unclear at present but it is speculated that it acts as a transport protein.

Urata and colleagues found angiotensin receptors located on myocardial membranes and on adrenergic nerve fibres.[207] This is of particular interest since the cardiac effects of ANG II partly depend on modulation of sympathetic neurotransmission. The same group identified a neutral serine protease, generating ANG II from ANG I in cardiac membrane preparations of human heart, which is not blocked by ACE inhibitors but by soybean trypsin inhibitor and other serine protease inhibitors.[88] According to the authors, the heart chymase has a high specificity for the conversion of ANG I to ANG II and does neither degrade ANG II, bradykinin or vasoactive intestinal peptide, nor generate ANG II directly from angiotensinogen.

Physiological actions of ANG II in the heart

- ANG II has both direct actions on the heart which affect rate, contractility and cell growth[210] and indirect actions on the heart that include central nervous actions such as stimulation of sympathetic outflow to the heart.[211]
- ANG II has positive inotropic and chronotropic effects on the heart,[212,213] but due to the concomitant systemic vasoconstriction and subsequent baroreceptor stimulation this effect cannot be observed in the intact animal. The inotropic effect may be mediated by activation of a voltage-sensitive slow calcium channel independent of the β-adrenergic system and cyclic adenosine monophosphate accumulation.[214]
- ANG II exerts opposing actions on heart rate: reflex slowing of the heart due to baroreceptor stimulation following the increase in blood pressure and an increase in heart rate directly and by various sympathoadrenal mechanisms. ANG II may also be involved in arrhythmogenesis via a direct action on cardiomyocytes.[210]

Kidney

Localization of the components of the RAS

The mRNA for renin, angiotensinogen and ACE has been found in various locations in the kidney using different techniques like Northern blot or in situ hybridization histochemistry. Ingelfinger et al.[215] found renin mRNA mostly at the vascular pole of the glomerulus and lesser amounts in the distal tubule and the glomerular tuft, whereas angiotensinogen mRNA was predominantly located to the proximal tubule.[216] The authors did not find renin mRNA in the proximal tubule, even though the presence of immunoreactive renin has been reported for this site. The immunoreactive renin at the proximal tubule may be taken up from the interstitium or originate from tubular filtrate. ACE mRNA, as well as the protein, occur in the tubular brush border of the proximal tubule.[217–219] ANG II has been measured in the glomerulus filtrate near Bowman's capsule and in the early, mid and late proximal tubule.[220] The concentration of the peptide was thousandfold higher in these renal compartments than in the circulation, indicating a local synthesis in the proximal tubule.[220]

Effects of ANG II on the kidney ANG II is known to modulate glomerular haemodynamics and tubular reabsorption. In addition, the peptide has growth-promoting effects on mesangial[221,222] and tubular cells.[223] ANG II induces dose-dependent contractions of isolated glomeruli and cultured mesangial cells,[223] and causes vasoconstriction mainly in the efferent arterioles. As a result of these effects, significant changes in single nephron glomerular plasma flow, single nephron glomerular filtration rate, intraglomerular capillary pressure and ultrafiltration coefficient are induced.[224] In the proximal convoluted tubule, ANG II directly enhances sodium and water reabsorption.[225,226] In addition, the peptide has been demonstrated to enhance the sensitivity of the vascular elements that mediate tubuloglomerular feedback-induced alterations in single nephron haemodynamics.[227] Thus, there is a functional synergy between tubular and vascular effects of ANG II, resulting in a concerted action towards increased renal electrolyte and water conservation. Via these actions ANG II may participate importantly in various physiologic actions of the kidney, including the renal response to change in dietary sodium intake and in intracellular volume, as well as in pathophysiological processes such as hypertension.

Renal ANG II receptors ANG II receptors in the kidney show a characteristic pattern of distribution. A high density of ANG II binding is seen in the glomeruli and in the inner zone of the outer medulla, corresponding to vasa recta bundles. A more moderate density of ANG II binding is found in the outer cortex, corresponding to the proximal convoluted tubules, and throughout the inner zone of the outer medulla in the interbundle areas.[118] Glomerular ANG II binding is highest in superficial and midcortical glomeruli and lowest in juxtamedullary glomeruli. Incidentally, a similar gradient has been described for the distribution of renin within the kidney.

Glomerular ANG II receptors can be modulated by several factors. For example, high dietary sodium,[228] renal denervation,[229] and old age[230] are associated with high glomerular ANG II receptor density, whereas low dietary sodium[231] or experimental diabetes mellitus[232] go along with a low glomerular ANG II receptor density. Recent autoradiographic studies using receptor subtype-specific ligands[118,233] revealed that renal ANG II receptors belong predominantly to the AT_1 receptor subtype. The distribution of renal ANG II receptors correlates well with the intrarenal sites at which the peptide has its major physiological effects.

Pathophysiology related to the kidney: pathophysiological implications for the kidney in association with heart failure In experimental heart failure, the excretory function of the kidneys for salt and water may be impaired. Ichikawa and coworkers[234] investigated the role of ANG II in the altered renal function of congestive heart failure. Infarcted animals were characterized by a 50% reduced glomerular plasma flow rate, a threefold increased efferent arteriolar resistance and an increased single nephron filtration fraction compared to control animals. Intravenous infusion of teprotide, an ACE inhibitor, led to the return of glomerular plasma flow rate, single nephron filtration fraction and efferent arteriolar vascular resistance to or toward the levels found in control animals.

Hodsman and coworkers studied neurohormonal responses to chronic myocardial infarction in rats.[235] They reported a marked fivefold increase in plasma atrial natriuretic peptide compared to control animals. Since the authors did not find sodium retention and only little, if any, stimulation of the RAS documented by unchanged plasma renin activity in infarcted animals, they suggest that the high circulating atrial natriuretic peptide levels may prevent or limit salt and water retention, either directly or indirectly, by inhibiting the RAS.

The renal angiotensinogen mRNA levels of rats with compensated heart failure is increased to 150% of control,[236] whereas renal renin mRNA, plasma renin activity and plasma ANG II levels remain unchanged in these animals. Chronic treatment of these animals with an ACE inhibitor normalizes the elevated angiotensinogen mRNA levels in the kidney, possibly by inhibiting the positive feedback mechanism of ANG II on renal angiotensinogen mRNA.

Genetic hypertension The spontaneously hypertensive rat (SHR) is a model of genetic hypertension. In a recent study SHR and normotensive Wistar-Kyoto (WKY) rats were subjected to normal and low sodium diets and the levels of renal renin and angiotensinogen mRNA were determined. In this study, angiotensinogen mRNA was lower in SHR than in WKY under normal conditions. Mild sodium depletion caused an increase in angiotensinogen mRNA in WKY but not in SHR, and the renin mRNA levels were unaltered in both rat strains.[54]

Severe sodium depletion induced by a low sodium diet and treatment with furosemide elicits a greater increase in renal renin mRNA in SHR than in WKY, although there are no differences in plasma renin activity between the two strains.[237] The pathophysiological role of this differential regulation is still unclear, but the reduced renal angiotensinogen mRNA in SHR compared to WKY may be compensated by the increased renin expression in severe sodium depletion. Kost and Jackson[238] reported enhanced renal responses to ANG II in SHR mediated by an AT_1 receptor. They found a significantly greater decrease in renal blood flow and glomerular filtration rate and a significantly enhanced renal vascular resistance in response to intrarenal ANG II infusions in anaesthetized SHR compared to WKY.

Transgenic mRen-2 rats (TGR) The TGR, a new animal model of genetic hypertension, carrying an additional mouse renin gene, was first described in 1990 by Mullins et al.[239] From the age of four weeks, these animals develop hypertension reaching maximum levels of systolic blood pressure of 180–210 mmHg in male animals at nine weeks of age. Although the pathogenesis of hypertension in the TGR still remains to be elucidated, several studies point to involvement of the renin-angiotensin-aldosterone system, the sympathetic nervous system and sex hormones in this hypertension model.[240,241]

Both the ACE inhibitor, ramipril, and the AT_1-receptor antagonist, losartan, elicited an enhanced response with respect to blood pressure and heart rate in TGR compared to Sprague-Dawley (SD) rats.[240,242] Basal plasma catecholamines were found to be elevated in TGR compared to normotensive SD rats.[240] With increasing age, TGR develop end-organ damage; at eight months, they show severe glomerular and tubular lesions, as demonstrated by morphometric studies.[243]

Renal artery stenosis and renovascular hypertension In experimental models of renovascular hypertension it was observed that, immediately after clipping one renal artery, renin mRNA of the ischaemic kidney increased and renin mRNA of the contralateral kidney decreased.[244,245] In a study with male Sprague-Dawley rats that were made hypertensive either by coarctation of the aorta between the renal arteries, or by clipping the left renal artery, only a transient elevation of renin mRNA in the ischaemic kidney was observed,[245] whereas another study with female 2-kidney–1-clip (2K1C) Wistar rats reported the renin mRNA content of the ischaemic kidney to be still elevated fourfold compared to the age-matched control after 20 weeks of clipping.[244]

Vascular wall

The endothelium plays an important role in the maintenance of normal vascular homeostasis. It produces vasodilator substances such as prostacyclin (PGI_2) and endothelium-derived relaxing factor (EDRF), as well as vasoconstrictive agents such as ANG II and endothelin. Endothelial dysfunction impairing the balance of production of endothelial vasodilators and vasoconstrictors, including ANG II,[246] may contribute to the pathogenesis of hypertension.[247–249]

Several authors have shown the presence of the protein components of the RAS in the different layers of the arterial wall: angiotensinogen and renin have been found in the endothelium, medial and adventitial layers,[250–254] whereas ACE is confined to the endothelium and, to a very limited extent, to the adventitial layer.[61–63,65] A localization of ACE in the medial layer is still controversial. A study by Velletri and Bean,[255] using a radioenzymatic technique, reported the distribution of ACE activity in the medial (42%) and adventitial (58%) layer. A major drawback of the study, however, is the lack of ACE activity in the intimal layer. Pipili et al.[256] measured ACE activity in intact and endothelium-denuded aortic rings. Although this study points to the presence of ACE at an extra-endothelial site, the exact localization, the medial or adventitial layer, or both, remains unclear. Indirect evidence for an extra-endothelial conversion of ANG I to ANG II in the vascular wall was provided by studies measuring the contraction of endothelium-denuded aortic rings after addition of ANG I in rabbit aorta[257] and in rat aorta.[258,259] It is possible that conversion of ANG I to ANG II occurs primarily in the adventitial layer, for example the vasa vasorum.[65]

Two recent studies from our laboratory indicate that most of the ACE is located on the luminal side of the vascular endothelium. Conversion of ANG I to ANG II was demonstrated in the lumen of isolated thoracic rabbit aorta, an effect that can be completely blocked either by removal of the endothelium or by co-administration of the ACE inhibitor ramiprilat.[64] In addition, an accumulation of 3H-ramiprilat was observed in the endothelial layer of isolated rabbit aorta.[260]

Theoretically, ANG II generating enzymes, other than ACE, may participate in vascular ANG II production. A chymostatine-sensitive ANG II generating enzyme (CAGE), distributed mainly in the adventitia, has been reported.[92] Furthermore, alternative pathways for ANG I to ANG II conversion, for example a consecutive action of carboxypeptidases to generate ANG II, may be involved in vascular ANG II production.

The actions of ANG II in the vascular bed include the regulation of local vascular tone and blood flow. Additionally, ANG II may contribute to the development of vascular hypertrophy and to vascular responses to inflammation and injury.[12] ANG II molecules, generated by ACE at the luminal side of the endothelium, may either enter the circulation to act in an endocrine fashion, or affect receptors on neighbouring cells, acting in a paracrine fashion.

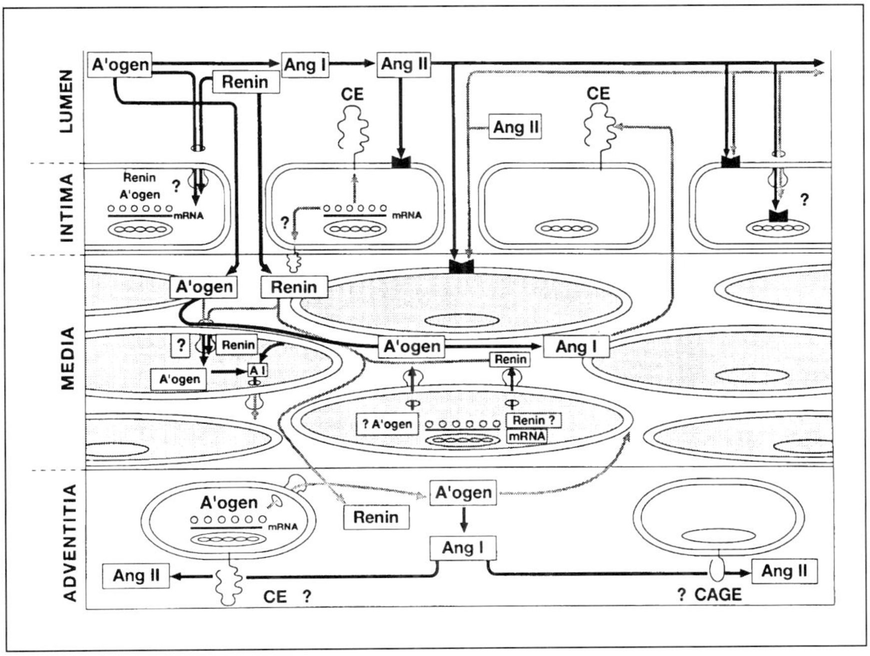

Figure 1.4
The various pathways through which ANG II van be formed locally in the vascular wall. Note that most of the ANG II molecules are generated by ACE at the luminal site of the endothelial cells. However, there might be alternative pathways as shown for the adventitial layer

Some ANG II molecules may also act on their original endothelial cell: this has been termed autocrine action.[148] There is an intimate relationship between the endocrine and para-autocrine action of angiotensin peptides (Fig. 1.4).

Pathophysiology of the renin-angiotensin and kallikrein-kinin systems

Julie Blasioli and Duncan J Campbell

Pathophysiology of the renin-angiotensin system

Pathophysiological conditions where the RAS is thought to be important include, among others, congestive heart failure, primary or renal hypertension and glomerulosclerosis.

Hypertension

Whereas the RAS is clearly implicated in the pathogenesis of hypertension associated with high renin states such as renal artery stenosis and renin-secreting tumours (see below), there is continuing controversy concerning its role in the pathogenesis of essential hypertension.

Recent linkage and association studies suggest that molecular variants of the angiotensinogen gene contribute to the inheritance of essential hypertension.[41] The mechanism by which these genetic variants increase blood pressure is not known, but it is of note that plasma angiotensinogen levels show a significant positive correlation with blood pressure in man[262] and, as mentioned above, increased plasma angiotensinogen levels are associated with an increase in renal vascular resistance.[263]

Any hypertensive process necessarily requires a change in the relationship between renal perfusion pressure and urinary excretion of salt and water;[264] such an alteration in renal function may be a primary or contributory cause of the hypertension. It may be argued that the appropriate response to a rise in blood pressure is a fall in renal renin secretion; thus, in essential hypertension a 'normal' plasma renin level is inappropriately elevated for the prevailing renal perfusion pressure.

Hypertensive subjects with either high, normal, or low circulating renin levels may show a reduction of blood pressure in response to inhibitors of the RAS. Given the multifactorial nature of blood pressure regulation, such a hypotensive effect should be interpreted with caution, and should not be taken as definitive evidence for a pathogenic role of the RAS in the hypertensive process. Many different hypotheses exist concerning the role of the RAS in the pathogenesis of essential hypertension. Laragh[265] has proposed that essential hypertension may result from nephron heterogeneity, in which a subpopulation of ischaemic nephrons hypersecretes renin and retains sodium, and the excess renin acts to impair adaptive natriuresis of neighbouring normal nephrons. Others have proposed that essential hypertension may result from increased ANG II formation at specific tissue sites such as the vasculature, adrenal or brain, but direct evidence in support of such hypotheses is lacking.

Renin-dependent hypertension

After kidney artery stenosis there is a decrease in blood pressure within the kidney, which is sensed by intrarenal baroreceptors leading to an enhanced secretion of renin. With increased renin concentrations, ANG II also increases and hypertension develops. Another form of renin-dependent hypertension is due to renin-producing tumours of the juxtaglomerular apparatus.

Left ventricular hypertrophy

The principal cardiac adaptive response to hypertension is concentric left ventricular hypertrophy (LVH). This hypertrophic response is not solely caused by blood pressure, but other factors such as the RAS and the sympathetic nervous system participate in its development. The hypertrophied heart is vulnerable to ischaemia, dysrhythmias and ventricular tachycardia, eventually leading to sudden death.

Heart failure

Like the sympathetic nervous system, the RAS is activated in patients with advanced congestive heart failure. Ventricular remodelling and compensatory hypertrophy after myocardial infarction may be related to an increased cardiac production of ANG II. Plasma renin activity is only modestly increased in patients with asymptomatic left ventricular dysfunction, but is markedly elevated once overt symptoms ensue and during phases of decompensation. Agents designed to block ANG II production can have a dramatic influence on the syndrome of heart failure. In this context, the non-vasodilator properties of ACE inhibitors may be as important as the vasodilator effects of these agents.

Increased expression of ACE in the septum, the free wall and the apex of the hypertrophied ventricle after experimental aortic stenosis has been reported by Schunkert et al.[266] The authors investigated the physiological significance of intracardiac conversion of ANG I to ANG II in the presence of chronic pressure overload. Their results indicate an increased intracardiac activation of ANG II in the presence of pressure overload, which impairs diastolic relaxation. It has recently been shown that treatment of SHR with antihypertensive doses of the ACE inhibitor ramipril effects a regression of cardiac hypertrophy, an improved myocardial metabolism and an increased cardiac capillary length density.[267] The latter may result from the potentiation of the kinin effect by ramipril. However, decreased ANG II levels after ramipril may also explain the capillary angiogenesis since ANG II has antiproliferative effects on cardiac microvascular endothelial cells from SHR.[167]

ANG II has been implicated in cardiac hypertrophy that is associated with hypertension. It stimulates protein synthesis and cell growth in embryonic chick heart cells[268] as well as in neonatal rat myocytes and adult rat cardiomyocytes.[269] These effects can be blocked by the angiotensin AT_1-receptor antagonist, losartan. Recently, in a rat model of cardiac hypertrophy induced by clamping the abdominal aorta, treatment with an ACE inhibitor prevented the increase in left ventricular mass without having an effect on cardiac afterload.[270,271] However, since this antihypertrophic effect of the ACE inhibitor could be prevented by HOE 140, a specific bradykinin B_2-receptor antagonist, it was most probably mediated by bradykinin.[271]

Chronic renal failure and glomerulosclerosis

In chronic renal failure the RAS seems to function properly but the response of renin secretion to volume changes is attenuated.

Glomerulosclerosis is characterized, morphologically, by expansion of the mesangium, which is due to mesangial cell hypertrophy and matrix accumulation. Most of the information about the activity of the RAS in patients with chronic renal failure is derived from measurements of renin and angiotensin in venous plasma under defined conditions of posture and fluid balance. It is important to emphasize that such measurements may not accurately reflect local generation of ANG II within the kidney. The highest renin activity is found amongst those patients with lesions of the renal microvasculature, while hyperreninaemia is less frequently seen in patients with glomerulonephritis. ANG II plays a direct role in the regulation of glomerular haemodynamics. The hyperfiltration of the glomeruli in experimental or human diabetes mellitus or high protein intake is linked to vasodilatation of the afferent arteriole. This may be a consequence of selective afferent insensitivity to ANG II. Metabolic factors or increased local prostaglandin synthesis may mediate this event. These views are backed by a vast body of micropuncture studies employing ACE inhibitors, which, under most circumstances, restore normal or proper glomerular haemodynamics.[272] ANG II enhances proteinuria and mesangial uptake of macromolecules and it may thereby contribute to the sclerotic evolution that invariably follows perturbed local microcirculation. The deleterious effect of ANG II on the glomerular filtration barrier may be mediated by altering concentrations of proteoglycans (specifically heparan sulphate) in the glomerular basement membrane. The latter alters its charge and, in turn, its protein-sieving characteristics. A separate haemodynamic effect of ANG II is to increase intraglomerular pressure, which can also cause proteinuria and lead to glomerulosclerosis.[273] ANG II also causes mesangial cell hypertrophy as recently demonstrated by Anderson et al. in quiescent, subconfluent cultured adult murine mesangial cells.[222] The effect seems to be mediated by the AT_1 receptor since it can be partially blocked by losartan.

Pathophysiology of the kallikrein-kinin system

Bradykinin (BK) has potent effects on many different aspects of physiology, including the regulation of blood pressure and of renal and cardiac function. Although the V_{max} for ANG I and BK are similar, ACE shows a 35-fold higher affinity for BK with a K_m of 0.85 μM, compared with a K_m of 30 μM for ANG I.[274] Consequently, one might predict that doses of ACE inhibitors which inhibit ANG I conversion may have little effect on BK degradation. In fact, this is not the case. Accumulating evidence indicates that BK mediates some of the antihypertensive and cardioprotective effects of ACE inhibitors. This evidence is due in part to the recent development of HOE 140, a highly potent, specific and long-acting antagonist of the type 2 BK receptor.[275] Moreover, some of the effects of neutral endopeptidase (NEP) 24.11 inhibitors may also be mediated by effects on BK metabolism.[276] The inference from these studies is that normal endogenous levels of BK make important contributions to vascular and cardiac physiology, and that inhibitors of ACE and NEP 24.11 produce increases in these endogenous levels of BK in plasma and tissue.

Components of the kallikrein-kinin system (KKS)

Kallikreins, kininogens and kinin peptides

Kinins are a group of potent vasodilator oligopeptides that contain BK within their structure. They act mainly as local tissue hormones, since they circulate at very low concentrations and are rapidly hydrolysed.[276,277] Kinins are released from high molecular weight precursors called kininogens by a group of serine proteases named kininogenases, which includes the kallikrein family of proteases. In man there are two forms of kininogen: high molecular weight (HMW) and low molecular weight (LMW) kininogen, which are derived by differential splicing of the primary transcript from a single gene. Proteolytic cleavage of two susceptible peptide bonds in the interior part of a kininogen molecule releases the effector peptide. Depending on the cleavage specificity of the protease, a number of different kinins can be generated (Fig. 1.5). A further complexity is that a proportion of HMW kininogen may be hydroxylated on the proline^3 (Hyp^3) of the BK sequence; thus, both hydroxylated and non-hydroxylated BK peptides may be derived from HMW kininogen.

Although many enzymes have kininogenase activity, such as trypsin, uropepsin, and plasmin, the two main kininogenases are plasma kallikrein and tissue (glandular) kallikrein. Plasma kallikrein, also known as Fletcher factor, releases kinins only from HMW kininogen, also known as Fitzgerald factor. Plasma kallikrein is found mainly in the inactive form (prekallikrein) and, together with HMW kininogen and Hageman factor, is involved in the intrinsic blood clotting mechanism through the activation of factor XI. This system is also involved in the activation of plasminogen to plasmin and in the organism's response to injury and inflammation. It is not known whether plasma kallikrein, through the release of BK, is involved in the regulation of blood pressure and blood flow. Patients with a deficiency of either plasma kallikrein (Fletcher trait) or HMW kininogen (Fitzgerald trait)

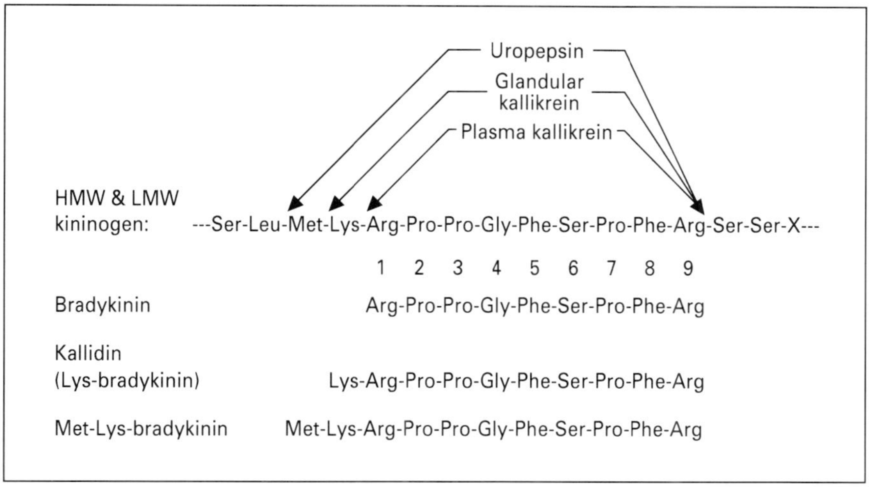

Figure 1.5
Schematic outline of kinin peptide formation from high molecular weight (HMW) and low molecular weight (LMW) kininogen. HMW kininogen may have a hydroxyl group on the Pro^3 of these peptides (Hyp^3).

have normal amounts of kinins in blood and urine, suggesting that neither plasma kallikrein nor HMW kininogen are essential components of the KKS,[278] but rather that tissue kallikrein and LMW kininogen play the dominant role in this system.

A further element of complexity is the presence of a kallikrein binding protein in plasma and tissues.[279] This binding protein is specific for tissue kallikrein and inhibits kallikrein activity. Thus, the binding protein may be an important regulatory element in tissue kinin production.

Kinin receptors

BK binds to G-protein-coupled receptors that activate phospholipase C or phospholipase A_2 and increase synthesis of inositol trisphosphate (IP_3) or arachidonic acid.[280] BK also stimulates an increase in intracellular calcium, probably secondary to IP_3 formation. There are two recognized BK receptors, type 1 (B_1) and type 2 (B_2),[281] although multiple B_2 receptors have been described. The rat uterine B_2 receptor has recently been cloned and sequenced, and shown to be expressed in multiple tissues, including the kidney and heart.[282] B_1 receptors are not present in normal tissues but are synthesized de novo during tissue injury, stress, or certain pathological conditions.[276] For the B_1 receptor, BK-(1–8) is a more potent agonist than BK-(1–9); by contrast, BK-(1–8) is inactive on the B_2 receptor.

Metabolism of kinin peptides

Kinins are rapidly inactivated by peptidases called kininases that are found in blood and tissues (Fig. 1.6). The relative importance of these different enzymes to kinin metabolism in different tissues is unknown. ACE inhibitors increase circulating kinin levels and urinary kinin excretion.[276,283] Our own studies indicate that ACE has an important role in BK metabolism in rat kidney, in that ACE inhibition caused an increase in renal BK levels.[277]

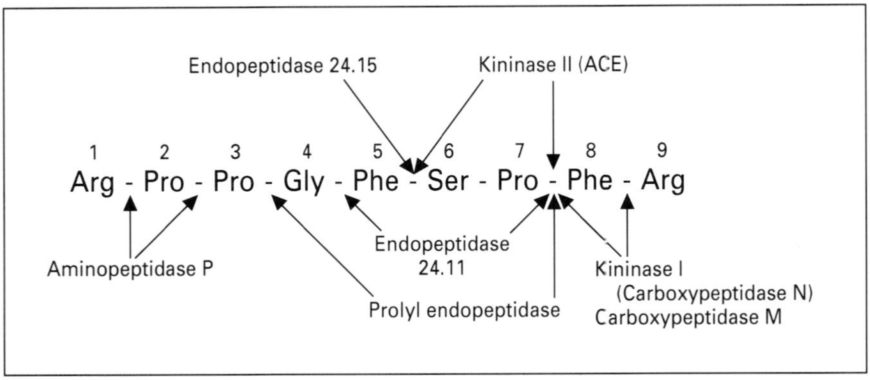

Figure 1.6
Sites of cleavage of bradykinin by enzymes which may play a role in bradykinin metabolism. Derived from Campbell et al.[277]

Operation of the kallikrein-kinin system

The actions of bradykinin peptides are summarized in Table 1.3.

Kinins and the cardiovascular system

Kinins have important actions on both blood vessels and the heart. By far the most important haemodynamic effect of BK in vivo is the hypotensive vasodilatation produced by stimulation of endothelial B_2 receptors of arteries and arterioles, with subsequent endothelial release of nitric oxide and prostaglandins, including prostacyclin (PGI_2). Evidence for a role for endogenous kinins in blood pressure regulation is the hypertensive effect of BK antagonists in normotensive[284,285] and hypertensive rats.[286,287] In addition, intravenous administration of BK antagonists or kinin antibodies reverses the hypotensive effect of ACE inhibition in hypertensive rats.[286,288–290]

Rat aortic smooth muscle cells in culture express kallikrein, kininogen, and kininase activity,[291] and kallikrein mRNA has been identified in rat vasculature and in cultured vascular smooth muscle cells,[292] thus providing evidence for the local vascular formation of BK. ACE inhibitors prevent neointima formation in response to endothelial injury; this effect is largely prevented by HOE 140, indicating a role for kinins in mediating this antiproliferative effect of ACE inhibition.[293]

The renal actions of kinins are important, given the dominant role of the kidney in blood pressure regulation. Kinins produce a diuresis and natriuresis and renal vasodilatation. Aprotinin (a non-specific serine protease inhibitor which inhibits renal kallikrein) and BK antagonists decrease renal blood flow and glomerular filtration rate. Moreover, BK antagonists partially reverse the effects of ACE inhibitors on renal papillary blood flow, urine flow and sodium excretion.[294,295] These findings suggest that endogenous kinins play a role in the regulation of renal function and contribute to the renal effects of ACE inhibitors. In addition, kinins stimulate the renal release of prostaglandins, including PGE_2 and PGI_2, which contribute to the vasodilator and natriuretic effects of kinins. Moreover, PGI_2 stimulates renin release. Mineralocorticoids increase urinary kallikrein excretion and renal kallikrein levels, and sodium restriction has similar effects, probably mediated by increased aldosterone levels. The renal KKS acts to ameliorate mineralocorticoid hypertension.[296]

Cellular actions	release of second generation mediators:	
	phospholipase C activation: phosphoinositide hydrolysis and calcium mobilization	
	phospholipase A2 activation: arachidonic acid and prostaglandin release	
	platelet activating factor	
	leukotrienes	
	substance P	
	acetylcholine	
	noradrenaline (sympathetic nerves)	
Blood vessels	endothelium:	nitric oxide release
		prostaglandin release
	smooth muscle:	variable endothelial dependence of vasodilatation
		prostaglandin release
		some vessels may show vasoconstriction
Heart	coronary vasodilatation and increased blood flow	
	improved cardiac function and metabolism	
	decreased reperfusion injury	
	decreased arrhythmias due to ischaemia	
Kidney	vasodilatation and increased blood flow	
	renin release	
	diuresis and natriuresis	
Pituitary	vasopressin release	
Adrenal	catecholamine release from adrenal medulla	
Non-vascular smooth muscle		
	uterus:	constriction
	duodenum:	relaxation
	bronchial smooth muscle:	biphasic effect with bronchodilatation at low doses and bronchoconstriction at high doses; asthmatics show bronchoconstriction
Central nervous system		
	behavioural response	
	sympathetic activation	
Metabolic	increased glucose utilization	
Inflammation	pain (stimulation of C-fibres, substance P release)	
	axon-mediated reflex flare	
	vasodilatation, increased capillary permeability, oedema, cytokine release	

Data derived from Regoli et al.,[281] Cherry et al.,[306] and Boola KD et al.[307]

Table 1.3
Actions of bradykinin peptides

The kallikrein-kinin system and hypertension

Several lines of evidence implicate the KKS in the pathogenesis of hypertension.[276] Patients with essential hypertension have reduced urinary kallikrein excretion, and some of these patients do not show a normal increase in kallikrein excretion in response to low sodium intake, despite an apparently normal aldosterone

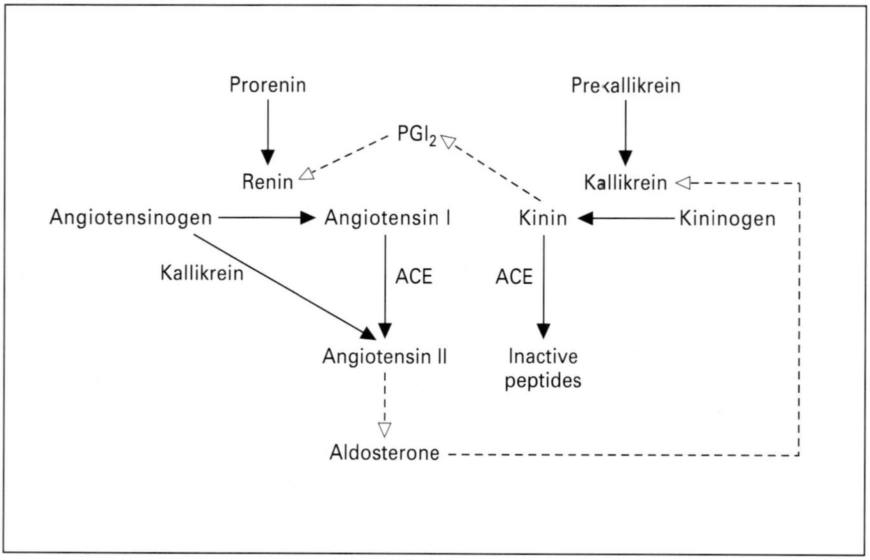

Figure 1.7
Schematic outline of the renin-angiotensin and kallikrein-kinin systems, and some of the interactions between these two systems. Shown in dotted lines are the stimulation of aldosterone secretion by angiotensin II, the stimulation of renal kallikrein production by aldosterone, and the stimulation of renin secretion by kinin, as mediated by prostacyclin (PGI_2). Other interactions are described in the text.

response.[297] Furthermore, in young normotensive individuals, urinary excretion of kallikrein has been found to be lower in those with hypertensive parents than in those with normotensive parents.[298] This finding is of interest, given that normotensive offspring of hypertensive parents have increased renal vascular resistance despite reduced plasma renin activity,[299,300] and is consistent with the suggestion that a genetically determined decrease in renal kallikrein activity might contribute to the pathogenesis of essential hypertension.

The kallikrein-kinin system and the heart

The main effect of BK on the normal heart is coronary vasodilatation. However, the situation is different in the ischaemic myocardium. BK has important protective effects on the reperfused ischaemic rat myocardium, with increased coronary blood flow, reduced infarct size, reduced reperfusion arrhythmias, reduced lactate dehydrogenase, creatine kinase, and lactate levels in the coronary effluent, and increased left ventricular pressure, and myocardial tissue levels of glycogen, ATP and creatine phosphate. These effects of BK are blocked by BK antagonists and partially blocked by indomethacin,[301] and may be mediated by BK-stimulated PGI_2 and nitric oxide release. Moreover, coronary occlusion results in increased BK levels in coronary sinus blood.[302]

ACE inhibitors reproduce the effects of BK in the ischaemic myocardium, and these effects of ACE inhibitors are blocked by BK antagonists or indomethacin,[303] thus providing strong evidence that ACE inhibitors produce increased BK levels in the myocardium. Furthermore, these BK-mediated effects of ACE inhibitors are over and above any effect of these drugs on ANG II levels.

Another important action of ACE inhibitors is the prevention or the reversal of cardiac hypertrophy, an effect which is blocked by HOE 140, again suggesting a role for endogenous BK in mediating this antihypertrophic effect of ACE inhibitors.[304]

The kallikrein-kinin system and the renin-angiotensin system

The KKS interacts directly with the RAS (Fig. 1.7). Plasma kallikrein is implicated in the activation of prorenin, in that patients with Fletcher factor deficiency have low levels of active renin and markedly elevated levels of prorenin.[305] Moreover, kallikrein and other serine proteases are able to cleave angiotensinogen to release ANG II directly. ACE is responsible for both the conversion of ANG I to ANG II, and the degradation of BK. Other sites of interaction between the two systems include the stimulation of renin secretion by PGI_2, the release of which is stimulated by BK, and the stimulation of urinary kallikrein excretion and renal kallikrein levels by ANG II and mineralocorticoids. These interactions play an important physiological role in that the tendency for stimulation of the RAS to reduce renal blood flow is counterbalanced by the effects of ANG II and aldosterone to stimulate renal kallikrein, BK, and prostaglandin levels and thus produce local vasodilatation and maintain renal blood flow, even in the face of high local ANG II levels in the kidney.

References

1 Tigerstedt R, Bergmann PG. Niere und Kreislauf. *Scand Arch Physiol* 1898; **8**:223–71.
2 Goldblatt H, Lynch J, Hanzal RF, Summervill WW. The production of persistent elevation of systolic blood pressure by means of renal ischemia. *J Exp Med* 1934; **59**:347–79.
3 Blalock A, Levy SE. Studies of the ethiology of hypertension. *Ann Surg* 1937; **106**:826–47.
4 Tobian L. Interrelationship of electrolytes, juxtaglomerular cells, and hypertension. *Physiol Rev* 1960; **40**:280–312.
5 Page IH, Helmer OM. Angiotensin activator, renin, and angiotonin inhibitor and the mechanism of angiotonin tachyphylaxis in normal, hypertensive and nephrectomized animals. *J Exp Med* 1940; **71**:485–519.
6 Braun-Menendez E, Fasciolo JC, Leloir JF, Munoz JM. The substance causing renal hypertension. *J. Physiol* 1940; **98**:283–98.
7 Page IH, McSwain B, Knapp GM, Andrus WD. Origin of renin activator. *Am J Physiol* 1941; **135**:214–35.
8 Skeggs LT, Kahn JR, Shumway NP. The preparation and function of the hypertensive converting enzyme. *J Exp Med* 1956; **103**:295–99.
9 Gross F. Renin und Hypertensin, physiologische und pathophysiologische Wirkstoffe. *Klin Wochenschr* 1958; **36**:693–705.
10 Unger Th, Badoer E, Ganten D, Lang RE, Rettig R. Brain angiotensin: pathways and pharmacology. *Circulation* 1988; **77 [suppl. I]**:I-40–I-54.
11 Campbell DJ. Circulating and tissue angiotensin systems. *J Clin Invest* 1987; **79**:1–6.
12 Dzau VJ. Circulating versus local renin-angiotensin system in cardiovascular homeostasis. *Circulation* 1988; **77 [suppl. I]**: I-4–I-13.
13 Blundell T, Sibanda BL, Pearl L. Three-dimensional structure, specificity and catalytic mechanism of renin. *Nature* 1983; **304**:273–5.
14 Dhanaraj V, Dealwis CG, Frazao C, et al. X-ray analyses of peptide-inhibitor complexes define the structural basis of specificity for human and mouse renins. *Nature* 1992; **357**:466–72.
15 Tang J, James MNG, Hsu IN, Jenkins A, Blundell TL. Structural evidence for gene duplication in the evolution of acid proteases. *Nature* 1978; **271**:618–22.
16 Morris BJ. Human renin protein and gene structures: present and future targets for a renin blockade in treatment of hypertension. *J Hypertens* 1989; **7(2)**:S9–S14.
17 Skott O, Jensen BL. Cellular and intrarenal control of renin secretion. *Clin Sci* 1993; **84**:1–10.
18 Soubrier F, Panthier JJ, Corvol P, Rougeon F. Molecular cloning and nucleotide sequence of a human renin cDNA. *Nucleic Acid Res* 1983; **11**:7181–90.
19 Hardman JA, Hort YJ, Catanzaro DF, et al. Primary structure of the human renin gene. *DNA* 1984; **3**:457–67.
20 Miyazaki H, Fukamizu A, Hirose S, et al. Structure of the human renin gene. *Proc Natl Acad Sci USA* 1984; **81**:5999–6003.
21 Burnham CE, Hawelu-Johnson CL, Frank BM, Lynch KR. Molecular cloning of rat renin cDNA and its gene. *Proc Natl Acad Sci USA* 1987; **84**:5605–9.
22 Fukamizu A, Nishi K, Cho T, et al. Structure of the rat renin gene. *J Mol Biol* 1988; **201**:443–50.
23 Rougeon F, Chambraud B, Foote S, Panthier JJ, Nageotte R. Molecular cloning of a mouse submaxillary gland renin cDNA fragment. *Proc Natl Acad Sci USA* 1981; **78**:6367–71.
24 Mullins JJ, Burt DW, Windass JD, McTurk P, George H, Brammar WJ. Molecular cloning of two distinct renin genes from the DBA/2 mouse. *EMBO J* 1982; **1**:1461–66.
25 Burt DW, Beecroft LJ, Mullins JJ, et al. Mouse renin gene structure, evolution and function. In: Kosta V, ed. Aspartic proteinases and their inhibitors. Berlin: Walter de Gruyter & Co. 1985: 355–77.

26 Holm I, Ollo R, Panthier JJ, Rougeon F. Evolution of aspartyl proteases by gene duplication: the mouse renin gene is organized in two homologous clusters of four exons. *EMBO J* 1984; **3**:557–62.

27 Panthier JJ, Holm I, Rougeon F. The mouse Rn locus: S allele of the renin regulator gene results from a single structural gene duplication. *EMBO J* 1982; **1**:1417–21.

28 Piccini N, Knopf JL, Gross KW. A DNA polymorphism, consistent with gene duplication, correlates with high renin levels in the mouse submaxillary gland. *Cell* 1982; **30**:205–13.

29 Hsueh WA, Carlson EJ, Israel-Hagman M. Mechanism of acid-activation of renin: role of kallikrein in renin activation. *Hypertension* 1981; **3 [suppl. I]**:I22–I29.

30 Leckie BJ, McGhee NK. Reversible activation-inactivation of renin in human plasma. *Nature* 1980; **288**:702–4.

31 Sealey JE, Atlas SA, Laragh JH. Prorenin and other large molecular weight forms of renin. *Endocr Rev* 1980; **1**:365–92.

32 Sealey JE, Rubattu S. Prorenin and renin as separate mediators of tissue and circulating systems. *Am J Hypertens* 1989; **2**:358–66.

33 Hackenthal E, Paul M, Ganten D, Taugner R. Morphology, physiology and molecular biology of renin secretion. *Physiol Rev* 1990;**70**:1067–1116.

34 Gomez RA, Lynch KR, Chevalier RL, et al. Renin and angiotensinogen gene expression and intrarenal renin distribution during ACE inhibition. *Am J Physiol* 1988; **254**:F900–F906.

35 Ménard J, Bouhnik J, Clauser E, Richoux JP, Corvol P. Biochemistry and regulation of angiotensinogen. *Clin Exp Hypertens* 1983; **A5**[7&8]:1005–19.

36 Ohkubo H, Kageyama R, Ujihara M, Hirose T, Inayama S. Cloning and sequence analysis of cDNA for rat angiotensinogen. *Proc Natl Acad Sci USA* 1983; **80**:2196–200.

37 Clouston WM, Evans BA, Haralambidis J, Richards RI. Molecular cloning of the mouse angiotensinogen gene. *Genomics* 1988; **2**:240–48.

38 Gaillard I, Clauser E, Corvol P. Structure of human angiotensinogen gene. *DNA* 1989; **8**:87–99.

39 Kageyama R, Ohkubo H, Nakanishi S. Primary structure of human preangiotensinogen deduced from the cloned cDNA sequence. *Biochemistry* 1984; **23**:3603–9.

40 Campbell DJ, Bouhnik J, Ménard J, Corvol P. Identity of angiotensinogen precursors in rat brain and liver. *Nature* 1984; **308**:206–8.

41 Jeunemaitre X, Soubrier F, Kotelevtsev YV, et al. Molecular basis of human hypertension: role of angiotensinogen. *Cell* 1992; **71**:169–80.

42 Campbell DJ, Habener JF. Cellular localization of angiotensinogen gene expression in brown adipose tissue and mesentery: quantification of messenger ribonucleic acid abundance using hybridization in situ. *Endocrinology* 1987; **121**:1616–26.

43 Campbell DJ, Habener JF. Angiotensinogen gene is expressed and differentially regulated in multiple tissues of the rat. *J Clin Invest* 1986; **78**:31–39.

44 Gomez RA, Lynch KR, Chevalier RL, et al. Renin and angiotensinogen gene expression in maturing rat kidney. *Am J Physiol* 1988; **254**:F582–F587.

45 Cassis LA, Saye JA, Peach MJ. Location and regulation of rat angiotensinogen messenger RNA. *Hypertension* 1988; **11**:591–96.

46 Lynch KR, Peach MJ. Molecular biology of angiotensinogen. *Hypertension* 1991; **17**: 263–69.

47 Klett C, Hellmann W, Suzuki F, et al. Induction of angiotensinogen mRNA in hepatocytes by angiotensin II and glucocorticoids. *Clin Exp Hypertens* 1988; **A10**:1009–22.

48 Ayers CR. Plasma renin activity and renin substrate concentration in patients with liver disease. *Circ Res* 1967; **20**:594–98.

49 Carretero O, Gross F. Renin substrate in plasma under various experimental conditions in the rat. *Am J Physiol* 1967; **213**:695–700.

50 Rasmussen S, Nielson MD, Giese J. Captopril combined with thiazide lowers renin substrate concentration: implications for methodology in renin assays. *Clin Sci* 1981; **60**:591–93.

51 Unger Th, Yukimura T, Marin-Grez M, Lang RE, Rascher W, Ganten D. SA446, a new orally active converting enzyme inhibitor: antihypertensive action and comparison with captopril in stroke-prone spontaneously hypertensive rats. *Eur J Pharmacol* 1982; **78**:411–20.

52 Unger Th, Ganten D, Lang RE, Schölkens BA. Is tissue converting enzyme inhibition a determinant of the antihypertensive efficacy of converting enzyme inhibitors? Studies with the two different compounds, HOE498 and MK421, in spontaneously hypertensive rats. *J Cardiovasc Pharmacol* 1984; **6**:872–80.
53 Unger Th, Moursi M, Ganten D, Hermann K, Lang RE. Antihypertensive action of the converting enzyme inhibitor perindopril (S9490-3) in spontaneously hypertensive rats: comparison with enalapril (MK421) and ramipril (HOE498). *J Cardiovasc Pharmacol* 1986; **8**:276–85.
54 Pratt RE, Zou WM, Naftilan AJ, Ingelfinger JR, Dzau VJ. Altered sodium regulation of renal angiotensinogen mRNA in the spontaneously hypertensive rat. *Am J Physiol* 1989; **256**:469–74.
55 Bernstein KE, Martin BM, Bernstein EA, Linton J, Striker L, Striker G. The isolation of angiotensin-converting enzyme cDNA. *J Biol Chem* 1988; **263**:11021–24.
56 Alhenc-Gelas F, Soubrier F, Hubert C, Allegrini J, Corvol P. The peculiar characteristics of the amino acid sequence of angiotensin I-converting enzyme, as determined by cDNA cloning of the human endothelial enzyme. *J Cardiovasc Pharmacol* 1989; **14 [suppl. 4]**:S6–S9.
57 Roy SN, Kusari J, Soffer RL, Lai CY, Sen GC. Isolation of cDNA clones of rabbit angiotensin converting enzyme: identification of two distinct mRNAs for the pulmonary and the testicular isozymes. *Biochem Biophys Res Commun* 1988; **155**:678–84.
58 Soubrier F, Alhenc-Gelas F, Hubert C, et al. Two putative active centers in human angiotensin I-converting enzyme revealed by molecular cloning. *Proc Natl Acad Sci USA* 1988; **85**:9386–90.
59 Hilbert P, Lindpaintner K, Beckmann JS, et al. Chromosomal mapping of two genetic loci associated with blood-pressure regulation in hereditary hypertensive rats. *Nature* 1991; **353**:521–29.
60 Cambien F, Poirier O, Lecerf L, et al. Deletion polmorphism in the gene for angiotensin-converting enzyme is a potent risk factor for myocardial infarction. *Nature* 1992; **359**:641–44.
61 Caldwell PRB, Seegal BC, Hsu KC, Das M, Soffer RL. Angiotensin-converting enzyme: vascular endothelial localization. *Science* 1976; **191**:1050–51.
62 Ryan US, Ryan JW, Whitaker C, Chiu A. Localization of angiotensin converting enzyme (kininase II). II. Immunocytochemistry and immunofluorescence. *Tissue Cell* 1976; **8(1)**:125–46.
63 Wilson SK, Lynch DR, Snyder SH. Angiotensin-converting enzyme labeled with [3H]captopril. Tissue localizations and changes in different models of hypertension in the rat. *J Clin Invest* 1987; **80**:841–51.
64 Gohlke P, Bünning P, Unger Th. Distribution and metabolism of angiotensin I and II in the blood vessel wall. *Hypertension* 1992; **20**:151–57.
65 Rogerson FM, Chai SY, Schlawe I, Murray WK, Marley PD, Mendelsohn FAO. Presence of angiotensin converting enzyme in the adventitia of large blood vessels. *J Hypertens* 1992; **10**:615–20.
66 Skidgel RA, Schulz WW, Tam LT, Erdös EG. Human renal angiotensin I converting enzyme and neutral endopeptidase. *Kidney Int* 1987; **31 [suppl. 20]**:S45–S48.
67 Yamada H, Fabris B, Allen AM, Jackson B, Johnston CI, Mendelsohn FAO. Localization of angiotensin converting enzyme in rat heart. *Circ Res* 1991; **68**:141–49.
68 Rix E, Ganten D, Schüll B, Unger Th, Taugner R. Converting-enzyme in the choroid plexus, brain, and kidney: immunocytochemical and biochemical studies in rats. *Neurosci Lett* 1981; **22**:125–30.
69 Chai SY, McKinley MJ, Mendelsohn FAO. Distribution of angiotensin converting enzyme in sheep hypothalamus and medulla oblongata visualized by in vitro autoradiography. *Clin Exper Theory and Practice* 1987; **A9(2&3)**:449–60.
70 Correa FMA, Plunkett LM, Saavedra JM, Hichens M. Quantitative autoradiographic determination of angiotensin-converting enzyme (kininase II) binding in individual rat brain nuclei with 125I–351A, a specific enzyme inhibitor. *Brain Res* 1985; **347**:192–95.

71 Paul M, Printz M, Harms E, Unger Th, Lang RE, Ganten D. Localization of renin (EC 3.4.23) and converting enzyme (EC 3.4.15.1) in nerve endings of rat brain. *Brain Res* 1985; **334**:315–24.

72 Jackson B, Cubela RB, Sakaguchi K, Johnston CI. Characterization of angiotensin converting enzyme (ACE) in the testis and assessment of the in vivo effects of the ACE inhibitor perindopril. *Endocrinology* 1988; **123**:50–55.

73 Erdös EG. Angiotensin I converting enzyme and the changes in our concepts through the year. *Hypertension* 1990; **16**:363–70.

74 Ehlers MRW, Riordan JF. Angiotensin-converting enzyme: new concepts concerning its biological role. *Biochemistry* 1989; **28**:5311–18.

75 Hooper NM. Angiotensin converting enzyme: implications from molecular biology for its physiological functions. *Int J Biochem* 1991; **23**: 641–47.

76 Bernstein KE, Martin BM, Edwards AS, Bernstein EA. Mouse angiotensin-converting enzyme is a protein composed of two homologous domains. *J Biol Chem* 1989; **264**: 11945–51.

77 Wei L, Alhenc-Gelas F, Corvol P, Clauser E. The two homologous domains of human angiotensin I-converting enzyme are both catalytically active. *J Biol Chem* 1992; **266**:9002–8.

78 Johnston CI, Burrell LM, Perich R, Jandeleit K, Jackson B. The tissue renin-angiotensin system and its functional role. *Clin Exp Pharmacol Physiol* 1992; **19** [**suppl. 19**]:1–5.

79 Erdös EG. Angiotensin I converting enzyme. *Circ Res* 1975; **36**:247–53.

80 Unger Th, Gohlke P, Gruber MG. Converting enzyme inhibitors. In: Ganten D, Mulrow PJ, eds. Handbook of experimental pharmacology, vol. 93. Berlin, Heidelberg: Springer-Verlag. 1990: 377–481.

81 Langford KG, Shai SY, Howard TE, Kovac MJ, Overbeek PA, Bernstein KE. Transgenic mice demonstrate a testis-specific promotor for angiotensin converting enzyme. *J Biol Chem* 1991; **266**:15559–62.

82 Stanislas-Leguern G, Mordelet-Dambrine M, Dusser D, Huesca M, Chretien J, Huchon GJ. In vitro synthesis of angiotensin-converting enzyme by alveolar macrophages is increased in disseminated sarcoidosis. *Lung* 1986; **164**:269–77.

83 Mendelsohn FAO, Lloyd CJ, Kachel C, Funder JW. Induction by glucocorticoids of angiotensin converting enzyme production from bovine endothelial cells in culture and rat lung in vivo. *J Clin Invest* 1982; **70**:684–92.

84 Fyhrquist FY, Forslund T, Tikkanen I, Grönhagen-Riska C. Induction of angiotensin I-converting enzyme in rat lung with captopril (SQ 14225). *Eur J Pharmacol* 1980; **67**:473–75.

85 Unger Th, Schüll B, Hübner D, et al. Plasma-converting enzyme activity does not reflect effectiveness of oral treatment with captopril. *Eur J Pharmacol* 1981; **72**:255–59.

86 Dzau VJ. Vascular renin-angiotensin: a possible autocrine or paracrine system in control of vascular function. *J Cardiovasc Pharmacol* 1984; **6**:S377–S382.

87 Urata H, Konoshita A, Misono KS, Bumpus FM, Husain A. Identification of a highly specific chymase as the major angiotensin II-forming enzyme in the human heart. *J Biol Chem* 1990; **265**(**36**):22348–57.

88 Kinoshita A, Urata H, Bumpus FM, Husain A. Multiple determinates for the high substrate specificity of an angiotensin II-forming chymase from the human heart. *J Biol Chem* 1991; **266**(**29**):19192–7.

89 Weintroub B, Klickstein LB, Dzau VJ, Watt KWK. Granulocyte-angiotensin system. Identification of angiotensinogen as substrate of leukocyte cathepsin G. *Biochemistry* 1984; **23**:227–32.

90 Tang SS, Loscalzo J, Dzau VJ. Tissue plasminogen activator activates renin angiotensin in vitro. *J Vasc Med Biol* 1989; **1**:67–74.

91 Krivoy N, Schlüter H, Karas M, Zidek W. Generation of angiotensin II from human plasma by tissue kallikrein. *Clin Sci* 1992; **83**:477–82.

92 Okunishi H, Miyazaki M, Okamura T, Toda N. Different distribution of two types of angiotensin II-generating enzymes in the aortic wall. *Biochem Biophys Res Commun* 1987; **149**:1186–92.

93 Welches WR, Santos RAS, Chappell MC, Brosnihan KB, Greene LJ, Ferrario CM. Evidence that prolyl endopeptidase participates in the processing of brain angiotensin. *J Hypertens* 1991; **9**:631–38.
94 Kohara K, Brosnihan KB, Chappell MC, Khosla MC, Ferrario CM. Angiotensin (1–7): a member of circulating angiotensin peptides. *Hypertension* 1991; **17**:131–38.
95 Johnson H, Drummer OH. Hydrolysis of angiotensin I by peptidase in homogenates of rat lung and aorta. *Biochem Pharmacol* 1988; **37**:1131–36.
96 Ryan JW. Peptidase enzymes of the pulmonary vascular surface. *Am J Physiol* 1989; **257**:L53–L60.
97 Abhold RH, Sullivan MJ, Wright JW, Harding JW. Binding, degradation and pressor activity of angiotensin II and III after amino peptidase inhibition with amastatin and bestatin. *J Pharmacol Exp Ther* 1987; **242**:957–62.
98 Jacobsen J, Poulsen K. In vivo generation and elimination of angiotensin in the rat. *Clin Exp Pharmacol Physiol* 1990; **17**:445–51.
99 Admiraal PJJ, Danser AHJ, Sjoukje-Jong M, Pieterman H, Derkx FHM, Schalekamp MADH. Regional angiotensin II production in essential hypertension and renal artery stenosis. *Hypertension* 1993; **21**:173–84.
100 Bumpus FM, Catt KJ, Chiu AT, et al. Nomenclature for angiotensin receptors. A report of the nomenclature committee of the Council for High Blood Pressure Research. *Hypertension* 1991; **17**:720–22.
101 Brechler V, Jones PW, Levens NR, De Gasparo M, Bottari SP. Agonistic and antagonistic properties of angiotensin analogs at the AT2 receptor in PC12W cells. *Regul Pept* 1993; **44**:207–13.
102 Furuta H, Guo DF, Inagami T. Molecular cloning and sequencing of the gene encoding human angiotensin II type 1 receptor. *Biochem Biophys Res Commun* 1992; **183**:8–13.
103 Murphy TJ, Alexander RW, Griendling KK, Runge MS, Bernstein KE. Isolation of a cDNA encoding the vascular type-1 angiotensin II receptor. *Nature* 1991; **351**:233–36.
104 Lassegue B, Alexander RW, Clark M and Griendling KK. Angiotensin II-induced phosphatidylcholine hydrolysis in cultured vascular smooth-muscle cells. *Biochem J* 1991; **267**:19–252.
105 Pfeilschifter J. Angiotensin II B-type receptor mediates phosphoinositide hydrolysis in mesangial cells. *Eur J Pharmacol* 1990; **184**:201–2.
106 Bottari SP, Taylor V, King IN, Bogdal Y, Whitebread S, De Gasparo M. Angiotensin II AT2 receptors do not interact with guanine nucleotide binding proteins. *Eur J Pharmacol* 1991; **207**:157–63.
107 Bottari SP, King IN, Reichlin S, Dahlstroem I, Lyden N, De Gasparo M. The angiotensin AT_2 receptor stimulates protein tyrosine phosphatase activity and mediates inhibition of particulate guanylate cyclase. *Biochem Biophys Res Commun* 1992; **183**:206–11.
108 Smith TL, Yamamura HI, Lee L. Effect of ethanol on receptor-stimulated phosphatidic acid and polyphosphoinositide metabolism in mouse brain. *Life Sci* 1986; **39**:1675–84.
109 Tamura CS, Speth RC. Effects of angiotensin II on phosphatidylinositide hydrolysis in rat brain. *Neurochem Int* 1990; **17**:475–79.
110 Sumners C, Myers LM, Kalberg CJ, Raizada MK. Physiological and pharmacological comparisons of angiotensin II receptors in neuronal and astrocyte glial cultures. *Prog Neurobiol* 1990; **34**:355–85.
111 Tsuda T, Kawahara Y, Shii K, Koide M, Ishida Y, Yokoyama M. Vasoconstrictor-induced protein-tyrosine phosphorylation in cultured vascular smooth muscle cells. *FEBS Lett* 1991; **285 (No. 1)**:44–8.
112 Marrero MB, Paxton WG, Duff Jl, Berk BC. Angiotensin II stimulates tyrosine phosphorylation of phospholipase C-γl in vascular smooth muscle cells. *J Biol Chem* 1994; **269** (**No. 14**):10935–9.
113 Chiu AT, Herblin WF, McCall DE, et al. Identification of angiotensin II receptor subtypes. *Biochem Biophys Res Commun* 1989; **165**:196–203.
114 Chiu AT, Duncia JV, McCall DE, et al. Nonpeptide angiotensin II receptor antagonists. III. Structure-function studies. *J Pharmacol Exp Ther* 1989; **250**:867–74.
115 Whitebread S, Mele M, Kamber B, De Gasparo M. Preliminary biochemical characterization of two angiotensin II receptor

subtypes. *Biochem Biophys Res Commun* 1989; **163**(1):284–91.
116 Obermüller N, Unger Th, Culman J, Gohlke P, De Gasparo M, Bottari SP. Distribution of angiotensin II receptor subtypes in rat brain nuclei. *Neurosci Lett* 1991; **132**:11–15.
117 Steckelings U, Obermüller N, Bottari SP, Quadri F, Veltmar A, Unger Th. Brain angiotensin: receptors, actions and possible role in hypertension. *Pharmacol Toxicol* 1992; **70 [suppl. II]**:23–27.
118 Zhuo J, Song K, Harris PJ, Mendelsohn FAO. In vitro autoradiography reveals predominantly AT1 angiotensin II receptors in rat kidney. *Renal Physiol Biochem* 1992; **15**:231–39.
119 Saveedra JM. Brain and pituitary angiotensin. *Endocr Rev* 1992; **13**:329–80.
120 Simon M, Flügge, Fuchs E, Gröne HJ. Evidence for two angiotensin II receptor subtypes in human fetal and adult kidney demonstrated by the A II receptor antagonists losartan and PD 123177 [abstract]. *FASEB J* 1991; **5**:A870.
121 Schinke M, Doods HN, Ganten D, Wienen W, Entzeroth M. Characterization of rat intestinal angiotensin II receptors. *Eur J. Pharmacol* 1991; **204**:165–70.
122 Herblin WF, Diamond SM, Timmermans PBMWM. Localization of angiotensin II receptor subtypes in the rabbit adrenal and kidney. *Peptides* 1991; **12**:581–84.
123 Herblin WF, Chiu AT, McCall DE, et al. Angiotensin II receptor heterogeneity. *Am J Hypertens* 1991; **4**:299S–302S.
124 Dudley DT, Hubbel SE, Summerfelt RM. Characterization of angiotensin II (AT2) binding sites in R3T3 cells. *Mol Pharmacol* 1991; **40**:360–67.
125 Chang RSL, Lotti VJ. Selective ligands reveal subtypes of angiotensin receptors in rat vasculature and brain. *Pharmacologist* 1989; **31**:150.
126 Chang RSL, Lotti VJ. Angiotensin receptor subtypes in rat, rabbit and monkey tissues: relative distribution and species dependency. *Life Sci* 1991; **49**:1485–90.
127 Chappell MC, Diz DI, Jacobson DW. Pharmacological characterization of angiotensin II binding sites in the canine pancreas. *Peptides* 1992; **13**:313–8.
128 Eglen RM, Rapp JM, Leung E. Characterization of angiotensin II receptors in guinea-pig gastrointestinal tract [abstract]. *Faseb J* 1991; **5**:A869.
129 Entzeroth M, Hadamovsky S. Angiotensin II receptors in the rat lung are of the AII-1 subtype. *Eur J. Pharmacol* 1991; **206**:237–41.
130 Pucell AG, Hodges JC, Sen I, Bumpus FM, Husain A. Biochemical properties of the ovarian granulosa cell type 2 angiotensin II receptor. *Endocrinology* 1991; **128**: 1947–59.
131 Gibson RE, Thorpe HH, Cartwright ME, et al. Angiotensin II receptor subtypes in renal cortex of rats and rhesus monkeys. *Am J Physiol* 1991; **261**:F512–F518.
132 Dudley DT, Panek RL, Major TC, et al. Subclasses of angiotensin II binding sites and their functional significance. *Mol Pharmacol* 1990; **38**:370–77.
133 De Gasparo M, Whitebread, S, Mele M, et al. Biochemical characterization of two angiotensin II receptor subtypes in the rat. *J. Cardiovasc Pharmacol* 1990; **16 [suppl. 4]**:S31–S35.
134 Bauer PH, Chiu AT, Garrison JC. DuP 753 can antagonize the effects of angiotensin II in rat liver. *Mol Pharmacol* 1991; **39**:579–585.
135 Chang RSL, Lotti VJ. Two distinct angiotensin II receptor binding sites in rat adrenal revealed by new selective nonpeptide ligands. *Mol Pharmacol* 1990; **29**:347–51.
136 Elton TS, Stephan CC, Taylor GR, et al. Isolation of two distinct type 1 angiotensin II receptor genes. *Biochem Biophys Res Commun* 1992; **184**:1067–73.
137 Ye MQ, Healy DP. Characterization of an angiotensin type-1 receptor partial cDNA from rat kidney: evidence for a novel AT1b receptor subtype. *Biochem Biophys Res Commun* 1992; **185**:204–10.
138 Yoshida H, Kakuchi J, Guo DF, et al. Analysis of the evolution of angiotensin II type 1 receptor gene in mammals (mouse, rat, bovine, human). *Biochem Biophys Res Commun* 1992; **186**:1042–49.
139 Ernsberger P, Zhou J, Damon TH, Douglas JG. Angiotensin II receptor subtypes in cultured rat renal mesangial cells. *Am J Physiol* 1992; **263**:F411–F416.
140 Tsutsumi K, Saveedra JM. Heterogeneity of angiotensin II AT_2 receptors in the rat brain. *Mol Pharmacol* 1992; **41**:290–97.

141 Tsutsumi K, Saavedra JM. Characterization and development of angiotensin II receptor subtypes (AT1 and AT2) in rat brain. *Am J Physiol* 1991; **261**:R209–R216.
142 Viswanathan M, Tsutsumi K, Correa FMA, Saavedra JM. Changes in expression of angiotensin receptor subtypes in the rat aorta during development. *Biochem Biophys Res Commun* 1991; **179**:1361–67.
143 Ji H, Sandberg K, Catt KJ. Novel angiotensin II antagonists distinguish amphibian from mammalian angiotensin II receptors expressed in *Xenopus laevis* oocytes. *Mol Pharmacol* 1991; **39**:120–23.
144 Chang RSL, Chen TD, Faust KA, Lotti VJ. Two angiotensin II binding sites in rat brain revealed using (125I)Sar7-Ile8-Angiotensin II and selective nonpeptide antagonists. *Biochem Biophys Res Commun* 1990; **171**:813–17.
145 Chaki S, Inagami T. Identification and characterization of a new binding site for angiotensin II in mouse neuroblastoma Neuro-2A cells. *Biochem Biophys Res Commun* 1992; **182**:388–94.
146 Swanson GN, Hannesworth JN, Sardinia MF, et al. Discovery of a distinct binding site for angiotensin (3–8) a putative angiotensin IV receptor. *Regul Pept* 1992; **40**:409–19.
147 Sandberg K, Hong J, Clark AJL, Shapiro H, Catt K. Cloning and expression of a novel angiotensin II receptor subtype. *J Biol Chem* 1992; **267**:9455–58.
148 Unger Th, Gohlke P, Paul M, Rettig R. Tissue renin-angiotensin systems: Fact or fiction? *J Cardiovasc Pharmacol* 1991; **19** [**suppl.** 2]:S20–S25.
149 Campbell-Boswell M, Robertson AL. Effects of angiotensin II and vasopressin on human smooth muscle cells in vitro. *Exp Mol Pathol* 1981; **35**:265.
150 Geisterfer AAT, Peach MJ, Owens GK. Angiotensin II induces hypertrophy, not hyperplasia, of cultured rat aortic smooth muscle cells. *Circ Res* 1988; **62**:749–56.
151 Griffin SA, Brown WCB, MacPherson F, et al. Angiotensin II causes vascular hypertrophy in part by a non-pressor mechanism. *Hypertension* 1991; **17**:626–35.
152 Naftilan AJ, Pratt RE, Dzau VJ. Induction of platelet-derived growth factor A-chain and c-myc gene expression by angiotensin II in cultured rat vascular smooth muscle cells. *J Clin Invest* 1989; **83**:1419–24.
153 Taubman MB, Berk BC, Izumo S, Tsuda T, Alexander RW, Nadal-Ginard B. Angiotensin II induces c-fos mRNA in aortic smooth muscle. *J Biol Chem* 1989; **264**:526–30.
154 Kawahara Y, Sunako M, Tsuda T, Fukuzaki H, Fukumoto Y, Takai Y. Angiotensin II induces expression of the C-fos gene through protein kinase C activation and calcium ion mobilization in cultured vascular smooth muscle cells. *Biochem Biophys Res Commun* 1988; **150**:52–59.
155 Bock KD, Gross F. Renin and angiotensin tachyphylaxis. *Circ Res* 1961; **9**:1044–55.
156 Ames RP, Borkowski AJP, Sicinski ALMM, Laragh JH. Prolonged infusions of angiotensin II and norepinephrine on blood pressure, electrolyte balance, aldosterone and cortisole secretion in normal man and in cirrhosis with ascites. *J Clin Invest* 1965; **44**:1171–86.
157 Benjamin N, Collier JG, Webb DJ. Angiotensin II augments sympathetically-induced venoconstriction in man. *Clin Sci* 1988; **5** [**suppl. 4**]:337–40.
158 Peach MJ, Cline WH, Watts DT. Release of adrenal catecholamines by angiotensin II. *Circ Res* 1966; **19**:571–75.
159 Oelkers W, Brown JJ, Fraser R. Sensitization of the adrenal cortex to angiotensin II in sodium-deplete man. *Circ Res* 1974; **34**:69.
160 Paguet JL, Baudouin-Legros M, Marchel P, Meyer P. Enhanced proliferation activity of cultured smooth muscle cells from spontaneously hypertensive rats. *Am J Hypertens* 1989; **13**:295–304.
161 Khairallah PA, Robertson AL, Davila D. Effects of angiotensin II on DNA, RNA and protein synthesis. In: Genest J, Koiw E, eds. Hypertension. New York: Springer Verlag, 1972: 212–20.
162 Berk BC, Vekshtein V, Gordon HM, Tsuda T. Angiotensin II-stimulated protein synthesis in cultured vascular smooth muscle cells. *Hypertension* 1989; **13**:305–14.
163 Re RN. The cellular biology of angiotensin: paracrine, autocrine and intracrine actions in cardiovascular tissues. *J Mol Cell Cardiol* 1989; **21** [**suppl. V**]:63–69.

164 Ganten D, Schelling P, Flugel RM, Ganten U. Effect of angiotensin and an angiotensin antagonist on iso-renin and cell-growth in 3T3 mouse cells. *Int Res Commun Med Sci* 1975; **3**:327.
165 Dzau VJ, Gibbons GH, Pratt RE. Molecular mechanisms of vascular renin-angiotensin system in myointimal hyperplasia. *Hypertension* 1991; **18 [suppl. II]**:II100–II105.
166 Takeuchi K, Nakamura N, Cook, NS, Pratt RE, Dzau VJ. Angiotensin II can regulate gene expression by the Ap-1 binding sequence via a protein kinase C-dependent pathway. *Biochem Biophys Res Commun* 1990; **172(3)**:1189–94.
167 Metsärinne KP, Stoll M, Gohlke P, Paul M, Unger Th. Angiotensin II is antimitogenic in vitro for heart capillary endothelial cells. *Pharm Pharmacol Lett* 1992; **2**:150–52.
168 Stoll M, Gohlke P, Metsärinne KP, Unger Th. Capillary growth induced by ACE inhibitors: in vivo and in vitro studies. In: *ACE inhibitors, endothelial function and atherosclerosis*. Schölkens BA, Unger Th (eds). (Chichester: Media Medica, 1993).
169 Zimmerman BG. Adrenergic facilitation by angiotensin: does it serve a physiological function. *Clin Sci* 1981; **60**:343–48.
170 Malik KU and Nasjletti A. Facilitation of adrenergic transmission by locally generated angiotensin II in rat mesenteric arteries. *Circ Res* 1976; **38**:26–30.
171 Unger Th, Rascher W, Schuster C, et al. Central blood pressure effects of substance P and angiotensin II: Role of the sympathetic nervous system and vasopressin. *Eur J Pharmacol* 1981; **71**:33–42.
172 Unger Th, Becker H, Petty M, et al. Differential effects of central angiotensin II and substance P on sympathetic nerve activity in conscious rats. *Circ Res* 1985; **56**:563–75.
173 Phillips MI, Speakman EA, Kimura B. Levels of angiotensin and molecular biology of the tissue renin angiotensin systems. *Regul Pept* 1993; **43**:1–20.
174 Bickerton RK, Buckley JP. Evidence for a central mechanism in angiotensin-induced hypertension. *Proc Soc Exp Biol Med* 1961; **106**:834.
175 Ganten D, Lang RE, Lehmann E, Unger Th. Brain angiotensin: on the way to becoming a well-studied neuropeptide system. *Biochem Pharmacol* 1984; **33**:3523–28.
176 Deschepper CF, Bouhnik J, Ganong WF. Colocalization of angiotensinogen and glial fibrillary acidic protein in astrocytes in rat brain. *Brain Res* 1986; **355**:195.
177 Imboden H, Harding JW, Hilgenfeldt U, Celio MR, Felix D. Localization of angiotensinogen in multiple cell types of rat brain. *Brain Res* 1987; **410**:74–77.
178 Oldfield BJ, Ganten D, McKinley MJ. An ultrastructural analysis of the distribution of angiotensin II in the rat brain. *J Neuroendocrin* 1989; **1**:121–28.
179 Guillot FL, Augus KL. Angiotensin peptide regulation of fluid-phase endocytosis in brain microvessel endothelial cell monolayers. *J Cereb Blood Flow Metab* 1990; **10**:827–34.
180 Schelling P, Ganten U, Sponer G, Unger Th, Ganten D. Components of the renin-angiotensin system in the cerebrospinal fluid of rats and dogs with special consideration of the origin and the fate of angiotensin II. *Neuroendocrinol* 1980; **31**:297–308.
181 Trolliet MR, Phillips MI. The effect of chronic bilateral nephrectomy on plasma and brain angiotensin. *J Hypertens* 1992; **10**:29–36.
182 Gehlert DR, Gackenheimer SL, Reel JK, Lin H-S, Steinberg MI. Non-peptide angiotensin II receptor antagonists discriminate subtypes of 125I-angiotensin II binding sites in the rat brain. *Eur J Pharmacol* 1990; **187**:123–26.
183 Rowe BP, Grove KL, Saylor DL, Speth RC. Discrimination of angiotensin II receptor subtype distribution in the rat brain using non-peptidic receptor antagonists. *Regul Pept* 1991; **33**:45–53.
184 Leung KH, Smith RD, Timmermans PBMWM, Chiu AT. Regional distribution of the two subtypes of angiotensin II receptor in rat brain using selective non-peptide antagonists. *Neurosci Lett* 1991; **123**:95–98.
185 Lang RE, Rascher W, Heil J, Unger Th, Wiedemann G, Ganten D. Angiotensin stimulates oxytocin release. *Life Sci* 1981; **29**:1425–28.
186 Unger Th, Horst P-J, Bauer M, Demmert G, Rettig R, Rohmeiss P. Natriuretic action of central angiotensin II in conscious rats. *Brain Res* 1989; **486**:33–38.

187 Hogarty DC, Speakman EA, Puig V, Phillips MI. The role of angiotensin, AT1 and AT2 receptors in the pressor, drinking and vasopressin responses to central angiotensin. *Brain Res* 1992; **586**:289–94.

188 Stadler T, Veltmar A, Quadri F, Unger Th. Angiotensin II evokes noradrenaline release from the paraventricular nucleus in conscious rats. *Brain Res* 1992; **569**:117–22.

189 Veltmar A, Culman J, Qadri F, Rascher W, Unger Th. Involvement of adrenergic and angiotensinergic receptors in the paraventricular nucleus in the angiotensin II-induced vasopressin release. *J Pharmacol Exp Ther* 1992; **263**:1253–60.

190 Qadri F, Culman J, Veltmar A, Maas K, Rascher W, Unger Th. Angiotensin II-induced vasopressin release is mediated through alpha-adrenoceptors and angiotensin II AT1 receptors in the supraoptic nucleus. *J Pharmacol Exp Ther* 1993; **267**(**2**);567–74.

191 Stauss H, Unger Th. Role of angiotensin II in thirst-induced drinking. *FASEB J* 1991; **5**(**4**): A658 (abstract).

192 Beyer C, Rohmeiss P, Qadri F, Strauch M, Unger Th. Characterization and localization of the central osmoreceptor [abstract]. *FASEB J* 1992; **7**:A406.

193 Edling O, Qadri F, Culman J, Gohlke P, Unger Th. Angiotensin II release in the hypothalamic nucleus paraventricularis (PVN) after local stimulation with hyperosmolaric saline; an in vivo microdialysis study [abstract]. *FASEB J* 1993; **7**:A36.

194 Blume A, Lebrun C, Herdegen T, Unger Th. Angiotensin II induces receptor-mediated expression of immediate-early genes in the rat central nervous system [abstract]. *FASEB J* 1993; **7**(**3**):1427 (abstract).

195 Dostal DE, Rothblum KN, Chernin MI, Cooper GR, Baker KM. Intracardiac detection of angiotensinogen and renin: a localized renin-angiotensin system in neonatal rat heart. *Am J Physiol* 1992; **263**:C838–C850.

196 Lindpaintner K, Jin M, Niedermeyer N, Wilhelm MJ, Ganten D. Cardiac angiotensinogen and its local activation in the isolated perfused beating heart. *Circ Res* 1990; **67**:564–73.

197 Dzau VJ. Cardiac renin-angiotensin system: molecular and functional aspects. *Am J Med* 1988; **84** [**suppl.** 3**A**]:22–27.

198 Dzau VJ, Ellison KE, Brody T, Ingelfinger JR, Pratt RE. A comparative study of the distribution of renin and angiotensinogen messenger ribonucleic acids in rat and mouse tissues. *Endocrinology* 1987; **120**:2334–38.

199 Suzuki F, Ludwig G, Hellmann W, et al. Renin gene expression in rat tissues: a new quantitative assay method for rat renin mRNA using synthetic cDNA. *Clin Exp Hypertens* 1988; **A10**:345–59.

200 Paul M, Wagner D, Metzger R, et al. Quantification of renin mRNA in various mouse tissues by a novel solution hybridization assay. *J Hypertens* 1988; **6**:247–52.

201 Ekker M, Tronik D, Rougeon F. Extra-renal transcription of the renin genes in multiple tissues of mice and rat. *Proc Natl Acad Sci USA* 1989; **86**:5155–58.

202 Iwai N, Inagami T. Quantitative analysis of renin gene expression in extrarenal tissues by polymerase chain reaction method. *J Hypertens* 1992; **10**:717–24.

203 Lindpaintner K, Wilhelm MJ, Manwen J, et al. Tissue renin-angiotensin systems: focus on the heart. *J Hypertens* 1987; **5** [**suppl. 2**]:S33–S38.

204 Mebazzaa A, Chevalier B, Mercadier JJ, Echter E, Rappaport L, Swynghedauw B. A review of the renin-angiotensin system in the normal heart. *J Cardiovasc Pharmacol* 1989; **14** [**suppl. 4**]:16–20.

205 Jin M, Wilhelm MJ, Lang RE, Unger Th, Lindpaintner K, Ganten D. Endogenous tissue renin-angiotensin systems: from molecular biology to therapy. *Am J Med* 1988; **84** [**suppl. 3A**]:28–36.

206 Phillips MI, Speakman EA, Kimura B. Levels of angiotensin and molecular biology of the tissue renin angiotensin systems. *Regulatory Peptides* 1993; **43**:1–20.

207 Urata H, Healy B, Stewart RW, Bumpus FM, Husain A. Angiotensin II receptors in normal and failing human hearts. *J Clin Endocrinol Metab* 1989; **69**:54–66.

208 Grinstead WC, Young JB. The myocardial renin-angiotensin system: existence, importance, and clinical implications. *Am Heart J* 1992; **123**:1039–45.

209 Sen I, Rajasekaran AK. Angiotensin II-binding protein in adult and neonatal rat heart. *J Mol Cell Cardiol* 1991; **23**:563–72.

210 Baker KM, Booz GW, Dostal DE. Cardiac actions of angiotensin II: role of an intracardiac renin-angiotensin system. *Ann Rev Physiol* 1992; **54**:227–41.

211 Stauss H, Busch O, Unger Th. Positiv inotroper Effect von zentralem Angiotensin II. *Nieren und Hochdruckkrankheiten* 1992; **11**:634–36.

212 Kobayashi M, Furkawa Y, Chiba S. Positive chronotropic and inotropic effects of angiotensin II in the dog heart. *Eur J Pharmacol* 1978; **50**:17.

213 Freer RJ, Pappano AJ, Peach MJ, et al. Mechanisms for the positive inotropic effect of angiotensin II on isolated cardiac muscle. *Circ Res* 1976; **39**:178.

214 Peach MJ. Molecular actions of angiotensins. *Biochem Pharmacol* 1981; **30**:2745–51.

215 Ingelfinger JR, Fon EA, Ellison KE, Dzau VJ. Localization of the intrarenal renin angiotensin system (RAS) by in situ hybridization of renin and angiotensinogen (ANG-N) mRNAs [abstract]. *Kidney Int* 1988; **33**:269A.

216 Ingelfinger JR, Zuo WM, Fon EA, Elison KE, Dzau VJ. In situ hybridization evidence for angiotensinogen messenger RNA in the rat proximal tubule. An hypothesis for the intrarenal angiotensin system. *J Clin Invest* 1990; **85**:417–23.

217 Bruneval P, Hinglais N, Alhenc-Gelas F, et al. Angiotensin I converting enzyme in human intestine and kidney. Ultrastructural immunohistochemical localization. *Histochem* 1986; **85**:73–80.

218 Ikemoto F, Song G-B, Tominaga M, Kanayama Y, Yamamoto K. Angiotensin-converting enzyme in the rat kidney. Activity, distribution, and response to angiotensin-converting enzyme inhibitors. *Nephron* 1990; **55 [suppl. 1]**:3–9.

219 Ikemoto F, Song GB, Tominaga M, Kanayama Y and Yamamoto K. Angiotensin converting enzyme predominates in the inner cortex and medulla of the rat kidney. *Biochem Biophys Res Commun* 1987; **144**:915–21.

220 Seikaly MG, Arant BS Jr, Seney FD, Jr. Endogenous angiotensin concentrations in specific intrarenal fluid compartments of the rat. *J Clin Invest* 1990; **86**:1352–57.

221 Ray PE, Aguilera G, Kopp JB, Horikoshi S, Klotman PE. Angiotensin II receptor-mediated proliferation of cultured human fetal mesangial cells. *Kidney Int* 1991; **40**:764–71.

222 Anderson PW, Do YS, Hsueh WA. Angiotensin II causes mesangial cell hypertrophy. *Hypertension* 1993; **21**:29–35.

223 Wolf G, Neilson EC. Angiotensin II induces cellular hypertrophy in cultured murine proximal tubular cells. *Am J Physiol* 1990; **259**:F768–F777.

224 Mitchell KD, Navar LG. Influence of intrarenally generated angiotensin II on renal hemodynamics and tubular reabsorption. *Renal Physiol Biochem* 1991; **14**:155–63.

225 Cogan MG. Angiotensin II: a powerful controller of sodium transport in the early proximal tubule. *Hypertension* 1990; **15**:451–58.

226 Morduchowicz GA, Sheikh-Hamad D, Dwyer BE, Stern N, Jo OD, Yanagawa N. Angiotensin II directly increases rabbit renal brush-border membrane sodium transport: presence of local signal transduction system. *J. Membrane Biol* 1991; **122**:43–53.

227 Mitchell KD, Navar LG. Interactive effects of angiotensin II on renal hemodynamics and tubular reabsorptive function. *Kidney Int* 1990; **38**:S69–S73.

228 Bellucci A, Wilkes BM. Mechanism of sodium modulation of glomerular angiotensin receptors in the rat. *J Clin Invest* 1984; **74**:1593–1600.

229 Wilkes BM, Pion I, Sollott S, Michaels S, Kiesel G. Intrarenal renin-angiotensin system modulates glomerular angiotensin receptors in the rat. *Am J Physiol* 1988; **254**: F345–F350.

230 Wilkes BM. Increased angiotensin II receptor binding with age in isolated rat glomeruli. *Kidney Int* 1988; **34**:241–47.

231 Ray PE, Castrén E, Ruley EJ, Saavedra JM. Different effects of sodium or chloride depletion on angiotensin II receptors in rats. *Am J Physiol* 1990; **258**:R1008–R1015.

232 Wilkes BM. Reduced glomerular angiotensin II receptor density in diabetes mellitus in the rat: time course and mechanism. *Endocrinology* 1987; **120**:1291–98.
233 Sechi LA, Grady EF, Griffin CA, Kalinyak JE, Schambelan M. Distribution of angiotensin II receptor subtypes in rat and human kidney. *Am J Physiol* 1992; **262**:F236–F240.
234 Ichikawa I, Pfeffer JM, Pfeffer MA, Hostetter TH, Brenner BM. Role of angiotensin II in the altered renal function of congestive heart failure. *Circ Res* 1984; **55**:669–75.
235 Hodsman GP, Kohzuki M, Howes LG, Sumithran E, Tsunoda K, Johnston CI. Neurohumoral responses to chronic myocardial infarction in rats. *Circulation* 1988; **78**:376–81.
236 Schunkert H, Hirsch AT, Mankodi S, Talsness C, Dzau VJ, Ingelfinger JR. Renal angiotensinogen gene expression in experimental heart failure: effect of angiotensin converting enzyme inhibition [abstract]. *Clin Res* 1989; **37**:584A.
237 Kitami Y, Hiwada K, Kokubu T. Kidney renin gene expression in spontaneously hypertensive rats. *J Hypertens* 1989; **7**:727–31.
238 Kost CK Jr, Jackson EK. Enhanced renal angiotensin II subtype 1 receptor responses in the spontaneously hypertensive rat. *Hypertension* 1993; **21**:420–31.
239 Mullins JJ, Peters J, Ganten D. Fulminant hypertension in transgenic rat harbouring the mouse Ren-2 gene. *Nature* 1990; **344**:541–44.
240 Chung O, Rohmeiss P, Gretz N, Unger Th. The transgenic mRen2 rat. In: Gretz N, Strauch M, eds. Experimental and genetic rat models of chronic renal failure. Basel: Karger: 124–28.
241 Peters J, Münter K, Bader M, Hackenthal E, Mullins JJ, Ganten D. Increased adrenal renin in transgenic hypertensive rats, TGR(mREN2)27, and its regulation by cAMP, angiotensin II, and calcium. *J Clin Invest* 1993; **91**:742–47.
242 Chung O, Schips T, Qadri F, Ganten D, Mullins JJ, Unger Th. Cardiovascular characterization of a new hypertensive animal model. *J Hypertens* 1991; **9 [suppl. 6]**:S430.
243 Bachmann S, Peters J, Engler E, Ganten D, Mullins JJ. Transgenic rats carrying the mouse renin gene. Morphological characterization of a low-renin hypertension model. *Kidney Int* 1992; **41**:24–36.
244 Samani NJ, Godfrey NP, Major JS, Brammar WJ, Swales JD. Kidney renin mRNA levels in the early and chronic phases of two-kidney, one clip hypertension in the rat. *J Hypertens* 1989; **7**:105–12.
245 Moffett RB, McGowan RA, Gross KW. Modulation of kidney renin messenger RNA levels during experimentally induced hypertension. *Hypertension* 1986; **8**:874–82.
246 Kofir I, Dzau VJ. Endothelial renin-angiotensin pathway: evidence for intracellular synthesis and secretion of angiotensins. *Circ Res* 1987; **60**:422–28.
247 Shepherd JT, Katusic ZS. Endothelium-derived vasoactive factors: endothelium-dependent relaxation. *Hypertension* 1991; **18 [suppl. III]**:III76–III85.
248 Marshall JJ, Kontos HA. Endothelium-derived relaxing factors. A perspective from in vivo data. *Hypertension* 1990; **16**:371–86.
249 Lüscher TF, Boulanger CM, Dohi Y, Yang Z. Endothelium-derived contracting factors. *Hypertension* 1992; **19**:117–30.
250 Cassis LA, Lynch KR, Peach MJ. Localization of angiotensinogen messenger RNA in rat aorta. *Circ Res* 1988; **62**:1259–62.
251 Naftilan AJ, Zuo WM, Ingelfinger JR, Ryan T, Pratt T, Dzau VJ. Localization and differential regulation of angiotensinogen mRNA expression in the vessel wall. *J Clin Invest* 1991; **87**:1300–11.
252 Samani NJ, Swales JD, Brammar WJ. Expression of the renin gene in extra-renal tissues of the rat. *Biochem J* 1988; **253**:907–10.
253 Re R, Fallon JT, Dzau VJ, Quay SC, Haber S. Renin synthesis by canine aortic smooth muscle cells in culture. *Life Sci* 1982; **30**:99–106.
254 Eggena P, Krall F, Eggena MP, Clegg K, Fittingoff M, Barrett JD. Production of angiotensinogen by cultured rat aortic smooth muscle cells. *Clin Exper Hyper Theory and Practice* 1990; **A12(7)**:1175–89.
255 Velletri P, Bean BL. The effects of captopril on rat aortic angiotensin-converting enzyme. *J Cardiovasc Pharmacol* 1982; **4**:315–25.
256 Pipili E, Manolopoulos V, Catravas J, Maragoudakis M. Angiotensin converting enzyme activity is present in the endothelium-denuded aorta. *Br J Pharmacol* 1989; **98**:333–35.

257 Saye JA, Singer HA, Peach MJ. Role of endothelium in conversion of angiotensin I to angiotensin II in rabbit aorta. *Hypertension* 1984; **6**:216–21.
258 Andre P, Schott C, Nehlig H, Stoclet J-C. Aortic smooth muscle cells are able to convert angiotensin I to angiotensin II. *Biochem Biophys Res Commun* 1990; **173**:1137–42.
259 Egleme C, Cressier F, Wood JM. Local formation of angiotensin II in the rat aorta: effect of endothelium. *Br J Pharmacol* 1990; **100**:237–40.
260 Gohlke P, Unger Th, Bünning P. Distribution of the angiotensin converting enzyme inhibitor ramiprilat in the blood vessel wall. *Pharm Pharmacol Lett* 1992; **2**:66–69.
261 Jeunemaitre X, Soubrier F, Kotelevtsev YV, Lifton RP, Williams CS, Charru A, Hunt SC, Hopkins PN, Williams RR, Lalouel J-M, Corvol P. Molecular basis of human hypertension: role of angiotensinogen. *Cell* 1992; **71**:169–80.
262 Walker WG, Whelton PK, Saito H, Russell RP, Hermann J. Relation between blood pressure and renin, renin substrate, angiotensin II, aldosterone and urinary sodium and potassium in 574 ambulatory subjects. *Hypertension* 1979; **1**:287–91.
263 Hollenberg NK, Williams GH, Burger B, Chenitz W, Hoosmand I, Adams DF. Renal blood flow and its response to angiotensin II: an interaction between oral contraceptive agents, sodium intake, and the renin-angiotensin system in healthy young women. *Circ Res* 1976; **38**:35–40.
264 Guyton AC, Hall JE, Coleman TG, Manning RD, Jr. The dominant role of the kidneys in the long-term regulation of arterial pressure in normal and hypertensive states. In: Laragh JH, Brenner BM, eds. *Hypertension: Pathophysiology, Diagnosis, and Management.* Raven Press: New York, 1990: 1029–52.
265 Laragh JH. The renin system and four lines of hypertension research: Nephron heterogeneity, the calcium connection, the prorenin vasodilator limb, and plasma renin and heart attack. *Hypertension* 1992; **20**:267–79.
266 Schunkert H, Dzau VJ, Tang SS, et al. Increased rat cardiac angiotensin converting enzyme activity and mRNA expression in pressure overload left ventricular hypertrophy. *J Clin Invest* 1990; **86**:1913–20.
267 Unger Th, Mattfeldt T, Lamberty V, et al. Effect of early onset ACE inhibition on myocardial capillaries in SHR. *Hypertension* 1992; **20**:478–82.
268 Baker KM, Aceto JF. Angiotensin II stimulation of protein synthesis and cell growth in chick heart cells. *Am J Physiol* 1990; **249**:H610–18.
269 Baker KM, Dostal DE, Chernin MI, Wealand AL, Conrad KM. Angiotensin II-mediated cardiac hypertrophy in adult rats. *J Cell Biochem* 1991; (suppl.):15C (abstract).
270 Baker KM, Chernin MI, Wixson SK, Aceto JF. Renin-angiotensin system involvement in pressure-overload cardiac hypertrophy in rats. *Am J Physiol* 1990; **259**:H324–32.
271 Linz W, Schölkens BA. A specific B2-bradykinin receptor antagonist HOE 140 abolishes the antihypertrophic effect of ramipril. *Br J Pharmacol* 1992; **105**:771–2.
272 Zatz R, Dunn BR, Meyer TW, Rennke HG, Brenner BM. Prevention of diabetic glomerulopathy by pharmacological amelioration of glomerular capillary hypertension. *J Clin Invest* 1986; **77**:1925–30.
273 Raij L, Keane WF. Glomerular mesangium: its function and relationship to angiotensin II. *Am J Med* 1985; **79**:24–30.
274 Soffer RL. Angiotensin-converting enzyme and the regulation of vasoactive peptides. *Ann Rev Biochem* 1976; **45**:73–94.
275 Bao G, Qadri F, Stauss B, Stauss H, Gohlke P, Unger T. HOE 140, a new highly potent and long-acting bradykinin antagonist in conscious rats. *Eur J Pharmacol* 1991; **200**:179–82.
276 Carretero OA, Scicli AG. Local hormonal factors (intracrine, autocrine, and paracrine) in hypertension. *Hypertension* 1991; **18 [suppl. I]**:I-58–I-69.
277 Campbell DJ, Kladis A, Duncan A-M. Bradykinin peptides in kidney, blood, and other tissues of the rat. *Hypertension* 1993; **21**:155–65.
278 Carretero OA, Scicli AG. The glandular kallikrein-kinin system: role in blood flow and blood pressure regulation and its inter-

relationship with other vasoactive systems. In: Robertson JIS, ed. Handbook of hypertension, vol. 1: Clinical aspects of essential hypertension. Amsterdam: Elsevier Science Publishers, 1983: 324–47.

279 Chao J, Chai KX, Chen L-M, Xiong W, Chao S, Woodley-Miller C, Wang L, Lu HS, Chao L. Tissue kallikrein-binding protein is a serpin. 1. Purification, characterization, and distribution in normotensive and spontaneously hypertensive rats. *J Biol Chem* 1990; **265**:16394–401.

280 Burch RM, Axelrod J. Dissociation of bradykinin-induced prostaglandin formation from phosphatidylinositol turnover in Swiss 3T3 fibroblasts: evidence for G protein regulation of phospholipase A_2. *Proc Natl Acad Sci USA* 1987; **84**:6374–78.

281 Regoli D, Rhaleb N-E, Drapeau G, Dion S, Tousignant C, D'Orléans-Juste P, Devillier P. Basic pharmacology of kinins: pharmacologic receptors and other mechanisms. In: Abe K, Moriya H, Fujii S, eds. Kinins V. Part A. New York: Plenum Press, 1989: 399–407.

282 McEachern AE, Shelton ER, Bhakta S, Obernolte R, Bach C, Zuppan P, Fujisaki J, Aldrich RW, Jarnagin K. Expression cloning of a rat B_2 bradykinin receptor. *Proc Natl Acad Sci USA* 1991; **88**:7724–28.

283 Shimamoto K, Matsuki T, Iimura O. Angiotensin-converting enzyme inhibitors and the kallikrein-kinin system. *J Cardiovasc Pharmacol* 1990; **15[Suppl.6]**:S83–S90.

284 Benetos A, Gavras H, Stewart JM, Vavrek RJ, Hatinoglou S, Gavras I. Vasodepressor role of endogenous bradykinin assessed by a bradykinin antagonist. *Hypertension* 1986; **8**:971–74.

285 Carbonell LF, Carretero OA, Madeddu P, Scicli AG. Effects of a kinin antagonist on mean blood pressure. *Hypertension* 1988; **11[Suppl. I]**:I84–I88.

286 Waeber B, Aubert J-F, Vavrek R, Stewart JM, Nussberger J, Brunner HR. Role of bradykinin in blood pressure regulation of conscious spontaneously hypertensive rats. *J Hypertens* 1986; **4[Suppl. 6]**:S597–S598.

287 Seino M, Abe K, Nushiro N, Omata K, Kasai Y, Yoshinaga K. Contribution of bradykinin to maintenance of blood pressure and renal blood flow in anaesthetized spontaneously hypertensive rats. *J Hypertens* 1988; **6[Suppl. 4]**:S401–S403.

288 Benetos A, Gavras I, Gavras H. Hypertensive effect of a bradykinin antagonist in normotensive rats. *Hypertension* 1986; **8**:1089–92.

289 Carretero OA, Miyazaki S, Scicli AG. Role of kinins in the acute antihypertensive effect of the converting enzyme inhibitor, captopril. *Hypertension* 1981; **3**:18–22.

290 Bao G, Gohlke P, Qadri F, Unger T. Chronic kinin receptor blockade attenuates the antihypertensive effect of ramipril. *Hypertension* 1992; **20**:74–79.

291 Oza NB, Schwartz JH, Goud HD, Levinsky NG. Rat aortic smooth muscle cells in culture express kallikrein, kininogen, and bradykininase activity. *J Clin Invest* 1990; **85**:597–600.

292 Saed GM, Carretero OA, MacDonald RJ, Scicli AG. Kallikrein messenger RNA in rat arteries and veins. *Circ Res* 1990; **67**:510–16.

293 Farhy Rd, Ho K-L, Carretero OA, Scicli AG, Kinins mediate the antiproliferative effect of ramipril in rat carotid artery, *Biochem Biophys Res Commun* 1992; **182**:283–8.

294 Fenoy FJ, Scicli G, Carretero O, Roman RJ. Effect of an angiotensin II and a kinin receptor antagonist on the renal hemodynamic response to captopril. *Hypertension* 1991; **17**:1038–44.

295 Hajj-ali AF, Zimmerman BG. Kinin contribution to renal vasodilator effect of captopril in rabbit. *Hypertension* 1991; **17**:504–9.

296 Majima M, Katori M, Hanazuka M, Mizogami S, Nakano T, Nakao Y, Mikami R, Uryu H, Okamura R, Mohsin SSJ, Oh-ishi S. Suppression of rat deoxycorticosterone-salt hypertension by kallikrein-kinin system. *Hypertension* 1991; **17**:806–13.

297 Margolius HS, Geller R, Pisano JJ, Sjoerdsma A. Altered urinary kallikrein excretion in human essential hypertension. *Lancet* 1971; **2**:1063–65.

298 Berry TD, Hasstedt SJ, Hunt SC, Wu LL, Smith JB, Ash O, Kuida H, Williams RR. A gene for high urinary kallikrein may protect against hypertension in Utah kindreds. *Hypertension* 1988; **13**:3–8.

299 Van Hooft IMS, Grobbee DE, Derkx FHM, De Leeuw PW, Schalekamp MADH, Hofman A. Renal hemodynamics and the renin-angiotensin-aldosterone system in normotensive subjects with hypertensive and normotensive parents. *N Engl J Med* 1991; **324**:1305–11.
300 Textor SC, Turner ST. Renal vascular response to sodium loading in sons of hypertensive parents. *Hypertension* 1991; **17**:982–88.
301 Scholkens BA, Linz W, Konig W. Effects of the angiotensin converting enzyme inhibitor, ramipril, in isolated ischemic rat heart are abolished by a bradykinin antagonist. *J Hypertens* 1988; **6**[**Suppl. 4**]:S25–S28.
302 Matsuki T, Shoji T, Yoshida S, Kudoh Y, Motoe M, Inoue M, Nakata T, Hosoda S, Shimamoto K, Yellon D, Iimura O. Sympathetically induced myocardial ischemia causes the heart to release plasma bradykinin. *Cardiovasc Res* 1987; **21**:428–32.
303 Linz W, Martorana PA, Schölkens BA. Local inhibition of bradykinin degradation in ischemic hearts. *J Cardiovasc Pharmacol* 1990; **15**[**Suppl. 6**]:S99–S109.
304 Linz W, Schölkens BA. A specific B_2-bradykinin receptor antagonist HOE 140 abolishes the antihypertrophic effect of ramipril. *Br J Pharmacol* 1992; **105**:771–72.
305 Schalekamp MADH, Derkx FHM. Plasma kallikrein and plasmin as activators of prorenin: links between the renin-angiotensin system and other proteolytic systems in plasma. *Clin Sci* 1981; **61**:15–21.
306 Cherry PD, Furchgott RF, Zawadzki JV, Jothianandan D. Role of endothelial cells in relaxation of isolated arteries by bradykinin. *Proc Natl Acad Sci USA* 1982; **79**:2106–10.
307 Bhoola KD, Figueroa CD, Worthy K, Bioregulation of kinins: kallikreins, Kininogens, and kininases, *Pharmacol Rev* 1992; **44**:1–80.

2

Basic cardiovascular pharmacology of ACE inhibitors

J-F Giudicelli, A Berdeaux, C Richer, E Pussard and P Demolis

The renin-angiotensin system (RAS) plays a central role in the maintenance of blood pressure and electrolyte homeostasis. Its involvement in the pathogenesis of a number of cardiovascular disorders, including hypertension and congestive heart failure has, over the years, led to pharmacological attempts at manipulation. These include renin inhibitors and, more recently, the non-peptidic angiotensin II receptor antagonists but, to date, the most successful approach to interrupting the RAS has been that of the angiotensin I converting enzyme inhibitors (ACE inhibitors).

Angiotensin I converting enzyme (ACE) is a zinc metallopeptidase; it removes the histidyl-leucine C-terminal dipeptide of angiotensin I (ANG I), to form angiotensin II (ANG II), the active octapeptide of the RAS. By competition with ANG I for ACE, ACE inhibitors prevent the formation of ANG II. As ACE is also bradykininase II, ACE inhibitors also prevent bradykinin breakdown (Fig. 2.1). These events occur in the circulation as the RAS is a circulating endocrine system. However, the existence of functional 'tissue' RASs has also been demonstrated in a large number of organs, including those involved in cardiovascular regulation, such as the vascular wall, heart and kidney. These tissue RASs are postulated to act through the paracrine and autocrine effects of locally synthesized ANG II. ACE inhibitors not only affect the circulating RAS but, with the exception of a few organs, also inhibit most tissue RASs.

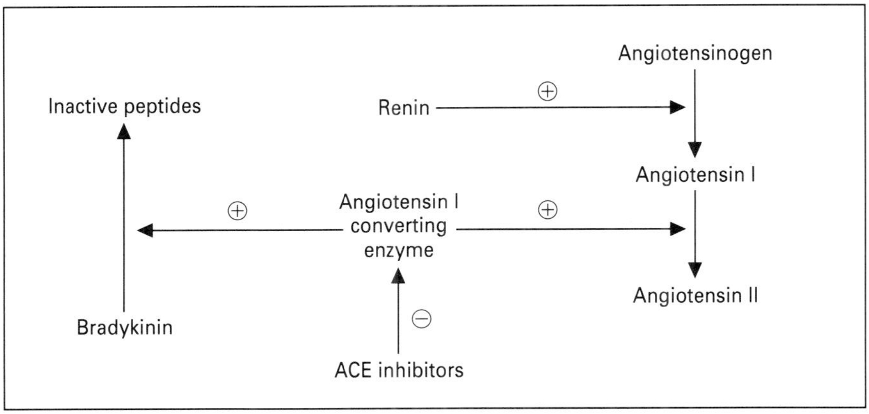

Figure 2.1
Effects of ACE inhibitors.

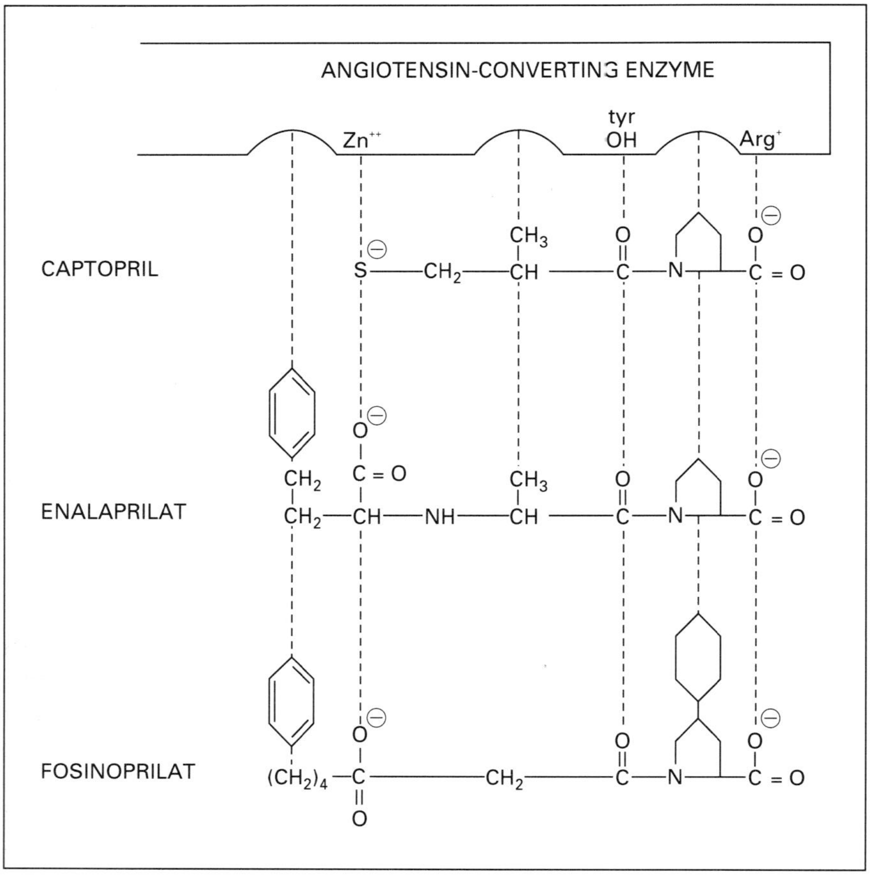

Figure 2.2
Hypothetical binding of competitive inhibitors to the active site of angiotensin converting enzyme.

Because of their efficacy in the treatment of major cardiovascular diseases and because of the paucity of their side-effects, ACE inhibitors now have a large therapeutic use, having emerged as a fast-growing new class of cardiovascular drugs. Hereafter, we will discuss their chemistry, the characteristics and the consequences of induced ACE-inhibition, the basic cardiovascular properties of ACE inhibitors in animals under both normal and pathological conditions, and in healthy subjects, and finally the pharmacokinetics of ACE inhibitors in humans.

Chemistry of ACE inhibitors

On the assumption that the natural substrate, ANG I, and ACE inhibitors bind in a similar fashion to ACE, a schematic model of the latter was initially developed by Ondetti et al.[1] (Fig. 2.2). That the active site of ACE contains a zinc ion which is pivotal in the catalytic process has over the years led to the synthesis of active-site directed ACE inhibitors which convert the catalytic interaction with the zinc ion into an effective binding interaction.

ACE inhibitors can be classified into three groups according to the nature of their zinc ligand (Fig. 2.2):

1 The sulfhydryl-containing ACE inhibitors, which bind to the zinc ion through their thiol function: these include captopril, its analogs with a free (rentiapril) or a masked (pivalopril, zofenopril) thiol function, and its prodrug (alacepril).
2 The carboxyalkyl dipeptidic ACE inhibitors, which bind to the zinc ion through one of their carboxylic functions. This is the largest class of ACE inhibitors, including enalapril, ramipril, perindopril, cilazapril, quinapril, trandolapril, benazepril and spirapril. In fact, these compounds are prodrugs, in other words monoester monoacid drugs, which are used for improved oral absorption; they have to be hydrolysed in vivo to yield the corresponding diacid active metabolites, for example enalaprilat, ramiprilat, perindoprilat, etc. It is the generated acid function that binds to the zinc ion. One exception in this group is lisinopril which possesses two carboxylic functions without the need for prior hydrolysis (and as such, its correct name should be lisinoprilat). Despite not being a prodrug, it is orally absorbed because of the addition of an aminobutyl side-chain.
3 The phosphorus-containing ACE inhibitors, which bind to the zinc ion through their phosphinic acid moiety. The latter, in the case of fosinopril, is obtained after in vivo hydrolysis of an ester which is used as a prodrug to improve oral absorption. In contrast, ceranapril is not a prodrug: despite a non-esterified phosphinic acid function, it is well absorbed because of the adjunction of an aminobutyl side-chain.

Besides these 'pure' ACE inhibitors, different types of hybrid molecules have also been described, combining either ACE inhibitory and nitrovasodilator-like properties (S-nitroso captopril), or ACE inhibitory and β-adrenoceptor blocking properties (BWA 575C), or ACE inhibitory and enkephalinase-inhibitory properties (glycoprilat, alatrioprilat).

ACE blockade

Characteristics of ACE blockade

Binding site of ACE inhibitors

Except for the directly active ACE inhibitors (captopril, lisinopril and ceranapril), the prerequisite for ACE blockade is the in vivo hydrolysis of prodrug esters to release their active ACE inhibitory moieties.

The primary structure of ACE has revealed the existence of two homologous domains, each of them containing a zinc atom. These two domains are both able to cleave ANG I and are both sensitive to ACE inhibitors. They have a similar affinity (K_m) for the substrate, but the catalytic constant (K_{cat}) of the N domain is much lower than that of the C domain.[2]

Binding of ACE inhibitors to the active sites of ACE is illustrated in Fig. 2.2. ACE inhibitors bind to the zinc atom through the sulfhydryl group for captopril and related drugs, through the carboxyl group released by prodrug hydrolysis for carboxyalkyl dipeptidic ACE inhibitors, and through the phosphinic acid residue for fosinopril and ceranapril. On the other hand, the terminal carboxyl residue of all ACE inhibitors interacts with the positive charge of an active arginine site of ACE. The heterocyclic ring next to the terminal carboxyl residue binds to a hydrophobic pocket of the enzyme structure. An additional hydrophobic binding with a 2-methyl residue restricts the conformation of the molecule. Except for

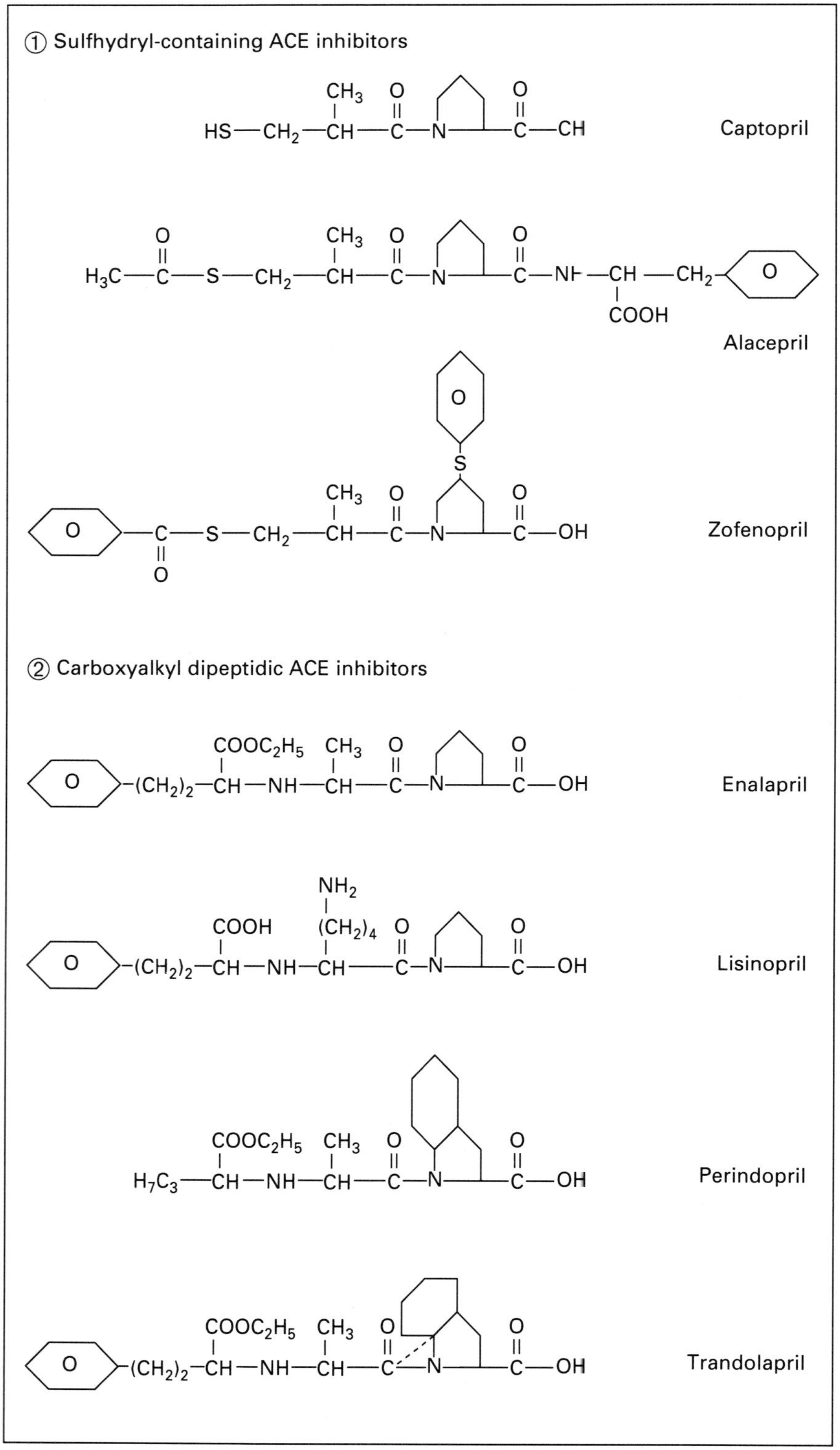

Figure 2.3
Chemical structures of ACE inhibitors.
(1) Sulfhydryl-containing ACE inhibitors
(2) Carboxyalkyl dipeptidic ACE inhibitors

Figure 2.3 cont. *Chemical structures of ACE inhibitors.*
(3) Phosphorus containing ACE inhibitors

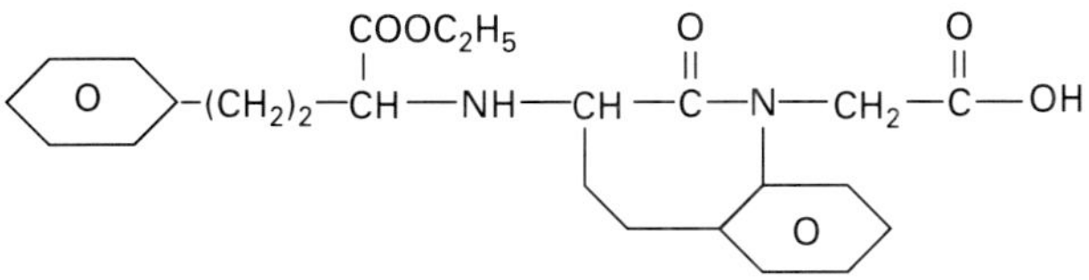

Benazepril

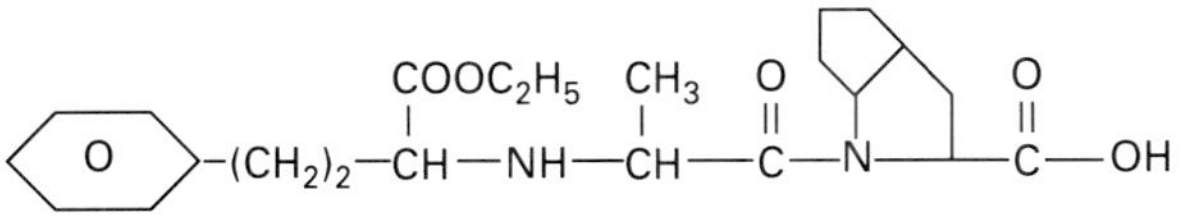

Ramipril

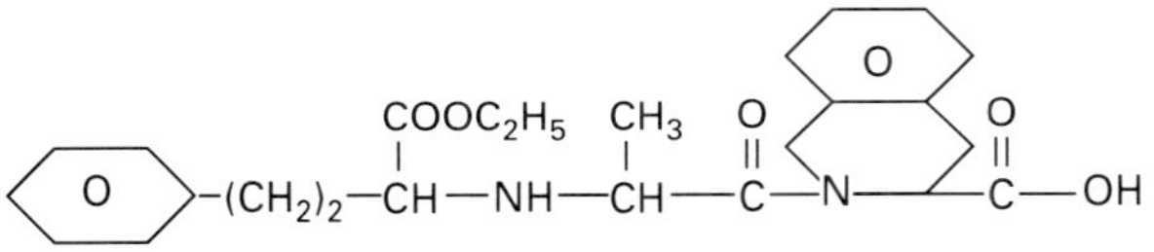

Quinapril

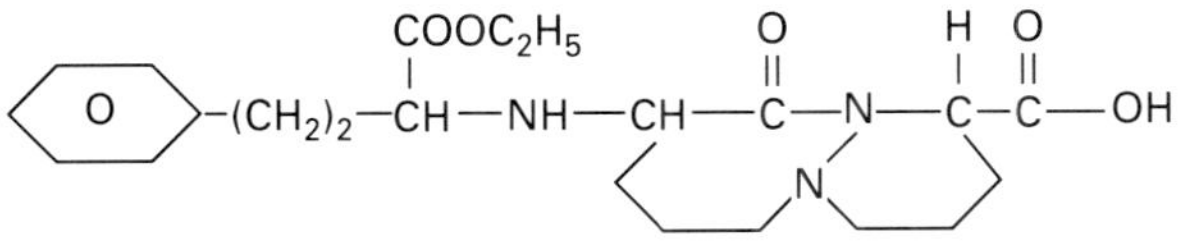

Cilazapril

③ Phosphorus containing ACE inhibitors

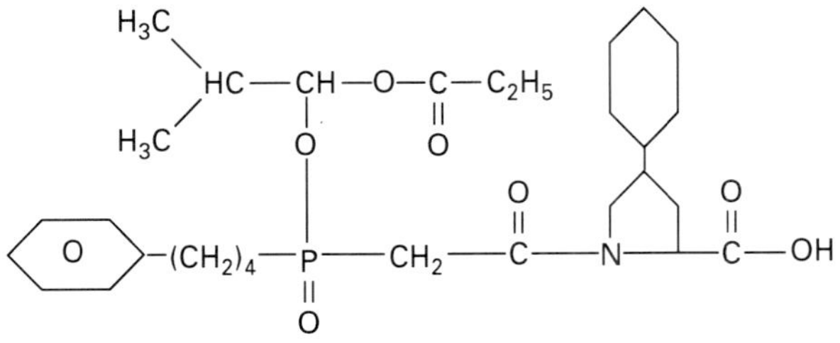

Fosinopril

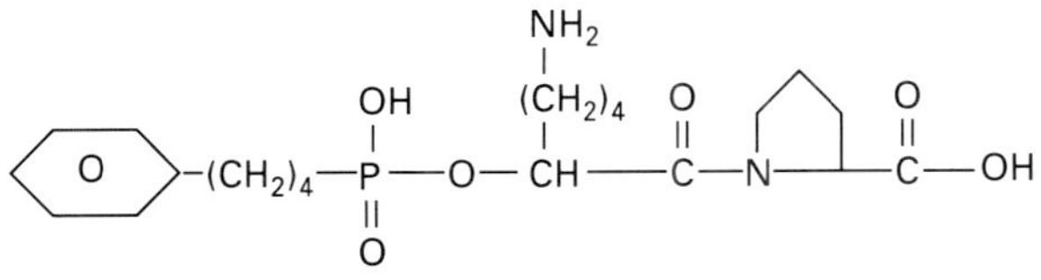

Ceranapril

captopril, the phenylalkyl side-chain introduces an additional site of interaction with the enzyme and increases the ACE inhibitors' potency. Finally, a hydrogen bond between a tyrosyl phenolic residue and a ketonic function of ACE inhibitors strengthens the interaction (Fig. 2.3).

ACE inhibitors interact with the active sites of both circulating ACE and membrane-bound ACE in vascular endothelial cells.

Specificity of ACE–ACE inhibitor interaction

The major consequence of ACE-inhibitor-induced ACE blockade is inhibition of ANG II formation and of bradykinin breakdown. However, as ACE has a broad substrate spectrum, ACE inhibitors can also potentially affect degradation of substance P, enkephalins and LHRH. Furthermore, ACE inhibitors are also potential inhibitors of other metallopeptidases such as enkephalinase, brain peptidyl dipeptidase and aminopeptidase, but blockade of these enzymes requires concentrations at least 100-fold greater than those inducing ACE inhibition.

Kinetics of ACE blockade

In plasma The competitive inhibition of circulating ACE differs in timing and duration according to the pharmacokinetic properties of ACE inhibitors. The onset of ACE blockade is related to the rate of hydrolysis of the prodrug and to the bioavailability of the active form. Thus, whereas maximal ACE blockade appears within 15 minutes with the directly active captopril, it occurs only after 2 hours with enalapril, ramipril and quinapril, after 2 to 4 hours with lisinopril and cilazapril, and after 6 hours with trandolapril. On the other hand, the strength of binding of the inhibitor to ACE determines the duration of ACE blockade. With the carboxyalkyl dipeptidic agents, plasma ACE activity begins to recover after 8 to 12 hours, but even 72 hours post-drug, it has not completely returned to baseline values. Finally, the terminal elimination half-life of ACE inhibitors is largely related to the dissociation rate of the ACE inhibitor–ACE complex (T1/2 ACE).[3]

There is a sigmoidal relationship between plasma ACE inhibitor concentrations and ACE inhibition in both normotensive subjects and hypertensive patients. In the latter, it has been shown that upon acute dosing, ACE inhibition of over 90% is required to induce a significant reduction in blood pressure.

In tissues The concept of the existence of independent RASs has emerged from the fact that ACE inhibitors reduce blood pressure irrespective of the pretreatment plasma renin values, and from the lack of any direct correlation between the antihypertensive effects of ACE inhibitors and the expected changes in the components of the circulating RAS, especially plasma ANG II levels. Accordingly, in most tissues, ACE activity is inhibited by ACE inhibitors and, despite different time courses and degrees which are likely to be related to the different drugs' local bioavailability, this inhibition is usually longer lasting than in the plasma. But the finding of inhibited tissue ACE activity is often difficult to interpret because biochemical measurements of the enzyme are usually performed in tissue homogenates, which contain vascular endothelium. It is well established that in the vascular RAS, ACE is only anchored in the endothelium and the major part of the enzyme, especially that containing the active sites, is floating within the vascular lumen. Thus, the measured tissue ACE inhibition may well represent endothelial rather than tissue inhibition, and the site of ANG II generation may well be the same for circulating and tissue RAS. If this were the case, there would be no need to evoke tissue penetration to explain ACE inhibitors' inhibitory effects on the paracrine and autocrine functions of ANG II.[4] There are, however, a few exceptions, such as the kidney, where an important part of ACE is localized in the tubular epithelial brush border cells, the brain and the testis. In the latter two organs, the existence of blood–tissue barriers is, and the lipophilicity of ACE inhibitors may be, of importance for tissue penetration and local ACE inhibition. Thus, following oral administration of ACE inhibitors, the effect in the brain is rapid but short-lasting with captopril, rapid and long-lasting with fosinopril, delayed and long-lasting with lisinopril and occurs only after large doses of enalapril. The ability of orally administered ACE inhibitors to block ACE in the cerebrospinal fluid parallels the drugs' liposolubility, ramipril being, in this respect, more potent than enalapril. Finally, the existence of a blood–tissue barrier prevents ACE inhibition in the testis.

Hormonal effects of ACE inhibition

Acute ACE inhibition results in dose-dependent blockade of ANG II production and of bradykinin breakdown. Suppression of ANG II-mediated feedback inhibition of renin results in an increase in plasma renin activity and in ANG I plasma levels. Aldosterone secretion is decreased but not completely suppressed as other stimuli of this secretion, adrenocorticotrophin and plasma potassium, are still operating. On the other hand, bradykinin accumulation tends to stimulate prostacyclin (PGI_2) synthesis.

During chronic ACE inhibitor treatment, plasma ANG II levels are decreased during the first 6 hours after drug intake but thereafter rise progressively despite maintained ACE inhibition.[5] This progressive return of plasma ANG II to baseline values is explained by the simultaneous existence of a small residual ACE activity and of feedback-induced high levels of its substrate, plasma ANG I.[6]

Assessment of ACE blockade

Measurement of ACE activity

In vitro ACE activity is assessed by measurement of the product resulting from hydrolysis of synthetic N-substituted tripeptides by plasma or tissue homogenates of ACE. Substrates used include Hip-Hist-Leu, carbobenzoxy-Phe-His-Leu and ^{3}H-Hip-Gly-Gly, and the reactions' product is measured using either a spectrophotometric assay of hippuric acid, a spectrofluorimetric determination of the complex formed between His-Leu and O-phtaldialdehyde, or a radioisotopic procedure. However, it has been shown that, depending upon the substrate, large differences can be observed in the determined. ACE inhibition values.

Ex vivo Intensity and time course of tissue ACE inhibition in animals pretreated with ACE inhibitors can be assessed ex vivo either by a radioinhibitor binding assay or by quantitation of the binding density of a labelled marker by autoradiography.[7]

Measurement of angiotensin I and II

Specific and highly sensitive radioimmunoassays are available for in vitro measurement of plasma ANG I, an index of RAS reactivity to ACE inhibition, and of plasma ANG II, an index of the actual status of the RAS. Finally, the plasma ANG II/ANG I ratio appears to be the best indicator of ACE blockade in vivo, as it relates the actual ANG II concentration to the prevailing level of ANG I, the native ACE substrate.[6]

Inhibition of ANG I responses
Functional ACE inhibition (onset, maximal blockade achieved and duration) is assessed by estimation of the suppression of ANG I effects. This estimation can be performed either in vitro (concentration-contractile response curves to ANG I before and after ACE inhibition in various isolated organs) or in vivo (dose-systemic pressor or regional vasoconstrictor response curves to ANG I before and after ACE inhibitors) in different animal species and in humans.

Comparison of ACE inhibitors

Potency
The relative potency of ACE inhibitors can be determined in vitro by comparison of the concentrations required to inhibit plasma ACE activity by 50%. Ester prodrug compounds appear to be much less potent than the corresponding diacid active forms. All carboxyalkyl dipeptidic drugs exhibit higher inhibitory potencies than the sulfhydryl and phosphorus containing ACE inhibitors.[7] Among the carboxyalkyl dipeptidic ACE inhibitors, the decreasing rank of order for ACE inhibition potency in human plasma is as follows: trandolaprilat > cilazaprilat > quinaprilat > ramiprilat > perindoprilat > enalaprilat.

Duration of action
The strength of drug binding to the active site of ACE determines not only the potency but also the duration of action of the drugs. As compared to captopril, the carboxyalkyl dipeptidic ACE inhibitors have a longer duration of action. Their longer terminal elimination half-life corresponds to the clearance of ACE-bound drug and reflects a higher affinity of the drug for the enzyme and/or a slower rate of dissociation of the inhibitor–ACE complex (T1/2 ACE). In addition, the potency contributes to the duration of the action since for a given plasma half-life, the duration of ACE blockade of a more potent inhibitor is longer.

Cardiovascular pharmacology of ACE inhibitors

ACE inhibitors in normotensive animals and subjects

Systemic and regional hemodynamics
In sodium replete normotensive animals (rats, cats, rabbits and dogs) as well as in normotensive subjects on a normal sodium diet, acute administration of ACE inhibitors results in a small, and often non-significant, decrease in blood pressure. Heart rate, myocardial contractility (dP/dt), cardiac output and stroke volume are usually not modified. Hence, the reduction in blood pressure is entirely related to a decrease in total peripheral resistance. This indicates that ACE inhibitors induce peripheral arteriolar vasodilatation. As could be expected, ACE inhibitor-induced vasodilatation and hypotensive effects are strongly enhanced in conditions of RAS stimulation, either by a low sodium diet or by diuretic-induced sodium depletion.

The absence of reflex tachycardia in response to the fall in blood pressure is a general property of ACE inhibitors after both acute ACE and chronic administration, which suggests an interaction between the inhibitors and the control of heart rate. On the one hand, acute ACE inhibitor administration, either to normotensive or hypertensive subjects, induces a rapid resetting of the baroreflex: the set point of the curve relating changes in RR interval to increases or decreases in systolic blood pressure following intravenous injections of phenylephrine and nitroglycerin respectively, is shifted to the left without any simultaneous change in the sensitivity or gain of the reflex.[8]

On the other hand, clinical studies of the vagally-mediated responses to the Valsalva manoeuvre, the diving test and the standing to lying test suggest that ACE inhibitors reinforce the vagal tone which buffers the expected hypotension-induced tachycardia.[9] This effect appears to be directly related to ACE inhibitor-induced ANG II suppression since infusion of a sub-pressor dose of ANG II abolishes the bradycardia induced by captopril during the diving test,[2] and is most likely due to the removal of ANG II vagolytic action at central and/or peripheral sites.

ACE inhibitor-induced vasodilatation appears to be heterogeneously distributed, preferentially affecting some vascular beds, especially those where ANG II exerts its most pronounced vasoconstrictor effects. Thus, in normotensive animals ACE inhibitors increase regional blood flow in the kidneys, the skin and the gastrointestinal tract, but not in the heart, the brain and the muscle. In healthy volunteers ACE inhibitors increase brachial blood flow more markedly than carotid blood flow and induce a redistribution of cardiac output towards the antebrachial musculocutaneous vascular territories. Finally, depending upon the vascular bed, dilatation may affect both the arterioles and the large arteries (forearm, coronaries), or mainly just the arterioles (carotid territory).

From a functional point of view, chronic ACE inhibitor administration to normotensive rats results in an improved carotid artery compliance.[10] In healthy volunteers also, the increase in brachial artery diameter with an unchanged or slightly decreased blood pressure indicates that ACE inhibitors improve the compliance of large arteries,[11] as confirmed by the simultaneously observed decrease in pulse pressure.

The mechanisms underlying ACE inhibitor-induced vasodilatation and decrease in blood pressure in normotension, but also in hypertension is discussed below.

Vital organs

Kidney

Renal blood flow and hemodynamics ACE inhibitors increase renal blood flow (RBF), both cortical and papillary, and decrease renal vascular resistance, in animals as well as in healthy volunteers. At the glomerular level, the vasodilating effect of ACE inhibitors is more pronounced on efferent than on afferent arterioles.[12] Intraglomerular pressure decreases, leading to a parallel reduction in filtration fraction, but glomerular filtration is maintained, as the renal blood flow increase compensates for the filtration fraction decrease. Most experiments using specific inhibitors tend to indicate that ANG II suppression rather than kinin-prostaglandin enhancement is responsible for the renal hemodynamic effects of ACE inhibitors.

Renal function The short-term effects of ACE inhibitors on renal function are mostly mediated by their hemodynamic actions. But, in addition to the increase in RBF, an increase in ultrafiltration rate due to enhanced glomerular capillary surface area also contributes to the stability of glomerular filtration rate, except in situations where the decrease in filtration fraction cannot be counterbalanced by an increase in RBF (for example in hypovolemia or renal artery stenosis). Urinary volume (UV) and sodium excretion (UNa) are increased after ACE inhibition and these actions are predominantly due to the suppression of the hemodynamic and direct antinatriuretic effects of ANG II. ANG II impairs the renal capacity to excrete a sodium load in normotensive dogs, and the sodium balance is equilibrated at higher levels of mean arterial pressure. Captopril reverses this phenomenon and allows the kidney to excrete the sodium load at lower blood pressure levels than in control animals. Finally, suppression of intrarenal ANG II tends to reduce the transglomerular passage of albumin by decreasing the mesangial uptake of macromolecules. All these effects occur at low doses, which do not necessarily affect plasma ACE activity.

The role of kinins and prostaglandins in the mechanism of ACE inhibitors on renal function is still a matter of debate. In conscious dogs, it has been shown that prostaglandins are responsible for the captopril-induced renal sympathoinhibition,[13] which could lead to enhanced urinary volume and sodium excretion. In hypovolemic rats, prostaglandins are probably also involved in the papillary blood flow increase observed after captopril administration and produce a pronounced increase in UV and UNa.

The long-term effects of ACE inhibitors on renal function are mainly related to the reduction of plasma aldosterone. However, administration of ACE inhibitors alone does not usually lead to hyperkalemia, except if the renal function is impaired or if there exists another cause for hyperkalemia.

Brain In normotensive animals and in healthy subjects,[14] ACE inhibitors usually have little if any effect on cerebral blood flow (CBF), upon either acute or long-term administration, even if there is a decrease of blood pressure.

In normal subjects, acute captopril does not change the CBF reactivity to carbon dioxide and does not modify the cerebral oxygen metabolism. More interestingly, ACE inhibitors, in contrast to other vasodilators, induce a downward shift of the lower limit of the CBF autoregulation plateau. As a result, the cerebral circulation exhibits a greater adaptability to large decreases in blood pressure, which allows the maintenance of normal CBF values in hypertensive patients during ACE inhibitor-induced blood pressure reduction. The mechanism of ACE inhibitor interaction with CBF is not yet clearly established.

Coronary arteries As ANG II is a potent coronary constrictor at the arteriolar level, a decrease in coronary resistance and an increase in coronary blood flow following ACE inhibition are to be expected. This is generally what is observed when drugs are administered in situations where the RAS is activated, for example myocardial ischemia or heart failure. However, when the resting RAS tone is low, as is the case in dogs and in the majority of other animal species on a normal sodium diet, or when ACE inhibitors are administered intracoronarily,[15] coronary resistance still decreases but coronary blood flow generally remains unchanged and even falls as the result of concomitant decrease in myocardial oxygen demand and/or fall in perfusion pressure. This clearly tends to indicate that the local RAS plays only a modest, if any role in arteriolar coronary vasodilatation, which is observed after intravenous ACE inhibitor administration and which thus appears to result mainly from the decrease in circulating ANG II synthesis.[15,16] Transmural distribution of coronary blood flow between the endocardium and the epicardium remains unaffected (unchanged endocardial /epicardial blood flow ratio) by ACE inhibitors.[16]

Regarding large coronary arteries, it has been shown that in conscious dogs on a normal sodium diet, IV enalaprilat increases the diameter of the epicardial coronary arteries, although it does not induce any concomitant dilatation of coronary arterioles nor any simultaneous increase in coronary blood flow. This clearly shows that enalaprilat exerts a direct dilating effect on large coronary conductance vessels since an indirect flow-mediated effect is excluded. The mechanism of this direct vasodilating effect appears to be the blockade of the tissue RAS within the wall of the large conductance vessels.[17] However, when enalaprilat is administered to the same dogs following RAS activation by sodium deprivation (furosemide), both large and small coronary vessels dilate and the increase in diameter of large coronary arteries is abolished when flow is maintained constant, clearly indicating that in those conditions a flow-mediated endothelium-dependent mechanism is involved. Although the absolute increase in large coronary artery diameter induced by ACE inhibitors is far smaller than that achieved by nitrate derivatives or calcium channel antagonists, the ability of ACE inhibitors to dilate large coronary arteries may contribute to the prevention of a steal phenomenon when these drugs are administered at hypotensive doses.

ACE inhibitors and hypertension

Hemodynamic effects

ACE inhibitors lower blood pressure in all models of experimental hypertension with the exception of the DOCA-salt rats. This antihypertensive effect is observed both in renin-dependent (2-kidney–1-clip Goldblatt, with aortic ligation between the two kidneys) and in non-renin-dependent (spontaneously hypertensive rat, SHR) models, but is usually more marked in the former after acute administration. Blood pressure is decreased after both acute and chronic administration and tolerance does not develop. After acute administration, the decrease in blood pressure is dose-dependent up to a given dose, beyond which it is the duration of the effect that is dose-dependent. The maximal decrease in blood pressure occurs earlier with non-prodrugs than with prodrugs. In addition, ACE inhibitors are effective both at preventing hypertension development and at reducing established hypertension. ACE inhibitors are almost unique among other antihypertensive drugs in that they exert long-lastingly persistent antihypertensive effects after treatment withdrawal in SHRs previously treated during the phase of genetic hypertension development (GHD).[18]

In all models of hypertension, available evidence indicates that the reduction (or the prevention of the rise) in blood pressure is due to a decrease (or the prevention of the rise) of total peripheral resistance since cardiac output is not simultaneously modified. No modification in heart rate is observed.

Thus, in hypertensive, as well as in normotensive animals, ACE inhibitors induce peripheral arteriolar vasodilation, which again is heterogeneous in its regional profile. In SHRs, vasodilation occurs preferentially in the kidney, in the gastrointestinal tract, in the splanchnic bed, in the skin and, to a lesser extent, in the brain, the heart and the skeletal muscle. Despite the decrease in blood pressure, that is, in perfusion pressure, blood flow increases in the kidneys, the gastrointestinal tract and the skin, but not in the liver, the muscle, the brain and the heart. The venous vascular bed is also dilated and cardiac preload is reduced by ACE inhibitors.

Mechanisms of vasodilatation In renin-dependent models of hypertension, the arteriolar vasodilating properties of ACE inhibitors are immediate and clearly related to inhibition of circulating ANG II synthesis and hence to suppression of ANG II direct (on vascular AT_1 receptors) and indirect (through potentiation of the sympathetic system) vasoconstrictor effects. However, in models with normal or low renin plasma levels, the vasodilating properties are somewhat delayed and appear to result at least partly from tissue RAS inhibition in the vascular wall. Blockade of vascular ANG II formation leads to suppression of local ANG II paracrine and autocrine effects,[4,19] including activation of smooth muscle cell AT_1 receptors and hence vasoconstriction, and facilitation of noradrenaline (norepinephrine) release from noradrenergic endings in the adventitia, etc.

There are a number of additional mechanisms which contribute to the vasodilating properties of ACE inhibitors:

- The prevention of kinin, especially bradykinin breakdown, not only in the plasma but also in the tissues, for example the arterial wall, where full kallikrein-kinin systems are present. Bradykinin, by stimulating the B_2-kininergic receptors present on endothelial cells, is a potent releaser of endothelium-derived relaxing factor (EDRF) and, through activation of phospholipase A_2 also stimulates the production of prostacyclin (PGI_2). ACE inhibitors potentiate the endothelium-dependent relaxation and the release of EDRF induced by bradykinin (trandolapril, perindopril, cilazapril). In addition, a number of studies performed in rats with aortic ligation-induced hypertension, with 2-kidney–1-clip hypertension, and more recently in SHRs have shown that bradykinin antagonists attenuate the vasodilating and antihypertensive effects of ACE inhibitors.[20] Furthermore, these effects are also blunted in SHRs pretreated with the nitric oxide synthesis inhibitor, LNMMA, clearly pointing out the role of bradykinin and of bradykinin-induced EDRF release in ACE inhibitor-induced vasodilatation.[20] In contrast, the contribution of PGI_2 to vasodilatation remains more controversial, but cannot be excluded.[20] It must be stressed that the role of diminished kinin degradation in ACE inhibitor-induced vasodilatation is probably more important in hypertension than in normotension.
- The flow-dependent release of EDRF. ACE inhibitor-induced rise in regional flow increases EDRF release and this contributes to both arteriolar and large artery dilatation.
- The potentiation of the vasodilating effects of endothelium-dependent relaxing agents other than bradykinin, for example, acetylcholine and purines.
- The interactions between ACE inhibitors and the sympathetic system. ACE inhibitors exert sympathoinhibitory effects, both pre- and postjunctionally,[21] resulting in diminished responses to sympathetic nerve stimulation and to direct α-adrenoceptor stimulation.

Hormonal effects

As previously mentioned, acute ACE inhibition results in the suppression of ANG II production and in the inhibition of bradykinin breakdown, but on prolonged treatment, circulating ANG II is not permanently decreased and its trough plasma levels increase progressively, whereas ACE remains inhibited. In addition, plasma aldosterone levels are partially decreased and no sodium and water retention develops.

In the long term, ACE inhibitors tend to decrease sympathetic nervous system activity, and plasma catecholamines levels are usually slightly reduced. ACE inhibitors reduce vasopressin plasma levels only when the latter are abnormally elevated.

Structural effects

Besides lowering blood pressure in hypertensive animals, ACE inhibitors also reverse (or prevent) the vascular and cardiac structural modifications that accompany hypertension.

Vessels At the vascular level, chronic ACE inhibitor administration prevents, reduces or reverses vascular wall hypertrophy and fibrosis. For example, in large arteries, ACE inhibitors decrease aortic media thickness and wall/lumen ratio and increase aortic nuclear density; they also reduce aortic collagen density.[10,22] Thus, these drugs oppose the vascular remodeling that accompanies hypertension, and this occurs both in renin-dependent (2-kidney–1-clip Goldblatt) and in non-renin-dependent (SHR) models of hypertension, and after both preventive and curative chronic administration. Furthermore, at least in SHR, the antihypertrophic and antifibrotic properties of ACE inhibitors persist for weeks after treatment withdrawal.[22–24]

In the small arteries, for example in mesenteric arterioles, chronic administration to SHRs also results in decreased media thickness and wall/lumen ratio and in increased lumen diameter.[22–24] These effects also persist for weeks after treatment withdrawal.[23,24] It is well known that ANG II is able to induce expression of proto-oncogenes and overexpression of growth factors (PDGF, $TGF_{\beta 1}$, endothelin)[19] through activation of the phosphoinositols pathway and hence to promote smooth muscle cell hypertrophy and increase collagen synthesis. Thus, it is most likely that the vascular antihypertrophic and antifibrotic properties of ACE inhibitors are directly linked to ANG II synthesis inhibition, especially within the vascular wall.

From a functional point of view, the potent and long-lasting effects of ACE inhibitors on vascular structures have a number of important consequences:

- The improvement (or the prevention of the deterioration) of vascular function. In both 2-kidney–1-clip Goldblatt rats and in SHRs, chronic ACE inhibitor administration results in a decrease in arteriolar and arterial wall stiffness and an increase in arteriolar and arterial compliance and thus an improved arterial damping function.[10,23] These effects persist after treatment withdrawal. As this decreased arterial compliance leads to an increase in systolic blood pressure and hence to augmented ventricular loading, and as the decrease in diastolic blood pressure may reduce coronary blood flow, ACE inhibitor administration will reduce cardiovascular morbidity and mortality in hypertension by improving arterial compliance.[25,26]
- ACE inhibitor-induced prevention of vascular wall hypertrophy development and of wall/lumen ratio increase suppresses the hyperresponsiveness to vasoconstrictor agents in SHRs,[27] an effect that persists in the long term after treatment withdrawal. Thus ACE inhibitor-induced prevention of vascular remodeling results in a long-lasting interruption of the so-called vascular pressor amplifier,[28] which is responsible for the perpetuation of hypertension.

Heart At the cardiac level, ACE inhibitors also prevent, reduce or reverse the myocardial alterations that develop in all experimental models of hypertension. In 2-kidney–1-clip Goldblatt rats and in SHRs, hypertrophic remodeling of the myocardium occurs, including myocyte growth and collagen accumulation. Myocyte growth and increase in left ventricular mass are mainly determined by ventricular loading, whereas interstitial and perivascular fibrosis of intramyocardial coronary arteries appear to be principally dependent upon circulating and/or tissue-derived ANG II and aldosterone levels.[29] ACE inhibitors oppose or reverse the adverse myocardial remodeling, as shown by reductions in heart weight/body weight ratio as well as in endocardial and epicardial collagen densities.[18,22,23] The mechanisms involved include both a reduction in left ventricle afterload and hence in systolic wall stress, and an inhibition of the direct growth stimulating properties of ANG II on cardiac myocytes and fibroblasts. The second mechanism is clearly independent from the first, since it has recently been shown that ramipril, in non-antihypertensive doses, prevents the development of aortic banding-induced left ventricular hypertrophy in rats. However, besides blockade of ANG II formation, the antimitogenic properties of the nitric oxide released by bradykinin may also contribute.[30]

From a functional point of view, ACE inhibitor-induced reversal (or prevention) of interstitial fibrosis opposes myocardial stiffening and abnormal mechanical behavior.[31] In addition, reversal (or prevention) of perivascular fibrosis and medial thickening of the intramural coronary arteries preserves coronary dilatation reserve and decreases the propensity to arrhythmogenesis. Thus ACE inhibitors, depending upon preventive or curative administration, exhibit cardioprotective or cardioreparative properties.[29] More importantly, ACE inhibitor-induced regression or prevention of left ventricular hypertrophy results in suppression during the treatment period of the so-called cardiac pressor amplifier.[28] However in SHRs, upon treatment withdrawal, left ventricular hypertrophy resumes a progressive but attenuated development, in parallel with the slow redevelopment of the rise in blood pressure.

It thus appears that, if the immediate systemic and regional hemodynamic actions of ACE inhibitors (resulting from circulating ACE inhibition) initiate their antihypertensive effects, the subsequent and simultaneous suppression of the vascular and cardiac pressor amplifier mechanisms (resulting from tissue ACE inhibition) is responsible for the maintenance of normal blood pressure.

ACE inhibitors and stroke

ACE inhibitors have been shown to afford cerebrovascular protection in the genetically stroke-prone spontaneously hypertensive rat (SHR-SP). In this special strain of rats, the animals develop a malignant hypertension, with blood pressure values reaching 250–260 mmHg. Non-genetic factors, such as replacing normal drinking water with a 1% sodium chloride solution, maintaining the animals on a low protein and low potassium diet and housing the animals individually, reinforce the blood pressure increase, accelerate the onset of stroke and decrease life span (less than 20 weeks in males). The animals develop cerebral infarcts and/or hemorrhage and histological observations reveal vascular alterations characterized by hyperplasia of intima and media cells, and fibrinoid necrosis. Myocardial infarcts, malignant nephroangiosclerosis and proteinuria develop simultaneously. Finally, the RAS which is initially depressed because of the ingestion of 1% sodium chloride, is secondarily markedly stimulated and vascular catecholamine content is consistently increased.

In this experimental model, ACE inhibitors lower blood pressure, prevent the development of proteinuria and exhibit cerebrovascular, cardiac and renal protection.[32] Moreover, they strongly decrease and/or prevent stroke-related mortality.[32] Although these beneficial effects have been shown to develop at low doses and to be independent of the drugs' antihypertensive effects, their mechanisms are not fully understood and could involve the reduction of ANG II formation and of its detrimental vascular effects, the increase in tissue kinins and a stimulation of prostaglandin synthesis.

ACE inhibitors and heart failure

ACE inhibitors have been investigated and shown to exert beneficial effects in almost all types of experimental models of heart failure, whether resulting from increased afterload (arterial hypertension), coronary flow impediment (following rapid ventricular pacing in dogs or coronary ligation in rats), volume overload (aortocaval fistula), or cardiomyopathy (either adriamycine-induced or hereditary, as in syrian hamsters). The beneficial effects of ACE inhibitors in heart failure usually develop at doses lower than those used in experimental hypertension.

Cardiac index is not similarly affected in the different experimental models, being increased in the volumetric loading model, unchanged in postischemic cardiac insufficiency or decreased after barometric loading, cardiac pacing, adriamycine or in the pre-edematous state in syrian hamsters. But whatever the model, the latest stage of the disease is always characterized by a depressed left ventricular performance with a fall in cardiac contractility (dP/dt) and a rise in left ventricular end-diastolic pressure (LVEDP). Congestive heart failure is also associated with neurohormonal changes in relation to activation of the RAS and of the sympathetic nervous system. Plasma aldosterone is not necessarily augmented, possibly because of the inhibitory effect of ANF, which is increased because of cardiac dilatation. From a structural point of view, cardiac hypertrophy, initially adaptative and beneficial, eventually leads to myocardial incompetence and ventricular dilatation. The viable myocardial mass tends to decrease and fibrosis develops.

Hemodynamic effects

Whatever the experimental model of heart failure, long-term treatment with ACE inhibitors (captopril, enalapril, trandolapril, quinapril) limits the development of ventricular dilatation and improves left ventricular function, as shown by increased dP/dt and ejection fraction index and decreased LVEDP values.[35] ACE inhibitors decrease blood pressure or prevent its normalization when it has previously been lowered by myocardial infarction,[33,34] thus resulting in decreased left ventricular loading. Simultaneously, ACE inhibitors increase diuresis which, together with venous dilatation, contributes to decrease preload.[34] When the cardiac index is decreased, ACE inhibitors normalize it. When the cardiac index is normal, ACE inhibitors improve its response to acute experimental ventricular loading.

Hormonal effects

Besides their classical effects on plasma renin, ANG I and ANG II levels, ACE inhibitors, because of the induced improvement in cardiac function, decrease or even normalize plasma ANF and, as a result, plasma and urinary cGMP values. Noradrenaline plasma levels are also decreased.

Structural effects

ACE inhibitors oppose the process of cardiac remodeling which is deleterious in the long term. Thus, these drugs decrease global cardiac hypertrophy, which affects both left and right ventricles as shown by a significant reduction in heart weight. Factors other than hemodynamic ones are involved in opposing the growth process, for example suppression of ANG II and its trophic role, and accumulation of kinins. Limitation of cardiac hypertrophy is associated with a reduction in myocardial fibrosis, both in the subepicardium and subendocardium, which contributes to increase ventricular compliance and to improve cardiac performance.[33,34] Limitation of fibrosis also occurs at the vascular level, especially in the coronary arteries, resulting in an increase in coronary compliance and dilatation reserve which contributes to better myocardial perfusion and improved survival.

Survival

As a result of their beneficial hemodynamic and structural effects, ACE inhibitors significantly improve survival rate and increase life expectancy in all experimental models of cardiac insufficiency.[32,34–36] In post-myocardial infarction heart failure, survival is increased even in animals with large infarcts. This is shown by the fact that, in contrast to the situation in untreated animals, survival is not correlated with infarct size.

ACE inhibitors and ischemic myocardium

Cardioprotection

Pioneering experimental[35] and clinical studies initially demonstrated that long-term inhibition of the RAS by ACE inhibitors prevents excessive ventricular dilatation, and ultimately cardiac failure, following myocardial infarction. Thus, in a large number of experimental models of ischemic myocardium, acute ACE inhibitor administration was shown to limit infarct size.[37] This beneficial effect was initially attributed to inhibition, in the ischemic myocardium, of the detrimental actions of ANG II (for example coronary vasoconstriction, positive inotropic effect, release of endogenous catecholamines through presynaptic activation) which reinforce the imbalance of myocardial oxygen supply/demand and exacerbate the ischemic process resulting from temporary or permanent coronary artery occlusion. However, other mechanisms have also been postulated to explain ACE inhibitor-induced limitation of infarct size. On the one hand, inhibition of oxygen free-radical-dependent reperfusion injury by sulfhydryl-containing ACE inhibitors (which exhibit antioxidant effects) has been proposed to explain their protective effects in the stunned myocardium.[38] However, the fact that ramipril, a non-sulfhydryl-containing ACE inhibitor exerts a similar protective action tends to question this mechanism. On the other hand, inhibition of bradykinin degradation by ACE inhibitors could also be involved in ischemic myocardium protection, as bradykinin improves cardiac function following ischemia in isolated rat hearts, and as HOE 140, a selective and potent bradykinin antagonist, abolishes ramipril-induced infarct size limitation.

It must be stressed, however, that this cardioprotective effect of ACE inhibitors is not universally recognized. Numerous studies have failed to demonstrate that captopril can salvage the jeopardized myocardium following an ischemic insult.[39] Duration and severity of ischemia, occurrence of reperfusion, resting RAS tone and, above all, collateral blood flow could partly explain these conflicting data. Indeed it has recently been shown that captopril exacerbates the ultimate infarct size in dogs with high pretreatment collateral flow but reduces it in animals with previously poorly developed collateral circulation.[39]

Antiarrhythmic effects

It is now well established that ACE inhibitors can exert protective effects against reperfusion arrhythmias and that this property is independent of their ability to scavenge oxygen free radicals since all ACE inhibitors, whether they contain a sulphydryl group or not, reduce to a similar extent the incidence of fibrillation during reperfusion. Similarly, this protective effect is not linked to their peripheral vasodilating properties as it is observed in both in vitro and in vivo preparations. At least two

ACE inhibitor	Ligand	Prodrug	t_{max} (h)	Bioavail-ability (%)	Protein binding (%)	Elimination route (%)	T1/2 ACE (h)	Cl_R (l/h)	V_D (l)
Captopril	Sulf.	N	1	70	30	K			
Alacepril	Sulf.	Y	1–2						
Zofenopril	Sulf.	Y	0.5–1.2	78	88	K			
Enalapril		Y	1						
Enalaprilat	Carb.		3–4	40	50	K	30–35		
Perindopril		Y	0.4–1.2	66				3–4	
Perindoprilat	Carb.		2–4	20	20	K(L?)	>30	6–10	9.3
Ramipril		Y	0.7		73				
Ramiprilat	Carb.		2.3	54	56	K	110	7	
Quinapril		Y	0.6	>37	97			111	
Quinaprilat	Carb.		1.5		97	K(L?)		13	
Benazepril		Y	0.5		97				
Benazeprilat	Carb.		1–2	28	95	K(L?)	21–22	1.54	8.7
Cilazapril		Y	0.8	77					25.4
Cilazaprilat	Carb.		1–2	52		K	37–50	10.8	20
Trandolapril		Y	0.5–1		80				
Trandolaprilat	Carb.		6–8	40–60	80–94	K		3.8	
Spirapril		Y	0.6						
Spiraprilat	Carb.		2			L(65) K(35)			
Lisinopril	Carb.	N	6–8	25	10	K	30–40	6.36	2
Fosinopril		Y		36					
Fosinoprilat	Phosp.		3	29	95	K(50) L(50)	12	1.02	9.8
Ceranapril	Phosp.	N	4	66	50	K	15		

Abbreviations. Sulf. sulfhydryl-containing ACE inhibitors; Carb. carboxyalkyl dipeptidic ACE inhibitors; Phosp. phosphorus-containing ACE inhibitors; N no; Y yes; K kidney; L liver.

Table 2.1
Pharmacokinetic properties of ACE inhibitors. Adapted from Burnier and Biollaz.[3]

mechanisms involved in the reduction of post-ischemic arrhythmias have been identified — inhibition of the formation of ANG II, and prolongation of the action of bradykinin. On the one hand, Schölkens et al.[40] have shown that addition of ANG I and ANG II after ischemia in isolated rat hearts increases the duration of ventricular fibrillation, whereas bradykinin reduces it. Conversely, inhibition of the formation of ANG II with either renin inhibitors or with ACE inhibitors (regardless of their chemical structure), and inhibition of ANG II effects by non-peptidic antagonists of ANG II AT_1 receptors reduce the incidence of ventricular fibrillation in rat hearts when they are administered before reperfusion at doses which do not alter coronary blood flow. This also indicates that ANG II produced during reperfusion is an important factor, directly involved in ventricular fibrillation. On the other hand, the deleterious effects of ANG II on global ischemic hearts are prevented by bradykinin perfusion in a concentration-dependent manner, and combination of subthreshold concentrations of bradykinin and of ramiprilat, which given alone are ineffective, results in a marked cardioprotective and antiarrhythmic effect in isolated rat hearts. That HOE 140 suppresses these beneficial effects further supports the view that bradykinin largely contributes to the cardioprotective and antiarrhythmic effects of ACE inhibitors. The mechanisms involved include bradykinin-induced preservation of high-energy phosphates, increase in glucose uptake, and release of prostacyclin and nitric oxide from the vascular endothelium.

Appendix

Pharmacokinetics of ACE inhibitors in humans (Table 2.1)

Most ACE inhibitors are prodrugs developed to circumvent poor absorption of the active compounds. For the carboxyalkyl dipeptidic ACE inhibitors (enalapril-type) and for fosinopril, the prodrug is a monoester monoacid molecule that is converted into the corresponding diacid active drug. Exceptions are captopril, a well-absorbed thiol molecule, and lisinopril and ceranapril which are diacid molecules, the absorption of which is improved by the addition of an aminobutyl side-chain.

Absorption of captopril and the prodrugs is usually rapid with T_{max} values between 0.5 and 1.5 hours. Absorption of lisinopril and ceranapril is slower, with T_{max} values of 6–8 and 4 hours respectively.

Conversion of prodrugs into their active metabolite is usually rapid, and occurs mainly in the liver but also in the plasma, kidney, etc. The T_{max} values of active diacid metabolites range between one and four hours, with the exception of trandolaprilat (6–8 hours).

After oral administration, bioavailability of active ACE inhibitor molecules ranges between 25% and 70%. In the plasma, protein binding of ACE inhibitors also varies greatly, ranging from 10% to 97%.

The majority of ACE inhibitors are eliminated, mostly in the form of the active molecules, by the kidney through glomerular filtration that may or may not be associated with tubular excretion. Spirapril is mostly cleared by the biliary tract (65%) and fosinopril is cleared equally by the kidney and the biliary tract.

The elimination half-life of captopril is short (0.7–1.9 hours) as is that of the prodrugs (range: 1–3 hours). In contrast, the decline in plasma concentrations of the diacid active metabolites is polyphasic, probably reflecting the binding to various components and ultimately the binding to circulating or fixed ACE (T1/2 ACE).

Severe renal impairment affects the pharmacokinetics of those of the ACE inhibitors that are cleared mostly by the kidney. When creatine clearance (Cl_{CR}) values are below 30 ml/min/1.73 m^2, plasma concentrations and the area under the curves (AUCs) of the active metabolites rise and dosage reduction is recommended. Hepatic impairment usually has little effect on ACE inhibitor pharmacokinetics. In congestive heart failure, clearance of ACE inhibitors is usually reduced. Finally, in elderly subjects, AUC and plasma levels of ACE inhibitors are often increased due to reduced clearance.

References

1 Ondetti MA, Rubin B, Cushman DW. Design of specific inhibitors of angiotensin-converting enzyme: a new class of orally active antihypertensive agents. *Science* 1977; **196**:441–46.

2 Wei L, Alhenc-Gelas F, Corvol P, Clauser E. The two homologous domains of human angiotensin I-converting enzyme are both catalytically active. *J Biol Chem* 1991; **266**:9002–8.

3 Burnier M, Biollaz J. Pharmacokinetic optimisation of angiotensin converting enzyme (ACE) inhibitor therapy. *Clin Pharmacokin* 1992; **22**:375–84.

4 Unger T, Gohlke P, Paul M, Rettig R. Tissue renin-angiotensin systems: fact or fiction? *J Cardiovasc Pharmacol* 1991; **18[Suppl 2]**:S20–S25.

5 Hodsman GP, Zabludowski JR, Zoccali C, Fraser R, Morton JJ, Murray GD, Robertson JIS. Enalapril (MK 421) and its lysine analogue (MK 521): a comparison of acute and chronic effects on blood pressure, renin-angiotensin system and sodium excretion in normal man. *Br J Clin Pharmacol* 1984; **17**:233–41.

6 Nussberger J, Juillerat L, Perret F, Waeber B, Bellet M, Brunner J, Ménard J. Need for plasma angiotensin measurements to investigate converting-enzyme inhibition in humans. *Am Heart J* 1989; **117**:717–21.

7 Fabris B, Chen B, Pupic V, Perich R, Johnston CI. Inhibition of angiotensin-converting enzyme (ACE) in plasma and tissue. *J Cardiovasc Pharmacol* 1990; **15[Suppl 2]**:S6–S13.

8 Giudicelli JF, Berdeaux A, Edouard A, Richer C, Jacolot D. The effect of enalapril on baroreceptor mediated reflex function in normal man. *Br J Clin Pharmacol* 1985; **20**:211–18.

9 Ajayi AA, Campbell BC, Meredith PA, Kelman AW, Reid JL. The effect of captopril on the reflex control of heart rate: possible mechanisms. *Br J Clin Pharmacol* 1985; **20**:17–25.

10 Levy BI, Michel JB, Salzmann JL, Azizi M, Poitevin P, Safar M, Camilleri JP. Effects of chronic inhibition of converting enzyme on mechanical and structural properties of arteries in rat renovascular hypertension. *Circ Res* 1988; **63**:227–39.

11 Richer C, Thuillez C, Giudicelli JF. Perindopril, converting enzyme blockade and peripheral arterial hemodynamics in the healthy volunteer. *J Cardiovasc Pharmacol* 1987; **9**:94–102.

12 Hall JE, Guyton AC, Jackson TE, Coleman TG, Lohmeier TE, Trippodo NC. Control of glomerular filtration rate by renin angiotensin system. *Am J Physiol* 1979; **236**:F252–F259.

13 Zucker IH, Chen JS, Wang W. Renal sympathetic nerves and hemodynamic responses to captopril in conscious dogs: role of prostaglandins. *Am J Physiol* 1991; **260**:H260–H266.

14 Demolis P, Chalon S, Annane D, Duhaze P, Giudicelli JF. Effect of an angiotensin converting enzyme inhibitor, ramipril, on intracranial circulation in healthy volunteers. *Br J Clin Pharmacol* 1992; **34**:224–30.

15 Noguchi K, Kato T, Ito H, Aniya Y, Sakanashi M. Effect of intracoronary captopril on coronary blood flow and regional myocardial function in dogs. *Eur J Pharmacol* 1985; **110**:11–19.

16 Berdeaux A, Bonhenry C, Giudicelli JF. Effects of four angiotensin I converting enzyme inhibitors on regional myocardial blood flow and ischemic injury during coronary artery occlusion in dogs. *Fundam Clin Pharmacol* 1987; **1**:201–12.

17 Holtz J, Busse R, Sommer O, Bassenge E. Dilation of epicardial arteries in conscious dogs induced by angiotensin-converting enzyme inhibition with enalaprilat. *J Cardiovasc Pharmacol* 1987; **9**:348–55.

18 Giudicelli JF, Freslon JL, Glasson S, Richer C. Captopril and hypertension development in the SHR. *Clin Exp Hypertens* 1980; **2**:1083–96.

19 Dzau VJ. Circulating versus local renin-angiotensin system in cardiovascular homeostasis. *Circulation* 1988; **77[Suppl 1]**:I4–I13.

20 Cachofeiro V, Sakakibara T, Nasjletti A. Kinins, nitric oxide, and the hypotensive effect of captopril and ramiprilat in hypertension. *Hypertension* 1992; **19**:138–45.

21 Antonaccio MJ, Kerwin L. Pre- and post-junctional inhibition of vascular sympathetic function by captopril in SHR. Implication of vascular angiotensin II in hypertension and antihypertensive actions of captopril. *Hypertension* 1981; **3[Suppl 1]**:154–62.

22 Richer C, Mulder P, Fornes P, Richard V, Camilleri JP, Giudicelli JF. Hemodynamic and morphological effects of quinapril during genetic hypertension development. *J Cardiovasc Pharmacol* 1991; **18**:631–42.

23 Freslon JL, Giudicelli JF. Compared myocardial and vascular effects of captopril and dihydralazine during hypertension development in spontaneously hypertensive rats. *Br J Pharmacol* 1983; **80**:533–43.

24 Christensen KL, Jespersen LT, Mulvany MJ. Development of blood pressure in spontaneously hypertensive rats after withdrawal of long term treatment related to vascular structure. *J Hypertens* 1989; **7**:83–90.

25 Scalbert E, Levy B, Devissaguet M, Safar ME. Vascular effects of perindopril: from experimental to clinical investigation. *J Cardiovasc Pharmacol* 1991; **18[Suppl 7]**:S25–S32.

26 Asmar RG, Pannier B, Santoni JP, Laurent S, London GM, Levy BI, Safar ME. Reversion of cardiac hypertrophy and reduced arterial compliance after converting enzyme inhibition in essential hypertension. *Circulation* 1988; **78**:941–50.

27 Adams MA, Bobik A, Korner PI. Enalapril can prevent vascular amplifier development in spontaneously hypertensive rat. *Hypertension* 1990; **16**:252–60.

28 Korner PI, Bobik A, Jennings GL, Angus JA, Anderson WP. Significance of cardiovascular hypertrophy in the development and maintenance of hypertension. *J Cardiovasc Pharmacol* 1991; **17[Suppl 2]**:S25–S32.

29 Weber KT, Brilla CG, Janicki JS. Signals for the remodeling of the cardiac interstitium in systemic hypertension. *J Cardiovasc Pharmacol* 1991; **17[Suppl 2]**:S14–S19.

30 Linz W, Scholkëns BA. A specific B_2-bradykinin receptor antagonist HOE 140 abolishes the antihypertrophic effect of ramipril. *Br J. Pharmacol* 1992; **105**:771–72.

31 Anversa P, Palackal T, Sonnenblick EH, Olivetti G, Meggs LG, Capasso JM. Myocyte cell loss and myocyte cellular hyperplasia in the hypertrophied aging rat heart. *Circ Res* 1990; **67**:871–85.

32 Stier CT, Benter IF, Ahmad S, Zvo H, Selig N, Roethel S, Levine S, Itskovitz H. Enalapril prevents stroke and kidney dysfunction in salt-loaded stroke-prone spontaneously hypertensive rats. *Hypertension* 1989; **13**:115–21.

33 Michel JB, Lattion AL, Salzmann JL, Cerol ML, Philippe M, Camilleri JP, Corvol P. Hormonal and cardiac effects of converting enzyme inhibition in rat myocardial infarction. *Circ Res* 1988; **62**:641–50.

34 Richer C, Mulder P, Fornes P, Domergue V, Heudes D, Giudicelli JF. Long-term treatment with trandolapril opposes cardiac remodeling and prolongs survival after myocardial infarction in rats. *J Cardiovasc Pharmacol* 1992; **20**:147–56.

35 Pfeffer MA, Pfeffer JM, Steinberg C, Finn P. Survival after an experimental myocardial infarction: beneficial effects of long-term therapy with captopril. *Circulation* 1985; **72**:406–12.

36 Sweet CS, Emmert SE, Stabilito II, Ribeiro LGT. Increased survival in rats with congestive heart failure treated with enalapril. *J Cardiovasc Pharmacol* 1987; **10**:636–42.

37 Ertl G, Kloner RA, Alexander W, Braunwald E. Limitation of experimental infarct size by an angiotensin converting enzyme inhibitor. *Circulation* 1982; **65**:40–48.

38 Przyklenk K, Kloner RA. Relationships between structure and effects of ACE inhibitors: comparative effects in myocardial ischaemic reperfusion injury. *Br J Clin Pharmacol* 1989; **28**:167–75.

39 De Lorgeril M, Ovize M, Delaye J, Renaud S. Importance of the flow perfusion deficit in the response to captopril in experimental myocardial infarction. *J Cardiovasc Pharmacol* 1992; **19**:324–29.

40 Schölkens BA, Linz W, Lindpainter K, Ganten D. Angiotensin deteriorates but bradykinin improves cardiac function following ischaemia in isolated rat hearts. *J Hypertens* 1987; **5[Suppl 5]**:S7–S9.

SECTION II

CURRENT CLINICAL PRACTICE

3

ACE inhibitors in hypertension

Michael Schachter

It is inevitable that this chapter will overlap to some extent with others in this book: on renovascular disease, on left ventricular hypertrophy and on adverse reactions, among others. However, it is hoped that there will not be too much irritating repetition. The focus of the chapter will therefore be on the use of angiotensin converting enzyme inhibitors (ACE inhibitors) in essential hypertension, especially in comparison with other classes of hypotensive agent or in combination with them, and to some extent the comparison of ACE inhibitors with one another. This aspect is particularly difficult, since at present no-one is sure which apparent differences, other than pharmacokinetic ones, are of real clinical importance.

Background and mechanisms

There are several plausible mechanisms by which angiotensin converting enzyme inhibitors (ACE inhibitors) may lower blood pressure. The relative importance of these mechanisms is controversial, even in experimental models, and still more so in human hypertension. The following are the principal modes of action according to the current, but not necessarily wholly correct, consensus.[1,2]

i) reduction in levels of circulating angiotensin II (ANG II) due to inhibition of converting enzyme in the endothelium, principally in the lung;

ii) inhibition of enzyme activity in the vascular wall and in the heart;

iii) increased levels of vasodilator kinins, such as bradykinin, which are also metabolized by ACE inhibitors;

iv) the reduction of ANG II-mediated sodium retention by effects on intrarenal blood flow and possibly also by counteracting the direct effects of ANG II on the proximal tubule;

v) reduction in the synthesis of aldosterone, hence also reducing sodium reabsorption (while increasing potassium retention);

vi) reduction in noradrenaline release from sympathetic nerve endings;

vii) possibly a central effect on baroreceptor function mediated by the brain's renin-angiotensin system (RAS);

viii) alteration of long-term structural changes in the vasculature leading to attenuation of the increased peripheral resistance.

There is little dispute that the first of these mechanisms can cause very profound acute falls in blood pressure, notably in patients who have high circulating renin levels. This can comprise several clinical groups:

- patients with renovascular hypertension, who can have greatly elevated plasma renin levels, regardless of sodium intake;
- patients who are sodium depleted, usually due to diuretic therapy;

- patients already treated with vasodilators, such as calcium channel blockers. The practical implications of ACE inhibitor therapy in these groups will be considered below.

By contrast, elderly patients with normal renal vasculature[3] and most black hypertensives from the USA or of Afro-Caribbean origin,[4] have relatively low renin levels in relation to their salt intake. It should therefore, in theory, be relatively easy to predict which hypertensive individuals are likely to benefit from chronic ACE inhibitor treatment, even as monotherapy. In fact this is only partly the case. While 'high-renin' patients are consistently good responders in terms of blood pressure lowering, similar responses are often or usually seen in low-renin patients: in the latter, the adoption of a low-sodium diet can enhance ACE inhibitor efficacy. The actual rate of response is very variable between studies, but these are often designed or analysed in substantially different ways.[5] However, it must be anticipated that a minority of patients will totally fail to respond to any ACE inhibitor, either acutely or chronically. At the moment, there is no wholly accurate way of predicting response in a given individual.

ACE inhibitors as monotherapy in essential hypertension

Captopril was the first ACE inhibitor to be used in antihypertensive therapy, usually in combination with a diuretic. Although initial studies from the late 1970s onwards confirmed the efficacy of the drug, they also produced alarming side-effects.[6] These included not only renal damage, presenting as proteinuria, deteriorating renal function or both, but also bone marrow dyscrasias and liver toxicity. Not surprisingly therefore, captopril was regarded as a second- or even third-line drug, to be used where other therapy had been unsuccessful. However, it was appreciated that these problems could be greatly improved by more careful patient assessment, particularly with respect to renovascular disease, and even more importantly by reducing captopril dosage as far as possible. This meant that 150 mg of captopril daily came to be considered as the maximum dose, only rarely to be reached, rather than a mean dose as in the early trials. Indeed, some studies found little or no benefit in raising captopril dosage above 37.5 mg per day.[1] At this level the drug is very well tolerated. Enalapril was the next ACE inhibitor to be clinically evaluated, again showing a relatively flat dose–response curve with regard to blood pressure lowering, with a range of 2.5–40 mg per day.[7,8] In fact, 10 mg was rarely exceeded. Similar dose–response relationships have been reported for numerous other ACE inhibitors: some of this data is summarized in Table 3.1, which is not intended to be comprehensive. The percentage of patients satisfactorily controlled, as defined by diastolic blood pressure of < 90 mmHg and a systolic pressure of < 160 mmHg, is generally quoted as 50–70%,[5] with a few reports indicating an even higher success rate.[9]

ACE inhibitors and diuretics

As indicated earlier, most of the earliest antihypertensive trials with ACE inhibitors used them in conjunction with diuretics, almost always thiazides and this has remained common clinical practice.[10–12] Such a combination is rational and relatively straightforward, on at least two counts. Firstly, the sodium-depleted high-renin state associated

Drug	Prodrug	Lipophilicity	t1/2 (h)	Daily dose (mg/day; frequency)
Captopril	No	Low	1–6	12.5–150 (1–2)
Cilazapril	Yes	Medium	9	1.25–5 (1)
Enalapril	Yes	Medium	8+	2.5–40 (1–2)
Fosinopril	Yes	High	12+	10–80 (1)
Lisinopril	No	Low	6–12	10–40 (1–2?)
Perindopril	Yes	Low	27+	2–8 (1)
Quinapril	Yes	Medium	2–3	5–40 (1–2)
Ramipril	Yes	Low	36+	1.25–10 (1)
Trandolapril	Yes	Low	16+	0.5–4 (1)

Table 3.1
Summary of data on ACE inhibitors used in the treatment of hypertension in the UK (early 1994). For prodrugs, the t1/2 values are for the active metabolite.

with diuretic therapy will increase the ANG II-dependence of the blood pressure and make it more susceptible to the removal of that hormonal stimulus. It should therefore make ACE inhibitors more effective in patients with low-renin hypertension, as in the case in blacks.[13] Secondly, it is widely accepted that diuresis is not the only or even the principal mode of action by which the thiazides lower blood pressure.[14]

The actual mechanism has still to be determined, but certainly raises the strong possibility of synergism with ACE inhibitors and possibly other classes of drugs. This allows the use of lower doses of both drugs than would otherwise be practicable, with consequent reduction of toxicity, especially since the adverse effects of the two types of drug are not additive. For instance, a further benefit of a combined regime is the maintenance of normal plasma potassium. The two classes of agents have opposing effects, ACE inhibitors tending to raise plasma potassium and thiazides tending to lower it. Given simultaneously, they will usually maintain normokalaemia. More speculatively, it has been suggested that ACE inhibitors can improve glucose tolerance in hypertensive patients, while there has long been evidence that thiazides have the opposite effect, at least at high doses.[15] The clinical significance of either change, in terms of cardiovascular morbidity and mortality, remains unproven but there is great interest in insulin resistance as a central feature of a metabolic syndrome of obesity, hypertension dyslipidaemia and non-insulin-dependent diabetes mellitus.[16] It is worth emphasizing that sodium restriction can further lower blood pressure, even in patients taking both ACE inhibitors and diuretics.

Drugs (daily doses)		Total numbers	Assessment	Reference
Captopril (100 mg)	Propranolol (160 mg)	425	C = P	18
Enalapril (10–40 mg)	Propranolol (80–240 mg)	46	E = P	19
Enalapril (10–40 mg)	Atenolol (50–100 mg)	180	E = 4	20
Perindopril (2–8 mg)	Atenolol (25–100 mg)	172	P = A	21
Lisinopril (10–40 mg)	Atenolol (50–100 mg)	144	L = A DBP L > A SBP	22

Table 3.2
Some comparative studies of ACE inhibitors and beta-blockers. All the trials were randomized and double-blind, with parallel

ACE inhibitors and other antihypertensive drugs: comparisons and combinations

Beta-adrenergic blockers

The ACE inhibitors have been more extensively compared with the beta-blockers than with any other class of hypotensive therapy. One main reason for this is historical: when the ACE inhibitors first emerged, β-blockers were very widely used throughout the world as first-line therapy for hypertension and were, in a sense, the principal 'competition'. There is also a parallel in terms of mechanism, even if an indirect one. Beta-blockers reduce renin release by a direct effect on the renal juxtaglomerular cells and will therefore lower ANG II levels.[17] Some of the comparative studies are listed in Table 3.2, for illustration rather than comprehensive coverage. In general, antihypertensive efficacy is similar for the two classes of drugs. It should be noted — and advocates of thiazide and β-blocker therapy rightly do so — that there are still no long-term outcome studies available using an ACE inhibitor in comparison with one or more other first-line agents. Trials of this kind are of course lengthy and very expensive, but it is likely that they will take place in Europe and the USA before the end of the decade.

One highly influential outcome of such comparisons concerns 'quality of life' (QOL) assessments. It is well-known, indeed notorious, that patient compliance is a major problem in the treatment of hypertension. Patients with mild to moderate elevation of blood pressure are usually asymptomatic. Treatment with, for instance β-blockers, can give rise to a wide range of adverse effects, such as lethargy, vivid dreams, impaired memory and sexual dysfunction. Most of these are not dangerous, but are nonetheless unpleasant. It is not too surprising that many patients take such medication irregularly, if at all. Several trials have indicated that one or more of the above side-effects are diminished or eliminated on ACE inhibitor therapy,[19,23]

Drugs (daily doses)		Total numbers	Assessment	Reference
Captopril (75 mg)	Nifedipine retard (40 mg)	19	C < N*	28
Captopril (100 mg)	Nitrendipine (20 mg)	24	C < N*	29
Enalapril (10–40 mg)	Nifedipine retard (20–80 mg)	128	E = N†	30
Lisinopril (20–80 mg)	Nifedipine retard (40–80 mg)	136	L = N	9

*The ACE inhibitor and calcium channel blocker were also used in combination in these studies, with better responses.
†This is the assessment for *seated* BP. The drugs were equivalent for standing BP.

Table 3.3
Some comparative studies of ACE inhibitors and calcium channel blockers. All studies were randomized and parallel but not necessarily double-blind.

though other studies have not confirmed this conclusion.[24,25] Nonetheless, this perspective has undoubtedly had a very significant role in the extraordinarily rapid increase in ACE inhibitor usage in populations where the relatively high acquisition costs are acceptable.

At first sight combination ACE inhibitor and β-blocker therapy would not seem particularly rational or advantageous. As noted previously, β-blockers lower plasma renin and hence would not be expected to enhance the effect of the ACE inhibitors — if anything, the reverse. The existing clinical evidence is contradictory and rather sketchy, with some studies showing unexpected additional hypotensive effects with combined therapy.[26] This situation is not uncommon in the treatment of hypertension, where 'irrational' combinations may work surprisingly well. However, the question needs further study if it is to be resolved.

Calcium channel blockers

The introduction of the calcium channel blockers (CCBs) considerably antedates the use of ACE inhibitors, although it is worth bearing in mind that these drugs were originally intended as anti-anginal agents rather than hypotensives. Comparative studies have been somewhat less extensive than with beta-blockers, but even so many combinations have been tried — not surprisingly, nifedipine has been the most widely investigated CCB in this context. A selection of these studies are reviewed above (Table 3.3). It will be seen that the hypotensive effect of the CCB was generally the greater, or there was no significant difference between the two agents. However, the CCB was often poorly tolerated with a high rate of discontinuation. It must be borne in mind that the patients in these trials were not selected on the basis of their renin-ANG II status, and the groups must therefore include patients where the ACE inhibitors may be expected to be less effective. For instance, in one of the trials cited, over half the patients were black.

Both ACE inhibitors and CCBs are at least in part 'vasodilators', but their mechanisms of action are quite different. Combining the types of agent is therefore reasonable, both in terms of efficacy and tolerability. Several trials have confirmed that this is the case in practice,[9,30] with apparent synergism in at least one study.[31] It is not clear whether the CCBs are likely to make previously resistant patients more sensitive to the CCBs, but it would be unsurprising if they did, given their renin-raising activity.

ACE inhibitors in acute and refractory hypertension

Although ACE inhibitors, particularly captopril, have been used for acute hypertensive emergencies, by both oral and parenteral routes,[32,33] they cannot be regarded as ideal for this purpose. Even though no serious mishaps occurred in the small groups studied, this is no grounds for complacency. Severe acute hypertension may be associated with renovascular disease or some other cause of extremely high plasma renin activity. Under those circumstances the fall in blood pressure may be precipitate and very difficult to control. Since safer alternatives exist, it is difficult to envisage the situation where the ACE inhibitors would be drugs of choice. One rare exception may be scleroderma crisis, where accelerated hypertension is accompanied by rapid decline in renal function.[34,35]

In severe chronic hypertension apparently refractory to vasodilator and beta-blocker therapy, the response to ACE inhibitors may be dramatic,[36] though the first dose effect must always be anticipated where renin activity is likely to be high. In other patients the addition of an ACE inhibitor may nonetheless be helpful, if less spectacular.[37] However, most clinicians will be looking after a small number of essential hypertensives who need three, four or five drugs of different classes for even imperfect blood pressure control. In these patients, very often with low or normal plasma renin levels, the ACE inhibitors have no special advantages. Such patients provide one justifiable impetus for the development of new hypotensive compounds.

ACE inhibitors and isolated systolic hypertension

The traditional emphasis on the importance of diastolic blood pressure has recently been undergoing modification, particularly in the elderly. Several recent clinical trials have drawn attention to the benefits of lowering high systolic blood pressures (usually those greater than 160 mmHg) in terms of reduction of cardiac and cerebrovascular events and in reducing total mortality in this group of patients.[38–40] It is generally accepted that no currently available antihypertensive agent is truly selective for systolic as opposed to diastolic blood pressure. There is also broad agreement that changes in the compliance of large, conduit arteries is a major factor in the pathogenesis of isolated systolic hypertension.[41] Reduced compliance, in other words increased rigidity, would allow increased pressure and therefore wall stress in the downstream resistance small arteries and arterioles. It is likely that this will cause remodelling or hypertrophy in these resistance vessels, though there is little direct evidence that this does occur. On the other hand, it is clear that ACE inhibitors do have a favourable effect on large arterial compliance, both in animal models and in man.[42,43] There is accompanying increase in brachial artery diameter and a reduction in the velocity of the pulse wave. This is not a property unique to these agents but is not necessarily common to vasodilators of other types. The clinical use of ACE inhibitors in this situation appears to be well-founded and their efficacy is not in doubt, despite the theoretical considerations discussed earlier.[44,45]

ACE inhibitors in special patient groups

The use of ACE inhibitors in several such groups is discussed in detail in other chapters in this book — for instance in diabetes, patients with renovascular disease and in blacks. Some others will be considered in this section.

Asthma and other chronic airways disease

ACE inhibitors have generally been considered safe in these patients and until very recently there was no reason to question the validity of this view,[46] despite the notorious tendency of these drugs to cause a dry irritating cough.[47] However, it now seems that they can occasionally cause bronchospasm and dyspnoea, and that patients with asthma may be more susceptible than the rest of the population.[48] This adverse reaction may occur in conjunction with cough, but this is not necessarily the case. In the series reported, cough actually occurred about 10 times as often as other respiratory side-effects. Nonetheless, use of ACE inhibitors in the patient with airways disease clearly requires some caution.

Coronary artery disease

The ACE inhibitors are not indicated as anti-anginal drugs, though some reports suggest an improvement in symptoms, perhaps mediated by improved left ventricular function.[49,50] In general, the effect is probably neutral and there is nothing to indicate that ACE inhibitors are likely to be harmful. Most clinicians would nevertheless consider β-blockers or calcium channel blockers as more likely first-line therapy in this group, except where contraindicated. The possible anti-atherosclerotic properties of ACE inhibitors are discussed in another chapter.

Peripheral vascular disease

Potentially beneficial effects of ACE inhibitors on medium to large calibre vessels have already been mentioned. However, it is unlikely that they will be apparent in severely atherosclerotic peripheral arteries in the lower limb. Not surprisingly, it appears that they are preferable to β-blockers in patients with claudication,[51] but this appears to represent maintenance of existing function rather than actual improvement.

Hypertension and heart failure

Although the association between acute left ventricular failure and hypertension is clear and represents one of the major hypertensive emergencies, the patient with chronic congestive heart failure is not commonly hypertensive. However, where the two conditions do coexist ACE inhibitors are a rational therapeutic choice. They should be beneficial for both, since there are several common pathophysiological mechanisms which may be influenced: sodium retention, peripheral vasoconstriction, and increased sympathetic activity. However, it must be borne in mind that these patients are often sodium- and volume-depleted due to previous diuretic therapy and there must be particularly close supervision in case of drastic hypotension.[52] Ideally, diuretics should be withdrawn for several days before introducing ACE inhibitors, but some patients are unable to tolerate this without distress. Of the other classes of antihypertensive drugs only the diuretics are unequivocally safe, though α-adrenergic blockers and calcium antagonists have a somewhat controversial status. It has also recently been suggested that β-blockers can help to alleviate refractory heart failure, but this is not orthodox therapy at present.

Hypertension in pregnancy

There is evidence for a teratogenic effect of ACE inhibitors affecting several systems, and animal studies also suggest that they may impair placental perfusion.[53,54] It is hardly surprising therefore that these drugs are considered to be strictly contraindicated in pregnancy-associated hypertension. It is also recommended that they should only be used in women of child-bearing age who are using effective contraception.

Which ACE inhibitor?

In early 1994 there were no fewer than nine ACE inhibitors licensed in the United Kingdom for the treatment of hypertension. This reflects several considerations. Firstly, hypertension represents an enormous market in the developed world. Secondly, the hypertensive population represents very disparate individuals, with inter-individual differences which are poorly understood and an enormously variable response to medication. It is therefore tempting to assume that there will always be niches, not just for different classes of drugs, but also for a variety of drugs within one class (cf. verapamil and the dihydropyridines). However, it is also increasingly necessary to persuade prescribers and purchasing authorities that a drug is not merely a 'me too'.

The ACE inhibitors differ from one another in numerous respects, as outlined below, but it should be emphasized that there is no convincing evidence that they have significantly different efficacy in lowering blood pressure. The issue is therefore whether the other differences are of real importance.

Pharmacokinetics

Some of the principal pharmacokinetic properties of currently used ACE inhibitors are summarized in Table 3.1. It is obvious that most of the drugs are pro-drugs and can be given once daily. This is widely considered to be useful, both in the avoidance of blood pressure fluctuations, and in the encouragement of compliance. Nonetheless, the relatively short-acting captopril is still very widely used, particularly when initiating ACE inhibitor therapy. The reason for this is partly historical, but in fact the limited duration of action is potentially an important advantage in case of severe 'first dose' hypotension. Another pharmacokinetic consideration relates to the route of elimination of the drug. In this context it has been suggested that fosinopril may be safer in patients with impaired renal function, since in these circumstances, hepatic metabolism of the drug assumes increasing importance.[55]

Inhibition of tissue ACE

It is now accepted that many tissues, notably the heart and at least some of the vasculature, possess an autonomous renin-angiotensin system. While all the ACE inhibitors reduce circulating levels of ACE, which is principally derived from the luminal surface of endothelial cells, tissue penetration differs very widely. This is related, at least in part, to the lipophilicity of the drug or its active metabolite but there are certainly other factors involved. For instance, fosinopril and quinapril are regarded as having good tissue penetration, while captopril and enalapril are less active in this respect.[56,57] How much does this matter? There is at present no compelling evidence that it does, though comparative outcome studies in hypertensives remain a welcome, if expensive, objective. However, it has been claimed in one study that fosinopril has a more beneficial effect on left ventricular function than captopril or lisinopril.[58] Other well-penetrating ACE inhibitors were not compared.

Non-renin-angiotensin effects of ACE inhibitors

It has long been known that ACE inhibitors have numerous actions unrelated to the renin-angiotensin system.[59] The most widely studied of these concern the inhibition of the breakdown of vasodilator kinins (notably bradykinin and substance P), though there is no consensus on the therapeutic relevance of this mechanism. Animal studies suggest that they may be important but one must always be cautious about extrapolation to hypertensive humans.[60] In any case, it is not clear that ACE inhibitors differ significantly in their ability to inhibit kinin breakdown. Another potentially important cellular event is the generation of vasodilator prostanoids such as prostacyclin. This is stimulated by bradykinin, and hence indirectly by all ACE inhibitors. However, captopril appears to have a direct effect on prostanoid synthesis and this appears to be additive with that of the kinins.[61] Again, we do not know whether this is of real importance in therapy.

Adverse effects

The above unique action of captopril is believed to be related to its free SH group, which is not present in the other currently used ACE inhibitors. This structural feature is probably responsible for some of the toxic reactions seen with high-dose captopril therapy: photosensitive rashes, proteinuria, neutropenia, taste disturbance and the rare occurrence of liver damage. This is strikingly similar to the adverse effects seen with penicillamine, which also has a free SH group. They are much less common with other ACE inhibitors, but it should be noted that dosage regimes for these agents have been more cautious than was originally the case with captopril.

ACE inhibitors in the management of hypertension

From the considerations discussed in this chapter and elsewhere in this book can we describe an 'ideal' hypertensive candidate for ACE inhibitor therapy? Yes, but at the usual risk of being overwhelmed by exceptions. Tentatively, the following lists the relevant qualifications:

- young or middle-aged male, or woman of child-bearing age with little or no risk of pregnancy;
- high-renin hypertension, with particular caution in patients with bilateral renovascular disease (see chapter 4);
- patient with diabetes, especially with microproteinuria or nephropathy;

- patient with dyslipidaemia of other origin;
- presence of peripheral atherosclerosis, but again with particular caution about renovascular disease;
- concurrent congestive heart failure.

It has also been asserted that the ACE inhibitors are especially advantageous in patients where it is 'particularly' important to preserve alertness, cognitive function and sexual function. One may be tempted to ask why this should be regarded as a special favour rather than a matter of course!

Conclusion and prospects

There can be no dispute that the ACE inhibitors have already proved to be extremely valuable additions to the therapeutic options available in hypertension. If anything, their versatility has exceeded expectations. There have also been welcome bonuses in terms of effects on left ventricular function, diabetic nephropathy and perhaps on the progression of atherosclerosis. At present the choice of individual ACE inhibitor from the present extensive (excessive?) list is largely a matter of habit and pharmacokinetics. It is too early to say whether the other apparent differences really matter, or whether others will emerge. After all, the recent TOMH study has shown that it is very difficult to differentiate between the major classes of antihypertensive agents in terms of efficacy and other parameters.[62]

Other components of the renin-angiotensin system are also therapeutic targets. Orally active renin antagonists and blockers of the angiotensin II AT_1 receptor are under development and both types of drug are effective hypotensives.[63] It remains to be seen whether they prove to have any advantages over the ACE inhibitors in terms of efficacy or tolerability: for instance, are they less likely to cause cough? It will also be interesting to note whether the pharmaceutical industry will consider it worthwhile to expend the enormous amount of time and money required to market these novel drugs in an exceptionally crowded and competitive field. For at least one angiotensin II antagonist, losartan, this now seems highly probable. Apart from any potential clinical benefits, these new agents will certainly enhance our understanding of the renin-angiotensin system in health and disease.

References

1 Unger T, Gohlke P, Gruber MG. Converting enzyme inhibitors. In: Ganten D, Mulrow PJ, eds. Handbook of experimental pharmacology vol. 93. Berlin: Springer Verlag, 1990.

2 Unger T, Gohlke P. Converting enzyme inhibitors in cardiovascular therapy: current status and future potential. *Cardiovasc Res* 1994; **28**:146–58.

3 Breckenridge A. Age-related effects of angiotensin converting enzyme inhibitors. *J Cardiovasc Pharmacol* 1988; **12[Suppl 8]**:S100–S104.

4 Savage DD, Watkins LO, Grim CE, Kumanyika SK. Hypertension in black populations. In: Latragh JH, Brenner BM, eds. Hypertension: pathophysiology, diagnosis and management. New York: Raven Press, 1990.

5 Williams GH. Converting-enzyme inhibitors in the treatment of hypertension. *New Engl J Med* 1988; **319**:1517–25.

6 Jenkins AC, Dreslinski GR, Tadros SS et al. Captopril in hypertension: seven years later. *J Cardiovasc Pharmacol* 1985; **7**:S96–S101.

7 Bergstrand R, Herlitz H, Johansson S et al. Effective dose range of enalapril in mild to moderate essential hypertension. *Br J Clin Pharmacol* 1985; **19**:605–11.

8 Salvetti A, Arzilli F. Chronic dose-response curve of enalapril in essential hypertensives. An Italian multicenter study. *Am J Hypertens* 1989; **2**:352–54.

9 Morlin C, Baglivo H, Boejinga JK et al. Comparative trial of lisinopril and nifedipine in mild to severe essential hypertension. *J Cardiovasc Pharmacol* 1987; **9[Suppl 3]**:S48–S52.

10 MacGregor GA, Markandu ND, Banks RA et al. Captopril in essential hypertension: contrasting effects of adding hydrochlorothiazide or propranolol. *BMJ* 1982; **284**:693–96.

11 Vlasses PH, Rotmensch HH, Swanson BN et al. Comparative antihypertensive effects of enalapril maleate and hydrochlorothiazide, alone and in combination. *J Clin Pharmacol* 1983; **23**:227–33.

12 Sassano P, Chatelier G, Billaud et al. Comparison of increase in the enalapril dose and addition of hydrochlorothiazide as second-step treatment of hypertensive patients not controlled by enalapril alone. *J Cardiovasc Pharmacol* 1989; **13**:314–19.

13 Pool JL, Gennari J, Goldstein R et al. Controlled multicenter study of antihypertensive effects of lisinopril, hydrochlorothiazide and lisinopril plus hydrochlorothiazide in the treatment of 394 patients with mild to moderate essential hypertension. *J Cardiovasc Pharmacol* 1987; **9[Suppl 3]**:S36–S42.

14 Kramer HJ. Diuretics. In: Ganten D, Mulro PJ, eds. Handbook of experimental hypertension, vol. 93. Berlin: Springer Verlag, 1990.

15 Pollare T, Lithell H, Berne C. A comparison of the effects of hydrochlorothiazide and captopril on glucose and lipid metabolism in patients with hypertension. *N Engl J Med* 1989; **321**:868–73.

16 DeFronzo RA, Ferrannini E. Insulin resistance. A multifaceted syndrome responsible for NIDDM, obesity, hypertension, dyslipidaemia and atherosclerotic cardiovascular disease. *Diabetes Care* 1991; **14**:173–94.

17 Leonetti G, Mayer G, Morganti A et al. Hypotensive and renin suppressive activities of propranolol in hypertensive patients. *Clin Sci Mol Med* 1975; **48**:491–99.

18 Croog SH, Levine S, Testa MA et al. The effect of antihypertensive therapy on the quality of life. *N Engl J Med* 1986; **314**:1657–64.

19 Simon AL, Levenson JA, Bouthier JD et al. Comparison of oral MK-421 and propranolol in mild to moderate essential hypertension and their effects on arterial and venous vessels of the forearm. *Am J Cardiol* 1984; **53**:781–85.

20 Canadian Enalapril Study Group. Comparison of monotherapy with enalapril and atenolol in mild to moderate hypertension. *Can Med Assoc J* 1987; **137**:803–8.

21 Morgan TO. Australian Multicenter Study of perindopril compared to atenolol in the management of hypertension. *JAMA* 1990; **6[Suppl]**:18–22.

22 Beevers DG, Blackwood RA, Garnham S et al. Comparison of lisinopril versus atenolol for mild to moderate essential hypertension. *Am J Cardiol* 1991; **67**:59–62.
23 Lichter I, Richardson PJ, Wyke MA. Differential effects of atenolol and enalapril on memory during treatment for essential hypertension. *Br J Clin Pharmacol* 1986; **21**:641–45.
24 Herick AL, Waller PC, Bern KE et al. Comparison of enalapril and atenolol in mild to moderate essential hypertension. *Am J Med* 1989; **86**:421–26.
25 Steiner SS, Friedhoff AJ, Wilson BL, Wecker JR, Santo JP. Antihypertensive therapy and quality of life: a comparison of atenolol, captopril, enalapril and propranolol. *J Hum Hypertens* 1990; **4**:217–25.
26 Belz GG, Essig J, Erb K et al. Pharmacokinetic and pharmacodynamic interactions between the ACE inhibitor cilazapril and the beta-adrenoceptor antagonist propranolol in healthy subjects and hypertensive patients. *Br J Clin Pharmacol* 1989; **27**:317S–322S.
27 Singer DRJ, Markandu ND, Shore AC, MacGregor GA. Captopril and nifedipine in combination for essential hypertension. *Hypertension* 1987; **17**:798–803.
28 Gennari C, Nami R, Pavese G et al. Calcium-channel blockade (nitendipine) in combination with ACE inhibition (captopril) in the treatment of mild to moderate hypertension. *Cardiovasc Drug Ther* 1989; **3**:319–25.
29 MacLean D, Ramsay LE, Richardson PJ. Enalapril and nifedipine in the treatment of mild to moderate essential hypertension: a 6 month comparison. *Br J Clin Pharmacol* 1990; **30**:203–11.
30 Brouwer RML, Bolli P, Erne P et al. Antihypertensive treatment using calcium antagonists in combination with captopril rather than diuretics. *J Cardiovasc Pharmacol* 1985; **7**:S88–S91.
31 Gennari C, Nami R, Pavese G et al. Calcium-channel blockade (nitrendipine) in combination with ACE inhibition (captopril) in the treatment of mild to moderate hypertension. *Cardiovasc Drug Ther* 1988; **2**:211–18.
32 Biollaz J, Waeber B, Brunner HR. Hypertensive crisis treated with orally administered captopril. *Eur J Clin Pharmacol* 1983; **25**:145–49.
33 Padfield PL, Rademaker M, Atherden SM et al. Inhibition of the renin-angiotensin-aldosterone axis by low-dose captopril as a treatment for accelerated phase hypertension. *J Hypertens* 1985; **3[Suppl 3]**:S475–77.
34 Lopez-Ovejero JA, Saal SD, D'Angelo WA et al. Reversal of vascular and renal crises of scleroderma by oral angiotensin-converting-enzyme blockade. *N Engl J Med* 1979; **300**:1417–19.
35 Milsom SR, Nicholls MG. Successful treatment of scleroderma renal crisis with enalapril. *Postgrad Med J* 1986; **62**:1059–60.
36 White NJ, Rajagopalan B, Yahaya H, Ledingham JGC. Captopril and frusemide in severe drug-resistant hypertension. *Lancet* 1980; **ii**:108–10.
37 Swales JD, Bing RF, Heagerty AM et al. Treating refractory hypertension. *Lancet* 1982; **i**:894–96.
38 SHEP Co-operative Research Group. Prevention of stroke by antihypertensive drug treatment in older persons with isolated systolic hypertension. Final results of the Systolic Hypertension in the Elderly Program (SHEP). *JAMA* 1991; **265**:3255–64.
39 Dahlof B, Lindholm LH, Hansson L et al. Morbidity and mortality in the Swedish Trial in Old Patients with Hypertension (STOP-Hypertension). *Lancet* 1991; **338**:1281–85.
40 MRC Working Party. Medical Research Trial of treatment of hypertension in older adults: principal results. *BMJ* 1992; **304**:405–12.
41 Frishman WH. Epidemiology, pathophysiology and management of isolated systolic hypertension. *Am J Med* 1991; **90[Suppl 4B]**:14S–20S.
42 Asmar RG, Pannier B, Santoni JP et al. Reversion of cardiac hypertrophy and reduced arterial compliance after converting enzyme inhibition in essential hypertension. *Circulation* 1988; **78**:941–50.
43 Asmar RG, Journo HJ, Lacolley PJ et al. Treatment for one year with perindopril: effect on cardiac mass and arterial compliance in essential hypertension. *J Hypertens* 1988; **6[Suppl 3]**:S33–S39.
44 Mulinari R, Gavras I, Gavras H. Efficacy and tolerability of enalapril monotherapy in mild to moderate hypertension in older patients compared to younger patients. *Clin Ther* 1987; 9:678–89.

45 Suraniti S, Berrut G, Marre M, Fressinaud P. Antihypertensive efficacy and acceptability of perindopril in elderly hypertensive patients. *Am J Cardiol* 1993; **71**:28E–31E.
46 Kaufman J, Schmitt S, Barnard J, Busse W. Angiotensin-converting enzyme inhibitors in patients with bronchial responsiveness and asthma. *Chest* 1992; **101**:922–25.
47 Just PM. The positive association of cough with angiotensin-converting enzyme inhibitors. *Pharmacotherapy* 1989; **9**:82–87.
48 Lunde H, Hedner T, Samuelsson O et al. Dyspnoea, asthma and bronchospasm in relation to treatment with angiotensin converting enzyme inhibitors. *BMJ* 1994; **308**:18–21.
49 Gibbs JSR, Crean PA, Mockus L et al. The variable effects of angiotensin converting enzyme inhibition on myocardial ischaemia in chronic stable angina. *Br Heart J* 1989; **62**:112–17.
50 Ikram H, Low CJS, Shirlaw T, Webb CM, Richards AM, Crozier IG. Antianginal, hemodynamic and coronary vascular effects of captopril in stable angina pectoris. *Am J Cardiol* 1990; **66**:164–67.
51 Roberts DF, Tsao Y, McLoughlin GA, Breckenridge A. Placebo-controlled comparison of captopril, atenolol, labetalol and pindolol in hypertension complicated by intermittent claudication. *Lancet* 1987; **ii**:650–53.
52 Cleland JGF, Dargie HJ, McAlpine H, Ball SG, Robertson JIS, Ford I. Severe hypotension after first dose of captopril in heart failure. *BMJ* 1985; **291**:1309–12.
53 Boutroy MJ, Vert P, Huralt de Ligney B, Milton A. Captopril administration in pregnancy impairs fetal angiotensin converting enzyme activity and neonatal adaptation. *Lancet* 1984; **ii**:935–36.
54 Knott PD, Thorpe SS, Lamont CAR. Congenital renal dysgenesis possibly due to captopril. *Lancet* 1989; **i**:451.
55 Hui KK, Duchin KL, Kripalani KJ et al. Pharmacokinetics of fosinopril in patients with various degrees of renal function. *Clin Pharmacol Ther* 1991; **49**:457–67.
56 Gohlke P, Urbach H, Scholkens BA, Unger T. Inhibition of converting enzyme in the cerebrospinal fluid of rats after oral treatment with converting enzyme inhibitors. *J Pharmacol Exp Ther* 1989; **249**:609–16.
57 Johnston CI, Jandeleit K, Mooser V et al. Angiotensin-converting enzyme and its inhibition in the heart and blood vessels. *J Cardiovasc Pharmacol* 1992; **20[Suppl B]**:S6–S11.
58 Zusman RM. Angiotensin-converting enzyme inhibitors: more different than alike? Focus on cardiac performance. *Am J Cardiol* 1993; **72**:25H–36H.
59 Sunman W, Sever PS. Non-angiotensin effects of angiotensin-converting enzyme inhibitors. *Clin Sci* 1993; **85**:661–70.
60 Bao G, Gohlke P, Qadri F, Unger T. Chronic kinin receptor blockade attenuates the antihypertensive effect of ramipril. *Hypertension* 1992; **20**:74–79.
61 Zusman RM. Effect of converting enzyme inhibitors on the renin-angiotensin-aldosterone, bradykinin and prostaglandin systems: correlation of chemical structure and biological activity. *Am J Kidney Dis* 1987; **10**:113–23.
62 Neaton JD, Grimm RH, Prineas RJ et al. Treatment of mild hypertension study. Final results. *JAMA* 1993; **270**:713–24.
63 Foote EF, Halsenston CE. New therapeutic agents in the management of hypertension: angiotensin II receptor antagonists and renin inhibitors. *Ann Pharmacother* 1993; **27**:1495–1503.

4

ACE inhibitors in renal and renovascular disease

Graham W Lipkin and Anthony EG Raine

Introduction

Approximately 1 in 200 patients have renal hypertension or chronic renal failure. Furthermore renal and renovascular disease are the commonest cause of secondary hypertension. In the last 10 years major advances have been made in our understanding of the mechanisms underlying the inexorable progression of chronic renal failure, and there has been increased appreciation of the relatively high incidence of atherosclerotic renal arterial disease in the elderly, which may lead to both renovascular hypertension and renovascular chronic renal failure. Increasingly sensitive non-invasive diagnostic tests have been developed for these conditions. Drug therapy has become available which is capable of retarding both the progression of incipient diabetic nephropathy and also the progression of established diabetic renal failure to end-stage renal disease.

Angiotensin converting enzyme inhibitors have played a significant part in all of these developments. This chapter aims to assess the evidence for their role, to review the renal effects of ACE inhibition in health and renal disease and to consider the potential adverse renal effects of these drugs and their pharmacokinetics in renal failure.

The renin-angiotensin system and the kidney

A basic understanding of the physiology of the renin-angiotensin system in relation to the kidney is essential to the understanding of the beneficial and deleterious effects of ACE inhibitors in renal disease. Almost 100 years ago, Tigerstedt and Bergman demonstrated that extracts of rabbit kidney cortex contained a vasopressive substance, which they named renin.[1] Subsequent investigation established that renin is an enzyme which acts on a high molecular weight precursor arising from the liver, angiotensinogen, to produce biologically inactive angiotensin I (ANG I). Removal of two end-terminal amino acids by angiotensin converting enzyme located mainly on endothelial cells produces angiotensin II (ANG II) which exerts its biological activity by binding to AT_1 and AT_2 receptors. Angiotensin II directly affects peripheral vascular resistance by causing constriction of resistance vessels and acts indirectly by facilitation of sympathetic nerve transmission. Plasma volume is increased through enhancement by angiotensin II of proximal tubular sodium and water reabsorption, and indirectly through angiotensin II-induced release of aldosterone, which then acts to increase sodium reabsorption in the distal nephron. Renin is released from granules in specialized cells within the afferent glomerular arteriole, the juxtaglomerular apparatus. The number of cells producing renin varies according to dietary sodium content, treatment with ACE inhibitors and diuretic usage.

Renin release is influenced by several different control mechanisms.[2] Baroreceptors in the wall of the afferent glomerular arteriole respond to a reduction in renal arterial perfusion pressure by including a non-linear rise in renin release, which increases sharply when mean arterial pressure is reduced below 90 mmHg. Simultaneous renal sympathetic nerve stimulation increases the sensitivity of this mechanism. A specialized area of the thick ascending limb of loop of Henle, the macula densa, lies in close proximity to the juxtaglomerular apparatus. Renin release by the afferent arteriole is inversely related to saline (specifically chloride) delivery to the macula densa within the physiological range. In addition, the juxtaglomerular cells themselves are innervated by sympathetic nerves. Low-level renal nerve stimulation increases renin release, an effect which is blocked by beta-adrenoceptor antagonists.

Perturbations of endothelial factors, particularly nitric oxide and endothelin, may exist in renal disease. Endothelin appears to inhibit renin release in a number of preparations in vitro, although these are not confirmed by low-dose endothelin infusion studies in man. Nitric oxide may also mediate the inhibition of renin release signalled by the macula densa. Atrial natriuretic factor, the plasma concentration of which is increased in renal disease, acts to inhibit renin release directly.[3] Systemic and locally produced angiotensin II directly inhibits the transcription rate of the renin gene and release of renin. This is in addition to its promotion of aldosterone secretion, which acts indirectly to suppress renin by causing salt and water retention and a rise in blood pressure.

The effects of ACE inhibition on the kidney

Administration of captopril to healthy volunteers induces an increase in renal blood flow and a reduction in mean arterial blood pressure, the magnitude of which is increased by prior activation of the renin-angiotensin system by a low-salt diet.[4] This effect appears to be exaggerated in patients with essential hypertension. Its mechanism largely involves loss of the vasoconstrictor action of angiotensin II on the efferent arteriole, although angiotensin II also produces afferent arteriolar vasoconstriction.[5]

Normally, when hypertension is induced, the glomerular filtration rate is well maintained during reductions in renal perfusion pressure as far as 70 mmHg. An intact renin-angiotensin system is required for this autoregulation, as co-administration of ACE inhibitors leads to a marked reduction in GFR. In addition to the glomerular arteriolar effects of ANG II, which indirectly modulate GFR, angiotensin II binds directly to glomerular mesangial cells, enhancing their contractility and thus reducing glomerular permeability and the glomerular ultrafiltration coefficient. This may in part explain the ability of ACE inhibition to reduce glomerular proteinuria in patients with renal disease.[6]

In humans, ACE inhibition increases renin secretion and consequently circulating angiotensin I levels as a result of reduction in the normal angiotensin II-mediated negative feedback suppression of renin release. This effect is dependent upon the state of activation of the renin system[7] and has relevance to the use of ACE inhibitors in the diagnosis of renovascular disease. In addition, administration of an ACE inhibitor to normal volunteers results in an acute natriuresis, consistent with inhibition of aldosterone and angiotensin II.

Pharmacokinetics of ACE inhibitors in renal disease

Captopril, enalapril, lisinopril, cilazapril and ramipril all rely predominantly upon renal excretion (filtration and secretion).[8,9] In patients with renal impairment, therefore, the use of standard doses would be expected to result in high plasma concentrations, with an increased risk of side-effects. The half-life of captopril in normal subjects is 0.6 hours compared with 3.5 hours in mild chronic renal failure and 32 hours in haemodialysis patients. It has therefore been recommended that the maintenance dose of captopril in end-stage renal failure should be reduced by 75%.[10] Similarly, enalapril has a half-life of 7 hours in normal subjects and 35 hours in severe renal impairment and it has been recommended that the dose be reduced by 50% in patients with a GFR below 30 ml/min.[11] Thirty-five per cent of an administered dose of captopril is removed during a four hour haemodialysis, although the metabolites are more slowly removed. In contrast, 70% of enalapril is removed by four hours' haemodialysis, the dialysis clearance of enalapril being approximately 68 ml/min.

The altered pharmacokinetics of ACE inhibitors in renal disease have practical implications for haemodialysis patients. Immediately after a dialysis session, such patients may be volume depleted, and administration of an ACE inhibitor at that point can result in a prolonged hypotensive effect. Currently, the only available angiotensin converting enzyme inhibitor available with significant hepatic metabolism is fosinopril. In renal impairment, its hepatic clearance increases relative to renal excretion and therefore no dose adjustment is required in renal failure.[11]

In other aspects, the clinical significance of the differing pharmacokinetics of the ACE inhibitors is not established. However, care should be taken to adjust the dose of the renally excreted inhibitors to reduce excessive hypotensive and toxic side-effects.

Non-renin-angiotensin effect of ACE inhibitors

Angiotensin converting enzyme is involved in the breakdown of bradykinin, and in humans the vasodilatory effect of arterial bradykinin is enhanced by a sub-hypotensive infusion of enalaprilat.[12] In the kidney, bradykinin is a potent natriuretic and vasodilator agent. In anaesthetized rabbits, captopril-induced reductions in blood pressure and increases in renal blood flow may be abolished by the administration of a bradykinin antagonist.[13] The relative importance of reduction in intrarenal angiotensin II and accumulation of bradykinin in mediating the renal effects of ACE inhibition should become clearer with the availability of specific angiotensin II-receptor-antagonists such as losartin.

Renal side-effects of ACE inhibitors

Isolated reports have implicated ACE inhibitors as a cause of interstitial nephritis and glycosuria.[14] The risks of this effect must be small considering the millions of prescriptions issued for ACE inhibitors each year. Of more concern, in early reports captopril was associated with the development of membranous glomerulonephritis.[15] Investigation of these reports suggested that at least some patients may have had preceding membranous glomerulonephritis. The initial doses of captopril were far in excess of those used today and with use of lower doses these side-effects have considerably diminished.

As glomerular filtration rate falls in patients with renal disease, serum potassium tends to rise. Inhibition of aldosterone production by administration of ACE inhibitors further reduces potassium excretion and may precipitate life-threatening hyperkalaemia, especially when combined with potassium supplements or potassium-sparing diuretics. Recent evidence indicates that in patients without significant functioning renal tissue (dialysis/anephric patients), ACE inhibitors may still increase serum potassium levels. The explanation of this is uncertain but may involve diminished faecal excretion of potassium, since the bowel contains similar transport mechanisms to the distal nephron.

Other side-effects of ACE inhibitors tend to be more common in patients with renal disease, probably as a result of unexpected drug accumulation. Serious adverse effects such as neutropenia occurred, particularly in patients with renal failure.[16] An unpredicted side-effect has been noted between ACE inhibitors and one particular haemodialysis membrane. Anaphylactoid reactions have been reported in patients receiving angiotensin converting enzyme inhibitors whilst undergoing haemodialysis with the polyacrylonitrile AN69 dialyser. The mechanism of this interaction is unknown, but may involve bradykinin accumulation. ACE inhibitors are currently recommended to be avoided in patients being dialysed with such a membrane.

Renal failure induced by ACE inhibitors in renovascular disease

Angiotensin converting enzyme inhibitors may induce a severe decline in renal function when renal perfusion pressure is critically reduced. Acute renal failure requiring dialysis therapy has been reported relatively frequently since the introduction of ACE inhibitors in the early 1980s. It is associated especially with the presence of advanced bilateral renal artery stenosis or renal artery stenosis to a solitary functioning kidney or a transplanted kidney.[17,18]

Renovascular disease is the commonest cause of secondary hypertension, accounting for 1–5% of unselected hypertensives and around 10–30% of patients presenting with malignant phase hypertension.[19] It has also recently been recognized as a relatively common underlying aetiology of chronic renal failure associated with hypertension in the elderly (14% of patients over the age of 50).[20] Furthermore, moderate/severe atheromatous renal artery stenosis occurs in up to 50% of patients being investigated for peripheral or coronary vascular disease.[21]

The potential for development of acute renal failure in patients with bilateral renal artery stenosis has led some to recommend avoidance of ACE inhibitors as a first-line antihypertensive drug therapy. Alternatively, simple measures to reduce the risks in treating patients with these useful agents may be recommended. When assessing patients with hypertension or congestive cardiac failure, note should be taken of the clinical variables commonly associated with renal atheroma — age, smoking, ischaemic heart disease, peripheral vascular disease, and abdominal or generalized arterial bruits. Patients with malignant, severe or resistant hypertension should have their renal function checked before and within one week of starting ACE inhibitor therapy.

Biochemical screening and urinalysis are essential in order to identify renal disease in hypertensive patients before starting ACE inhibitor therapy. Patients receiving therapy should have regular clinical and biochemical monitoring to detect functional renal deterioration at an early stage. If the presence of renovascular disease is strongly suspected, ACE inhibitors should not be administered as antihypertensive treatment unless screening tests employing ACE inhibitor renography and/or angiography exclude this possibility.[22]

The reasons for renal functional deterioration associated with ACE inhibition have been clarified in animal studies. When renal perfusion is reduced, glomerular filtration rate is maintained by enhanced renin release which augments intrarenal angiotensin II formation, thereby increasing efferent arteriolar tone.[23] Inhibition of angiotensin converting enzyme, by blocking angiotensin II formation, causes efferent arteriolar vasodilation, which normally leads to an increase in renal plasma flow. However, if renal perfusion is limited by fixed stenoses of the renal arteries or arterioles, renal plasma flow cannot increase to compensate and hence glomerular perfusion pressure is reduced, resulting in a precipitous fall in glomerular filtration rate. This explanation would account for the development of acute renal failure in patients with established renal disease, who may subsequently turn out not to have major renal artery stenoses.[24]

These explanations are supported by further clinical observations. Reduction in blood pressure by non-ACE inhibitor therapy does not generally lead to the same fall in glomerular filtration rate.[17] Confirmatory evidence comes from animal studies in which equipotent blood pressure lowering with triple therapy in hypertensive rats with bilaterally constricted arteries did not impair kidney function, in contrast to ACE inhibition.[25] In addition, in some patients where critical advanced bilateral renal artery narrowing is present, blood pressure lowering per se may lower GFR by non-specific loss of perfusion pressure.[26]

In a recent retrospective analysis covering a six-year period, in 21 of 530 patients who presented with an acute uraemic emergency there was a close relationship to the introduction of ACE inhibitor therapy.[24] Although reversal of acute renal failure is the norm following withdrawal of ACE inhibitor therapy, prolonged dialysis support may be required and lack of recovery, which is probably related to renal arterial thrombosis, has been reported. Indeed in the series of Kalra et al. five out of 21 patients failed to regain renal function.[24]

It is essential that any patient who presents with reversible renal impairment following introduction of an ACE inhibitor should be subsequently investigated for the presence of renovascular disease, which may be amenable to correction by percutaneous transluminal renal angioplasty (PTRA) or surgical revascularization.

The enhancement of screening tests for renovascular disease by ACE inhibitors

Successful identification of patients with significant renovascular disease at an early stage offers the possibility of treatment which may improve hypertension and slow the progression of renal failure.[27,28] However, whilst renal artery stenosis is relatively common, the incidence of haemodynamically significant renovascular disease is much less prevalent. Screening tests for this condition must therefore have a high sensitivity and specificity, to detect functionally significant rather than purely anatomical renal artery narrowing.

Currently-available tests

Plasma renin activity, rapid sequence intravenous urography, and isotope dynamic renography have traditionally been used as screening methods for renovascular disease but lack adequate sensitivity and specificity and have therefore largely been abandoned.[29] Intravenous digital subtraction angiography is expensive and fails to visualize the renal artery in up to 20% of cases. It also exposes the patient to the risk of contrast nephropathy.[30]

Duplex Doppler ultrasound of the renal artery is well established in the diagnosis of transplant renal artery stenosis. With advances in technology and the advent of colour flow mapping to assist in renal artery localization it has been applied to the diagnosis of native renal artery stenosis. It is, however, very operator dependent, time consuming and fails to localize the renal artery in 20% of cases. It is therefore unlikely to become a routine screening test.[31] Indeed, the non-invasive tests so far described do not have adequate sensitivity and specificity to screen a condition with a relatively low incidence, even in a high-risk population.

Recent recognition of the importance of the renin-angiotensin system in mediating the compensatory renal haemodynamic changes in a hypoperfused kidney has led to the increasing use of ACE inhibitors in increasing the diagnostic accuracy of the existing screening tests.[32,33]

Intravenous urography

Until recently, rapid sequence intravenous urography was the mainstay of the non-invasive diagnosis of renovascular disease. The cooperative study of renovascular hypertension established three main criteria to distinguish patients with a positive test:

- discrepancy in renal size of greater than 1.5 cm;
- discrepancy in appearance time of calyceal contrast;
- discrepancy in contrast media concentration on late films.[34,35]

Two of the three positive diagnostic criteria are GFR-dependent and the sensitivity of the test may be increased by prior treatment with an angiotensin converting enzyme inhibitor, although this may heighten the risk of contrast nephropathy in patients with chronic renal failure. While enhancement of the IVU has been demonstrated in the case of the 2-kidney–1-clip Goldblatt hypertensive dog,[36] little clinical evidence is available to date.

Peripheral plasma renin activity

Plasma renin activity is usually higher in patients with renal artery stenosis than control subjects. Nevertheless, literature reviews have indicated that the sensitivity and specificity of this test for diagnosing renovascular disease are only in the region of 57% and 66% respectively.[32] An early study showed that administration of an ACE inhibitor resulted in marked increases in plasma renin activity in patients with renovascular hypertension.[37] The mechanism of the exaggerated rise in plasma renin activity may result from loss of angiotensin II-mediated suppression of renal renin release. In addition, the reduction in GFR associated with reduced sodium chloride delivery to the macula densa may itself stimulate renin release.

In a retrospective study, Muller et al. defined criteria for a positive captopril test. Using these criteria a sensitivity of 100% and specificity of

95% were achieved in diagnosing renovascular disease in patients with normal renal function, although sensitivity and specificity was considerably reduced in patients with renal failure.[38] Two recent prospective studies have come to opposing conclusions concerning this test.[39,40] Although the study of Postma[40] may have been affected by the inclusion of patients with insignificant renovascular disease, it has been generally concluded that the captopril-stimulated renin test is not suitable for routine screening.[41] In addition, the test is affected by other antihypertensive medication, posture and the state of sodium balance.

Renal vein renin

In animal models of renal artery stenosis, renin release is increased by the clipped kidney and is suppressed in the contralateral kidney. Some workers have therefore used the invasive measurement of renal vein renin sampling in the diagnosis of renovascular disease. Several studies have reported that the administration of an ACE inhibitor improves the diagnostic accuracy of this test,[42] but this has not been confirmed by a recent large study.[43]

ACE inhibitor induced changes in the isotope renogram

Isolated dynamic renography in the diagnosis of significant renovascular disease has largely been replaced by studies before and after administration of the ACE inhibitor captopril[29] with improved diagnostic accuracy.[44] This approach may provide evidence of the functional significance of renal artery stenosis. It relies on the fact that in the presence of a fixed renal artery stenosis, GFR is maintained by angiotensin II-mediated constriction of the efferent arteriole. Loss of angiotensin II following ACE inhibition leads to a marked fall in GFR. In unilateral renovascular disease, overall renal function is not affected.[45] However, the loss of GFR by one kidney may be detected by determining split renal function using the isotopes $^{99m}T_c$ DTPA, MAG 3, of ^{123}I-hippuran. The exact mechanism remains speculative as it is technically not possible to obtain reliable data using micropuncture studies in post-stenotic kidneys after ACE inhibition.

The diagnostic features of significant renal artery stenosis have been derived from assessment of the classical Goldblatt animal models of renal vascular disease — the 2-kidney–1-clip model, representing unilateral renal artery stenosis, and the 1-kidney–1-clip model, representing stenosis of a solitary functioning kidney or transplant kidney. Following ACE inhibition in the 2-kidney–1-clip Goldblatt hypertensive dog there is a reduction in GFR, urine flow, and renal cortical elimination of the isotope in the post-stenotic kidney.[46] Reports of the effects of ACE inhibition on renal haemodynamics in the 1-kidney–1-clip model are more variable.[47]

Initial studies of ACE inhibitor renography using either DTPA or hippuran indicated sensitivities of 80–100% and specificities of 72–100% for the diagnosis of significant renovascular disease.[44,48–50] The sensitivity of the test may be increased by the use of hippuran rather than DTPA.[51] Data from later studies including larger numbers of patients which compared captopril renography with angiographically proven significant renovascular disease demonstrated a sensitivity and specificity of roughly 85%.[52–55] The largest of these studies, which include 424 patients, 230 of whom had renal artery stenosis greater than 70% on angiography, demonstrated a sensitivity and specificity of the test of 80% and 84% respectively.

The reasons for the differences in sensitivity and specificity noted in the different studies remains unexplained. There are, however, a number of factors which may be relevant. Definition of significant renal artery stenosis on angiography remains controversial: the only true gold standard definition of renovascular hypertension or renovascular renal failure is the improvement in blood pressure or renal function in patients following revascularization. Certainly, the definition of a significant degree of angiographic stenosis varies between studies.[54,55] Indeed the multicentre study[54] indicated a better sensitivity when more severe narrowing of the renal artery was present.

In addition, sensitivity of renography is reduced in patients with renal impairment[55] and the number of patients with chronic renal failure differs between the different studies. There is as yet no consensus as to which radiopharmaceutical should be used: DTPA is a GFR marker, whereas hippuran and MAG 3 reflect renal blood flow, as they are both filtered and secreted. Furthermore, there is at present no standardization between study protocols in terms of continuation or withdrawal of non-ACE inhibitor medication, sodium balance, hydration and dose of captopril use, or the criteria applied for diagnosing a positive test.[32]

With regard to the latter point, some centres have used comparison of GFR changes within the same kidney both before and after captopril, whereas others have used comparison between kidneys both before and after captopril. Other centres have used derived renographic variables of time to peak activity and cortical retention time (transit time). Further studies are required to elucidate the optimal protocol, isotope and preparation of the patient.

From the practical perspective, ACE inhibitor renography is generally well tolerated but is not without risks. Administration of a relatively large dose of an ACE inhibitor to patients with renin-dependent hypertension may result in a precipitous fall in blood pressure, particularly if patients are volume depleted as a consequence of large doses of loop diuretics. Patients should be closely monitored following administration of captopril with blood pressure measurement every 15 minutes for the first hour and there should be facilities for lying the patient in a supine position and for administering intravenous saline if necessary.

This test is undoubtedly an advance in the diagnosis of renal vascular disease, although diminished accuracy in the presence of renal failure (that is, patients at particular risk of contrast nephropathy) is a major potential limitation. The functional basis of the test may well enable the prediction of the response to revascularization.[29,44]

ACE inhibition, proteinuria, renal hypertension and the progression of chronic renal failure

Following the loss of a critical amount of functioning renal mass, renal function inexorably declines even though the initial injurious cause may no longer exist.[56] The progressive nature of renal disease and the fact that histological examination of the end-stage kidney provides no clue to the underlying aetiology, suggests that common mechanisms may contribute to renal damage in different kidney diseases. Our current understanding of the mechanisms that may underline this process have largely been derived from animal models of chronic renal failure, in particular the subtotal renal ablation model in rats. Brenner's original hypothesis suggested that progressive glomerulosclerosis resulted from glomerular hyperfiltration in the remaining nephrons as a consequence of diminished renal mass.[57] In the light of studies involving the use of ACE inhibitors and dietary protein restriction, this theory has now been refined to suggest that intraglomerular hypertension is more important.[58] It is suggested that intraglomerular hypertension leads to a sequence of events culminating in progressive glomerulosclerosis. In animal models of essential hypertension, the glomeruli are protected from systemic blood pressure by preglomerular vasoconstriction, but in chronic renal failure this autoregulation is lost, exposing the glomeruli to elevated systemic pressure with resulting renal damage.[58]

It is well known that control of malignant hypertension arrests and can reverse the progression of renal impairment. Considerable evidence now indicates that in the presence of renal failure, systemic hypertension increases the rate of decline of renal function.[58] Current theories of progression of renal disease predict that control of systemic hypertension and of intraglomerular hypertension may be more effective in controlling the rate of development of glomerulosclerosis than control of systemic hypertension alone. In this respect, ACE inhibitors may have an advantage over other antihypertensive agents.

Control of systemic hypertension in animal models of renal failure slows or arrests the progressive decline in renal function seen with placebo treatment alone.[59] Although few in number, clinical studies in diabetic and non-diabetic renal disease confirm this observation in man.[60–62]

Are ACE inhibitors superior to other antihypertensive agents in retarding progression of chronic renal failure?

Animal studies

In animal models of diabetic and non-diabetic renal disease, treatment with ACE inhibitors is more effective in retarding the progression of renal failure and in reducing proteinuria than 'conventional' antihypertensive agents, despite similar control in systemic blood pressure.[63,64] In these studies, rats rendered uraemic by subtotal nephrectomy were allocated to receive treatment with enalapril or reserpine, hydralazine and hydrochlorthiazide. Systemic blood pressure was controlled equally in both groups, but only the ACE inhibitor group demonstrated reduction of intraglomerular hypertension. This was associated with a considerable reduction in the development of progressive proteinuria and glomerular sclerosis compared with the non-ACE inhibitor treated group. Similar interventions in rats rendered hyperglycaemic by streptozotocin demonstrated equivalent results.[64] It is generally assumed that the reduced glomerular pressure induced by ACE inhibitors is a consequence of the loss of angiotensin II-mediated efferent glomerular arteriolar vasoconstriction.

Efficacy of antihypertensive drug therapy in retarding the progression of chronic renal failure in humans

In evaluating clinical intervention studies in progressive renal disease, certain criteria must always be applied. First, studies should have a prospective randomized double blind design. Secondly, glomerular filtration rate should be measured as the clearance of inulin or isotopic tracer. Many studies have used serum creatinine, creatinine clearance or the slope of reciprocal creatinine plot. These latter surrogate markers of GFR have been shown to be inaccurate in the setting of chronic renal failure.

Angiotensin converting enzyme inhibitors are undoubtedly effective in reducing blood pressure in patients with renal disease and have an important role in the control of blood pressure in dialysis patients who have renin-dependent hypertension which does not respond to fluid removal alone.[65] They are particularly effective in reducing proteinuria in diabetic and non-diabetic renal disease during both long- and short-term administration.[66,67] Generally, they appear to be superior to other antihypertensive agents in this respect.[7,68]

Non-diabetic renal failure

A recent review of both short-term (less than 12 months) and long-term (greater than 12 months) studies of ACE inhibition in non-diabetic renal disease concluded that the majority of studies in humans had not been adequately randomized and were of too short a duration or confounded by design difficulties to allow any conclusions concerning the effect of ACE inhibitors on the progression of renal disease in this setting.[66] Some studies, however, are worthy of note. Using ^{51}CR-EDTA clearance techniques, Rekola et al. demonstrated that patients with non-diabetic chronic renal failure who did not use ACE inhibitors as antihypertensive therapy exhibited a faster decline in renal function than did those patients treated with ACE inhibitors. Unfortunately, baseline renal function and the length of time of follow-up between the groups differed markedly.[69] In a two-year non-blinded study, Kamper et al. randomized non-diabetic patients with renal impairment to receive either enalapril or conventional therapy including β-blockers, diuretics and vasodilators. The decline in isotopically measured GFR was less in patients treated with enalapril than in those receiving conventional therapy. However, the number of patients who reached end-stage renal failure did not differ between the groups. In contrast, Zucchelli and colleagues randomized similar patients to receive captopril or nifedipine after a 12-month observation period with follow-up for a subsequent three years. The reduction in the progression rate induced by the ACE inhibitor was not different from that induced by the calcium antagonist.[71]

In conclusion, prospective long-term randomized double-blind trials are still required to assess the relative effectiveness of ACE inhibition compared to other antihypertensive agents in retarding the progression of non-diabetic renal disease.

Diabetic renal disease

Natural history Diabetic nephropathy develops in around 30% of patients with insulin-dependent diabetes and follows a predictable course of events. Following a variable period (minimum of five years) of glomerular hyperfiltration associated with supra-normal glomerular filtration rate, patients develop microalbuminuria (urinary albumin excretion rate of 30–300 mg/24 hours). This is associated with development of hypertension, left ventricular hypertrophy and a 100-fold increase in mortality.[72] Clinical proteinuria ensues (established nephropathy) and precedes an inexorable decline in renal function culminating in end-stage renal failure. Most intervention studies employing ACE inhibitors in an attempt to retard this process have so far been performed in type I diabetes.

Overt hypertensive diabetic nephropathy Bjorck et al studied 40 insulin-dependent diabetic patients with established diabetic nephropathy (GFR 46 ml/min/1.73 m^2) and proteinuria (1.5 g/24 hour). They were randomized to receive enalapril and frusemide, or metoprolol and frusemide, and followed for a mean period of 2.4 years. GFR declined by a mean of 2 ml/min per year in the group given enalapril and frusemide, compared to 5.6 ml/min per year in the metoprolol and frusemide group, despite equivalent blood pressure control. In addition, urinary albumin excretion was 60% lower in those patients treated with enalapril.[73] Important evidence of the beneficial effect of ACE inhibitors in retarding the progression of renal failure comes from a recent study which included 409 patients with type I diabetes. Treatment with captopril led to a 48% reduction in risk of doubling baseline creatinine, in comparison with the non-ACE inhibitor treated group, despite similar reductions in blood pressure in both groups. The effect was more pronounced in patients with more severe renal impairment (baseline serum creatinine greater than 177 µmol/l). Treatment with captopril was associated with a 50% reduction in the risk of death or end-stage renal failure.[74]

Thus, it appears that in patients with established diabetic nephropathy, the rate of progression of chronic renal failure is slowed when ACE inhibitor therapy is used. The risk of developing end-stage renal failure is also reduced, compared with patients not receiving an ACE inhibitor. In these patients, renal function must be monitored carefully soon after starting the ACE inhibitor and subsequently during the course of therapy to detect any rapid decline in renal function which may signify coexistent renovascular disease. A meta analysis of 100 trials in which patients received ACE inhibitors for hypertensive diabetic nephropathy recently concluded that these drugs can decrease proteinuria and preserve glomerular filtration rate in patients with diabetes, and that these effects occur independently of changes in systemic blood pressure.[67]

Normotensive early diabetic nephropathy
Studies of the use of converting enzyme inhibitors in normotensive insulin-dependent diabetic patients with microalbuminuria (urine albumin excretion 20–200 µg/min) and normal renal function have also yielded encouraging results. Treatment of normotensive microalbuminuric insulin-dependent diabetic patients with an ACE inhibitor reduces microalbuminuria and may also retard the progression to overt diabetic nephropathy.[75–77] In the study by Mathiesen et al. 7 out of 23 patients not receiving captopril progressed to diabetic nephropathy, whereas none of the 21 patients receiving captopril progressed to overt diabetic nephropathy.[77] It is interesting that this effect was achieved without difference in systemic blood pressures although there is some suggestion that 24-hour ambulatory blood pressure may have differed between the two groups. Studies comparing ACE inhibitors with other antihypertensive agents are awaited in patients with non-insulin dependent diabetic nephropathy.

Conclusion

The kidney is the major source of active plasma renin in the circulation. The renin-angiotensin system has profound effects upon renal function and renal haemodynamics. The therapeutic introduction of angiotensin converting enzyme inhibition has markedly increased the sensitivity and specificity of non-invasive screening tests for renal artery stenosis. Of special importance, ACE inhibitors are particularly effective in the control of hypertension in renal disease and in achieving a greater reduction in proteinuria than conventional agents. As yet, no convincing evidence exists that they are more beneficial in retarding the progression of non-diabetic renal disease than other agents.

Treatment of early diabetic nephropathy with angiotensin converting enzyme inhibitors may retard the progression to overt nephropathy. Moreover, treatment of insulin-dependent diabetics with established renal failure with ACE inhibitors appears to retard the progression of chronic renal failure. The data available to date suggest that these agents may be superior to other antihypertensive drugs in this effect, although results of further controlled studies are awaited. Nevertheless, use of ACE inhibitors in patients with renal impairment is not without potential hazards. Renovascular disease is more common in this setting than previously suspected and in this group of patients, ACE inhibitors may lead to a marked decline in renal function, which on occasion may be irreversible.

References

1 Tigerstedt R, Bergman PG. Niere und Krieslauf. *Scand Arch Physiol* 1989; **8**:223–71.

2 Skott O, Jensen BL. Cellular and intrarenal control of renin secretion. *Clin Sci* 1993; **84**:1–10.

3 Anderson JV, Donkier J, Payne NN, Beachern J, Slater JDH, Bloom SR. Atrial natriuretic peptides; evidence of action as a natriuretic hormone at physiological plasma concentrations in man. *Clin Sci* 1987; **72**:312.

4 Hollenberg NK, Meggs LG, Williams GH, Katz J, Garnic JD, Harrington DP. Sodium intake in renal responses to captopril in normal man and in essential hypertension. *Kidney Int* 1981; **20**:240–45.

5 Brunner HR. ACE inhibitors in renal disease. *Kidney Int* 1993; **42**:463–79.

6 Apperloo AJ, de Zeeuw D, Sleuter HE, de Jong PE. Differential effects of enalapril and atenolol on proteinuria and renal haemodynamics in non-diabetic renal disease. *BMJ* 1991; **3**:821–24.

7 Biollaz J, Durr J, Brunner HR. Escape from mineralocorticoid excess: the role of angiotensin II. *J Clin Endocrinol Metab* 1982; **54**:1187–93.

8 Salvetti A. Newer ACE inhibitors: a look at the future. *Drugs* 1990; **40**:800–28.

9 Kelly JG. Clinical pharmacokinetics of the newer ACE inhibitors. A review. *Clin Pharmacokin* 1990; **19**:177–96.

10 Duchin K, Pierides A, Heald A, Singvi S, Mommel AJ. Elimination kinetics of captopril in patients with renal failure. *Kidney Int* 1984; **25**:942–47

11 Sica DA. Kinetics of angiotensin converting enzyme inhibitors in renal failure. *J Cardiovasc Pharmacol* 1992; **20[Suppl 10]**:S13–D20.

12 Simmermam W. Non-angiotensin effects of angiotensin converting enzyme inhibitor. *Clin Sci* 1993; **85**:661–70.

13 Haji-Ali AF, Zimmerman BG. Kinin contribution to the renal vasodilator effect of captopril in rabbits. *Hypertension* 1990; **18**:504–9.

14 Donker AJM. Nephrotoxicity of angiotensin converting enzyme inhibitors. *Kidney Int* 1987; **31[Suppl 20]**:132–37.

15 Hoorntje SJ, Kallenberg CGM, Weening JJ, Donker AJM, The TH, Hoedemaeker PJ. Immune complex glomerulopathy in patients treated with captopril. *Lancet* 1980; **2**:1212–14.

16 Frohlich ED, Cooper RA, Lewis EJ. Review of the overall experience of captopril in hypertension. *Arch Int Med* 1984; **144**:1441–44.

17 Hricik DE, Browning PJ, Kopelman R, Goorno WE, Madias NE, Dzau VJ. Captopril induced functional renal insufficiency in patients with bilateral renal artery stenosis or renal artery stenosis in a solitary kidney. *N Engl J Med* 1983; **308**:373–76.

18 Curtis JJ, Luke RG, Whelchel JD, Jones P, Dustan HP. Inhibition of angiotensin converting enzyme in renal transplant recipients with hypertension. *N Engl J Med* 1983; **308**: 377–81.

19 Peters AM. Renal artery stenosis. Renovascular hypertension and predicting the blood pressure response to renal revascularisation. *Nuc Med Comm* 1990; **11**:1–5.

20 Scoble JE, Maher, Hamilton G, Dick R, Sweeney P, Moorhead JF. Atherosclerotic renovascular disease causing renal impairment — a case for treatment. *Clin Nephrol* 1989; **31**:119–22.

21 Choudri AH, Cleland JGF, Rolands PS, Tran TL, McCarty M, Al Kautoubi MAO. Unsuspected renal artery stenosis in peripheral vascular disease. *BMJ* 1990; **301**:119.

22 Lipkin GW, Raine AEG. Hypertension and the kidney; recent advances. *Hospital update* (In press).

23 Kastner PR, Hall J, Gallatin AC. Control of glomerular filtration rate; role of intrarenally formed angiotensin II. *Am J Physiol* 1984; **246**:F897–F906.

24 Kalra PA, Mamtora H, Holmes AM, Waldek S. Renovascular disease and renal complications of angiotensin converting enzyme therapy. *Q J Med* 1990; **282**:1013–18.

25 Helmchen U, Grown HJ, Korchertzej et al. Contrasting renal effects of different antihypertensive agents in hypertensive rats with bilaterally constricted arteries. *Kidney Int* 1982; **Suppl 12**:S198–S205.
26 Textor FC, Novick AC, Tarazi RC, Klimas V, Vidt DG, Pohl M. Critical perfusion pressure for renal function in patients with bilateral atherosclerotic renal vascular disease. *Ann Intern Med* 1985; **102**:308–14.
27 Ramsay LE, Waller PC. Blood pressure response to percutaneous transluminal angioplasty for renovascular hypertension: an overview of published series. *BMJ* 1990; **300**:569–72.
28 Schreiber MJ, Pohl MA. Preserving renal function by revascularisation. *Ann Rev Med* 1990; **41**:423–29.
29 Saddler MC, Black HR. Captopril renal scintigraphy: a clinician's perspective. *Nuc Med Comm* 1990; **4**:1–8.
30 Harvey RJ, Krumlovsky F, Deljerco F, Martin HG. Screening for renovascular hypertension. Is renal digital subtraction angiography the preferred non-invasive test? *JAMA* 1985; **254**:388–93.
31 Dresberg AL, Paushter DM, Lammert GK et al. Renal artery stenosis: evaluation with colour flow imaging. *Radiology* 1990; **177**:749–53.
32 Jonker GJ, Huisman RM, de Zeeuw W. Enhancement of screening tests for renovascular hypertension by angiotensin converting enzyme inhibition. *Nephrol Dial Transplant* 1993; **9**:798–807.
33 Wilcox CS. The use of angiotensin converting enzyme inhibitors for diagnosing renovascular hypertension. *Kidney Int* 1993; **44**:1379–90.
34 Bookstein JJ, Abrahams HL, Duenger RE et al. Radiological aspects of renovascular hypertension: II. The role of urography in unilateral disease. *JAMA* 1972; **22**:1223–30.
35 Bookstein JJ, Maxwell MH, Abrahams HL et al. Cooperative study of radiological aspects of renovascular hypertension: bilateral renovascular disease. *JAMA* 1977; **237**:1706–09.
36 Jonker GJ, de Zeeuw D, Huisman RM, Piers DA, Beekhuis H, van der Hem GK. Angiotensin converting enzyme inhibition improves diagnostic procedures for renovascular hypertension in dogs. *Hypertension* 1988; **12**:411–49.
37 Case DB, Atlas SA, Laragh JH. Reactive hyperreninaemia to angiotensin blockade identifies renovascular hypertension. *Clin Sci* 1979; **57**:313S–316S.
38 Muller FB, Sealey JE, Case DB et al. The captopril test for identifying renovascular disease in hypertensive patients. *Am J Med* 1986; **80**:633–44.
39 Frederickson ED, Wilcox CS, Bucci CM et al. A prospective evaluation of a simplified captopril test for the determination of renovascular hypertension. *Arch Int Med* 1990; **150**:569–72.
40 Postma CT, van der Stem PHM, Hoefnagels WHL, De Boo T, Thien T. The captopril test in detection of renovascular disease in hypertensive patients. *Arch Int Med* 1990; **150**:625–28.
41 McCarthy JE. The captopril test in renovascular hypertension. *Arch Int Med* 1990; **150**:493–94.
42 Tomoda F, Takata M, Ohashis C, Ueno H, et al. Captopril stimulated renal vein renin in hypertensive patients with or without renal artery stenosis. *Am J Hypertens* 1990; **3**:918–26.
43 Roubidoux MA, Dunnick NL, Klotman PE. Renal vein renins: inability to predict response to revascularisation in patients with hypertension. *Radiol* 1991; **178**:819–22.
44 Geyskes GG, Oei Y, Puyleart CBAJ, Dorhout Mees EJ. Renovascular hypertension identified by captopril induced changes in the renogram. *Hypertension* 1987; **9**:451–58.
45 Levenson DJ, Dzau VJ. Effects of angiotensin converting enzyme inhibition on renal haemodynamics in renal artery stenosis. *Kid Int* 1988; **31**:S173–S179.
46 Jonker GJ, Visscher CA, De Zeeuw W. Changes in renal function induced by ACE inhibition in the conscious two kidney, one clip Goldblatt hypertensive dog. *Nephron* 1992; **6**:226–31.
47 Nally JD, Clarke HS, Gupta BK et al. Captopril renography in two kidney and one kidney Goldblatt hypertension in dog. *J Nuc Med* 1987; **28**:1171–79.
48 Fommei E, Ghione S, Pall L et al. Renoscintigraphic captopril tests in the diagnosis of renovascular hypertension. *Hypertension* 1987; **10**:212–20.

49 Maher ER, Othman S, Frankel AH, Sweeney MP, Moorhead JF, Hilson HAW. Captopril enhanced $^{99m}T_c$ DTPA scintigraphy in the detection of renal artery stenosis. *Nephrol Dial Transplant* 1988; **3**:608–11.
50 Kremer Hovinga TK, de Jong PE, Piers DA, Beekhuis H, Van der Ham GK, de Zeeuw D. Diagnostic use of angiotensin converting enzyme inhibitors in radioisotope evaluation of unilateral renal artery stenosis. *J Nuc Med* 1989; **30**:605–14.
51 Sfakianakis GN, Bourgoignie JJ, Jaff D et al. Single dose captopril scintigraphy in the diagnosis of renovascular hypertension. *J Nuc Med* 1987; **28**:1383–92.
52 Dondi M, Franchi R, Levorati et al. Evaluation of hypertensive patients by means of captopril enhanced renal scintigraphy with DTPA. *J Nuc Med* 1989; **30**:615–21.
53 Mann SJ, Pickering TG, Sos TA et al. Captopril renography in the diagnosis of renal artery stenosis: accuracy and limitations. *Am J Med* 1991; **9**:32–40.
54 Fommei E, Mezzasalma L, Ghione S, Volterrani D, Oei Y, Hilson AJW, Carrieri F. The European captopril radionucleotide test multicentre study group. Preliminary results. Inspective renographic analysis. *Am J Hypertens* 1991; **4**:690S–697S.
55 McLean AG, Hilson AJW, Scoble JE et al. Screening for renovascular disease with captopril enhanced renography. *Nephrol Dial Transplant* 1992; **7**:211–15.
56 Klahr S, Schreiner G, Ichikawa I. The progression of renal disease. *N Engl J Med* 1988; **318**:1657–66.
57 Brenner BM, Mayer TW, Hostetter TH. Dietary protein intake and the progressive nature of kidney disease. *N Engl J Med* 1982; **307**:652–59.
58 El Nahas AM. Glomerulosclerosis: insights into pathogenesis and treatment. *Nephrol Dial Transplant* 1989; **4**:843–53.
59 Neugarten J, Kaminetsky B, Feiner H, Schacht RG, Liu DT, Baldwin DS. Nephrotoxic serum nephritis with hypertension: amelioration by antihypertensive therapy. *Kidney Int* 1985; **28**:135.
60 Mogensen CE. Long term antihypertensive treatment inhibiting progression of diabetic nephropathy. *BMJ* 1982; **285**:685–98.
61 Brazy PC, Stead WW, Fitzwilliam JF. Progression of renal insufficiency: role of blood pressure. *Kidney Int* 1989; **35**:670–74.
62 Parving HH, Anderson AR, Smidt UM, Hommel E, Mathiesen ER, Svendsen PA. Effective antihypertensive treatment in kidney function in diabetic nephropathy. *BMJ* 1987; **294**:1443.
63 Anderson S, Mayer TW, Rennke HG, Brenner BM. Therapeutic advantage of converting enzyme inhibitors in arresting progressive renal disease associated with systemic hypertension in the rat. *J Clin Invest* 1986; **77**:1993–2000.
64 Anderson S, Rennke HG, Cijrcia L, Brenner BM. Short and long term effects of antihypertensive therapy in the diabetic rat. *Kidney Int* 1989; **36**:526–36.
65 Raine AEG, Ledingham JGG. Treatment of hypertension in renal disease. In: Buhler FR, Laragh JH, eds. Handbook of hypertension. Amsterdam: Elsevier Scientific Publishers, 1990: 465–82.
66 Ter Wee PM, Epstein M. Angiotensin converting enzyme inhibitors in the progression of non-diabetic chronic renal disease. *Arch Int Med* 1993; **153**:1749–59.
67 Kasiske BL, Kalil RSN, Mar JZ, Lee AO, Liao M, Keane WF. Effect of antihypertensive therapy on the kidney in patients with diabetes: a META regression analysis. *Arch Int Med* 1993; **118**:129–38.
68 Bianchi S, Bigazzi R, Baldari G, Campese EM. Long term effects of enalapril and nicardipine on urinary albumin excretion in patients with chronic renal failure: a one year follow-up. *Am J Nephrol* 1991; **11**:131–37.
69 Rekola S, Berkstrand A, Bucht H. Deterioration rate in hypertensive nephropathy: comparison of the converting enzyme inhibitor and a betablocker agent. *Nephron* 1991; **51**:57–60.
70 Kamper AL, Strandgaard S, Leyssac PP. Effect of enalapril on the progression of chronic renal failure: a randomised controlled trial. *Am J Hypertens* 1992; **5**:423–30.
71 Zuccheli P, Zuccala A, Borghi I et al. Long term comparison between captopril and nifedipine in the progression of renal insufficiency. *Kidney Int* 1992; **42**:453–58.

72 Watkins PJ. Diabetic nephropathy. *Q J Med* 1990; **280**:767–69.
73 Bjorck S, Mulec H, Johsen SA, Norden G, Aurell M. Renal protective effect of enalapril in diabetic nephropathy. *BMJ* 1992; **304**:339–43.
74 Lewis EJ, Hunsickler G, Bain RP, Rohde RD. The effect of angiotensin converting enzyme inhibition on diabetic nephropathy. *N Engl J Med* 1993; **329**:1456–62.
75 Marre M, Le Blanch H, Surarez L, Guyenne TT, Menard J, Passa P. Converting enzyme inhibition and kidney function in normotensive diabetic patients with persistent microalbuminuria. *BMJ* 1987; **294**:1448–52.
76 Marre E, Chantellier G, Le Blanc H, Unguyene TT, Menard J, Passa P. Prevention of diabetic nephropathy with enalapril in normotensive diabetics with microalbuminuria. *BMJ* 1988; **297**:1092.
77 Mathiesen J, Hommel B, Giese J, Parving HH. Efficacy of captopril in postponing nephropathy in normotensive insulin dependent diabetics with microalbuminuria. *BMJ* 1991; **303**:81–95.

5

ACE inhibitors in the management of cardiac failure

Eckhard P Kromer and Günter AJ Riegger

Congestive heart failure (CHF) was defined by Braunwald as: 'the pathophysiological state in which, because of an abnormality of systolic and/or diastolic cardiac function, the heart fails to pump blood at a rate commensurate with the requirements of the metabolizing tissues, or else does so only from an elevated filling pressure'.[1] A second definition, by Cohn, states that: 'Chronic heart failure is a syndrome characterized by left ventricular dysfunction, reduced exercise tolerance, impaired quality of life and markedly shortened life expectancy'.[2]

Recently, some historic aspects of the definition of CHF have been reviewed by Lutz.[3]

Epidemiology

The exact worldwide prevalence of CHF is unknown, however, based on the available data it may be estimated in the range of 10 to 20 million subjects.[4] In 1971, The Framingham Study demonstrated arterial hypertension to be the most frequent precursor (75%) of CHF.[5] Two recently published megatrials showed that since then, coronary artery disease has become the most common cause of CHF, occurring in 67% of subjects compared to 43% with arterial hypertension.[6,7] Other causes for heart failure include idiopathic cardiomyopathy, myocarditis, end-stage valvular or congenital heart disease, and alcoholic cardiomyopathy. The mortality of subjects with CHF is high and ranges from 15% in unselected patients to 50% among those in New York Heart Association functional class IV in the first year.[4,8] Accordingly, the number of deaths due to CHF may be between one and two million per year.[7] Furthermore, it could be shown, that mortality may depend on sex,[5] etiology of CHF,[9] left ventricular volumes,[10] the severity of concomitant ventricular arrhythmias[11] and the degree of neurohormonal stimulation.[12–14]

Pathophysiology

After an initial insult to the myocardium, several compensatory mechanisms are activated in order to preserve cardiocirculatory function (Table 5.1). Depending on the severity of the insult, cardiac function may be stabilized or even normalized or may progressively deteriorate along with the development of CHF.

The role of changing end-diastolic fibre length (Frank–Straub–Starling's law) has long been recognized as an important mechanism of the heart to meet acute changes of afterload.[15] In heart failure, however, preload reserve may be easily exhausted

- in trying to meet the need for a sustained supernormal stroke volume (volume overload),
- in trying to restore stroke volume to normal after it has been depressed due to a sustained increase in afterload (pressure overload),

Table 5.1 *Compensatory mechanisms in congestive heart failure*

Ventricular dilatation to stretch sarcomeres (Frank–Straub–Starling's law)
Myocardial hypertrophy
Reprogramming of myocardial gene expression
Stimulation/alteration of vasoconstricting systems:
sympathetic nervous system
renin-angiotensin-aldosterone system
arginine-vasopressin
endothelin-1
Stimulation/alteration of vasodilating systems:
natriuretic peptids (ANP, BNP, CNP, urodilatin)
prostaglandins (E_2, I_2)
endothelium-derived relaxing factor
bradykinin
dopamine
Exercise deconditioning

- and/or in trying to overcome the depression in effective contractility due to a local (in case of myocardial infarction) or a diffuse (in case of cardiomyopathy) loss of contractile units. If there is concomitant depletion of cardiac noradrenaline (norepinephrine) stores and impairment of β-receptor function, the load cannot be matched by increasing inotropy and a state of 'afterload mismatch' will develop, inducing myocardial hypertrophy.[15–17]

Left ventricular hypertrophy may present with an increase of sarcomeres in parallel (pressure overload) or in series (volume overload). It may be associated with a decrease of coronary vasodilator response and an increase of myocardial stiffness inducing further diastolic and/or systolic dysfunction. Recent animal data clearly indicate, that the renin-angiotensin-aldosterone system may play an important role as a non-hemodynamic factor for myocardial hypertrophy.[18,19]

The decrease of stroke volume and the increase of filling pressure activate arterial and cardiopulmonary baroreflexes, which play an important role in the control of the heart and peripheral circulation. They may be found to be significantly depressed in patients with CHF, at least in the moderate and severe stages of disease.[20–22] These alterations in sympathetic and parasympathetic control may play a pivotal role in the observed stimulation of vasoconstricting and/or vasodilating neurohormonal systems in CHF.

Changes in the sympathetic nervous system have long been recognized. Patients with CHF demonstrate an increase of the central sympathetic outflow and of noradrenaline (norepinephrine) spillover from the sympathetic nerve clefts,[23,24] both of which are responsible for the elevation of plasma-noradrenaline (norepinephrine) levels.[25] Myocardial noradrenaline (norepinephrine) stores are depleted,[16,26] possibly due to reduced activity of synthesizing enzymes and/or the limitation of the transport of dopamine across the membrane of the noradrenergic vesicle.[27] Recent studies using molecular-biological techniques have demonstrated significant changes of the β-receptor pathway in patients with CHF:

- There is a reduction of β_1-receptor mRNA levels and number by 50–70% whereas mRNA levels and the number of β_2-receptors are unaltered.
- mRNA levels of β-receptor kinase (β-ARK$_1$ >> β-ARK$_2$) are elevated, paralleled by enhanced enzyme activity.[29]
- Along with receptor downregulation and uncoupling by β-arrestin there is an increase of $G_{i\alpha}$-protein.[30–32]

In contrast, desensitization of β-adrenoceptors in limb vessels does not occur in patients with CHF.[33] Whether autoantibodies against β-adrenoceptors play a significant role in the pathophysiology of CHF due to idiopathic dilated cardiomyopathy remains to be determined.[34] The role of the α-adrenergic nervous system in CHF is not completely understood, however, there is evidence that myocardial α_1-receptor density is unchanged, and that peripheral α_1-receptor function is unaltered, although the blood pressure response to exogenous noradrenaline (norepinephrine) is significantly blunted in CHF.[35–37]

The activation of the renin-angiotensin-aldosterone system in patients with CHF was first noted in 1941 and 1946.[38,39] There are several mechanisms involved in renal renin secretion, such as sympathetic nerve activity, an intrarenal baroreceptor, the amount of sodium chloride sensed by the macula densa, hormonal factors and electrolytes. However, in early stages of CHF the renal secretion of renin may be suppressed by atrial natriuretic peptide and/or by inhibitory effects of cardiopulmonary baroreceptors.[40–42] If plasma renin activity is found elevated in early heart failure, it may be secondary to diuretics.[43,44] However, even if plasma renin activity is within the normal range in early CHF, there is experimental evidence for local activation in the myocardium, which may be delineated from a significant elevation of mRNA for angiotensinogen and for angiotensin converting enzyme.[45,46] In severe CHF, progressive elevation of plasma renin activity and plasma angiotensin II levels may be observed, both associated with a very poor prognosis.[12,47] Preliminary data suggest that myocardial angiotensin $II_{1,2}$ receptor density is unchanged in CHF,[48] however these data must be confirmed and do not exclude possible subtype alterations.

In severe CHF the marked stimulation of the renin-angiotensin system and alterations of cardiopulmonary baroreceptors that normally exert inhibitory effects on arginine vasopressin release, may be responsible for the inadequate elevation of plasma arginine vasopressin.[42,49,50] This may play a pivotal role, particularly in hyponatremia.[51]

Endothelin, a newly discovered vasoconstrictor peptide, was found to be elevated up to 5-fold in CHF.[52] Recently, it has been shown that these plasma levels exert acute biological effects, however, the exact role in chronic CHF remains to be clarified.[53]

Atrial natriuretic peptide is normally produced in the atria and released due to stretch of the myoendocrine cells. In severe CHF, expression of the atrial natriuretic peptide gene has also been reported in ventricular tissue.[54] Its plasma levels correlate with the severity of cardiac impairment.[55] For patients with symptomless LV dysfunction after myocardial infarction, the level of circulating N-terminal ANP may serve as a very sensitive marker.[56] The possible beneficial effects of atrial natriuretic peptide in CHF may be due to a reduction of volume and load of the heart through its vasodilatory, diuretic and natriuretic properties.[57] Recently, 'brain natriuretic peptide', which is predominantly secreted from the ventricles, has also been found at elevated levels in patients with CHF.[58,59]

The role of prostaglandins in CHF is very complex. At least in severe CHF, plasma levels of vasodilating prostaglandins are significantly elevated, and inhibition of their synthesis by indomethacin may further depress cardiac function.[60] There is a complex interrelationship with other neurohormonal systems, for example inhibition of their synthesis by aspirin may significantly deteriorate renal function in mild to moderate CHF,[61] or may blunt the acute hemodynamic effects of enalapril in severe CHF.[62] Experimental data indicate that renal prostaglandins blunt the vasoconstrictor effects of noradrenaline (norepinephrine), angiotensin II and arginine vasopressin.[63–65] Additionally, prostaglandin E_2 inhibits Na-K-ATPase activity in the inner collecting duct, thereby increasing natriuresis.[66]

Endothelium-derived relaxing factor plays an important role in vasomotor control and is the most potent endogenous vasodilator known.[67] Recently, it was demonstrated that endothelium-dependent vasomotion is significantly impaired in CHF, which may contribute to the increased peripheral resistance.[68]

Circulating dopamine originates from noradrenergic neurons and from the adrenal medulla, where it is synthesized as a precursor of noradrenaline (norepinephrine) and adrenaline (epinephrine). A considerable amount is produced from the decarboxylation of plasmatic L-dopa in the tubular epithelium of the kidneys.[69] Dopamine has complex actions in the kidney and the circulation.[70] It mediates vasodilatation (preferentially in the renal vascular bed) either directly (vascular DA_1 receptors), or through the control of sympathetic tone (prejunctional DA_2 receptors). DA_2 receptors in the adrenal cortex exert a tonic inhibitory control on aldosterone secretion.[71] Dopamine also inhibits sodium reabsorption in the proximal convoluted tubule.[72] Thus, dopamine may exert beneficial effects in CHF,[73] however its exact pathophysiologic role has not yet been determined.

Another compensatory mechanism in CHF is exercise deconditioning. Immobility induces skeletal muscle atrophy.[74] Skeletal muscle oxidative enzymes decrease,[75] and there is a shift from type I to type IIB fibers (slow oxidative versus fast glycolytic). Deconditioning may possibly be protective; the early onset of glycolysis in working muscle can decrease pH and initiate a reflex that may be one of the signals compelling the CHF patient to stop exercising.[75]

As long as vasodilating systems can counterbalance the effects of vasoconstricting neurohormonal systems, cardiocirculatory function after an initial insult to the myocardium may not deteriorate. In case of a progressive reduction of ventricular function, however, vasoconstricting systems will overcome the effects of vasodilating systems. This leads to an increase of arteriolar and venous tone, and an increase of sodium and water retention, elevating the loading conditions of the failing heart. Additionally, the sodium and water content of the great vessels may increase[76] due to neurohormonal stimulation. As might be inferred from recent experimental data, noradrenaline (norepinephrine), angiotensin II and arginine vasopressin stimulate hypertrophy of smooth muscle cells in vitro.[77–79] The latter two mechanisms would alter the physical properties of the great vessels, thereby also increasing the pulsatile component of left ventricular afterload, which is not the case in early CHF.[80] As a result of the increased loading conditions, the already impaired ventricle will further deteriorate, inducing progressive neurohormonal stimulation. If this vicious circle is not stopped, it usually leads to death due to progressive pump failure. This concept was first suggested by McMichael in 1949.[81] Alternatively, the combination of local ischemia, electrolyte imbalance due to neurohormonal stimulation and enhanced sympathetic activity may induce ventricular arrhythmias and sudden cardiac death.[82]

With respect to these pathophysiological considerations, therapy of CHF should address three important points:

1. In order to prevent further deterioration of pump function, the hemodynamic load should be reduced.
2. because compensatory activation of neurohormonal systems plays such a crucial role for the progression of CHF, therapy should attenuate or at least not increase their unfavorable systemic and/or local effects.
3. Death due to cardiac arrhythmias, which accounts for about one-third of all cardiovascular deaths in CHF, for example 29.6% in the SOLVD trials, should be reduced.[6,83]

'Unloading' the failing heart

Initially the failing heart was unloaded by venesection and application of venous tourniquets, both means producing immediate relief of symptoms. Elevated venous pressure in CHF was first reduced by pharmacological means using aminophylline in 1946.[84,85] Later tetraethylammonium, veratrum viride, priscoline as the first sympatholytic drug and hexamethonium were investigated in patients with CHF.[86–90] In 1956 hydralazine, and in 1957 nitroglycerine were used for the acute treatment of heart failure.[91,92] Thereafter vasodilators were forgotten until 1971, when their use

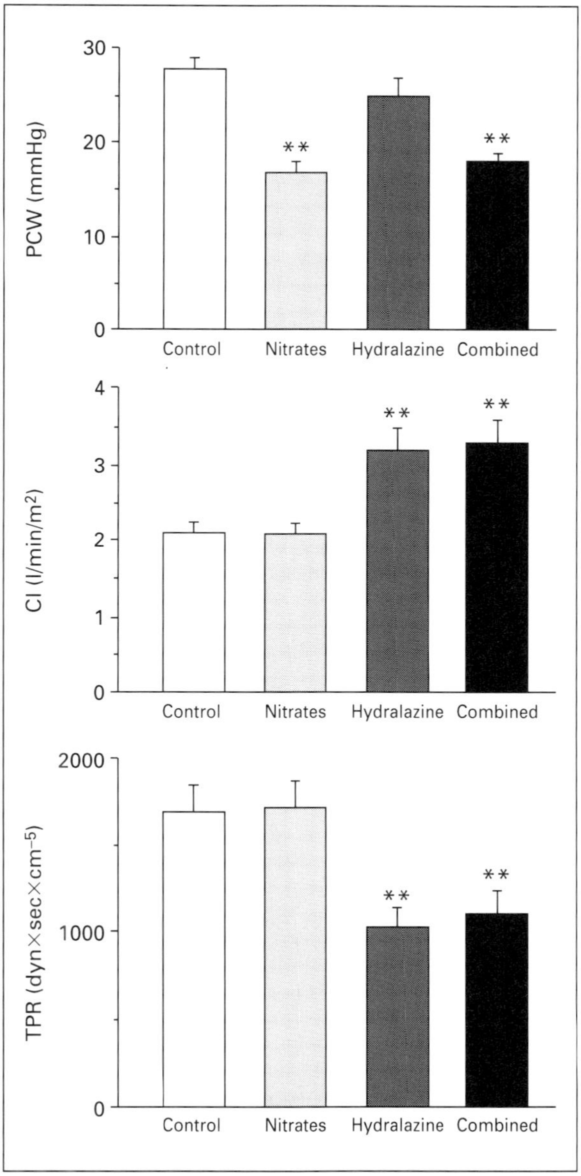

Figure 5.1
*Mean values with the standard error of the mean of hemodynamic variables after administration of individual doses of nitrates, hydralazine and the combination of both in patients with chronic heart failure. Adapted from Massie et al.[99] PCW = pulmonary capillary wedge pressure; CI = cardiac index; TPR = total peripheral resistance, ** = P < 0.001 versus control.*

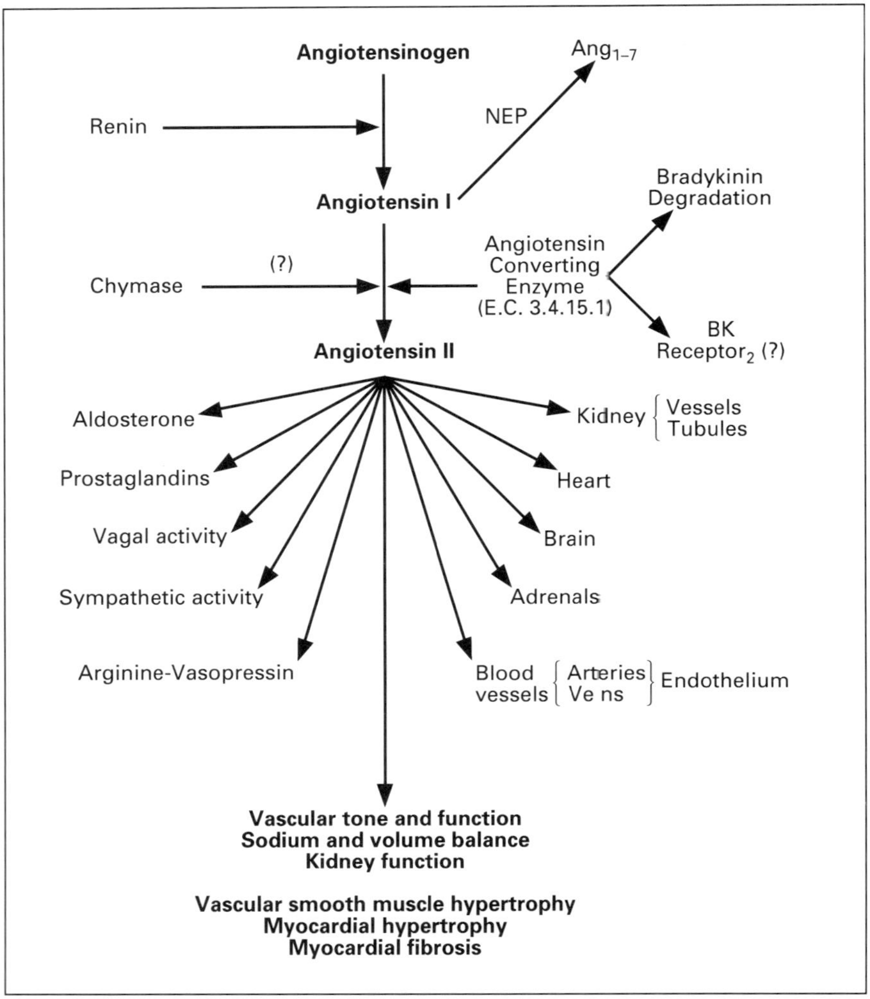

Figure 5.2
The renin–angiotensin–aldosterone system
BK_2 = bradykinin receptor type 2
Ang_{1-7} = Angiotensin 1–7
NEP = neutral endopeptidase

in the chronic treatment of CHF was extensively investigated.[93] Numerous studies using different designs investigated the effects of various direct vasodilators,[94] however the results of controlled long-term studies were disappointing, often due to the development of tolerance.[95–98] In this field the combination of nitrates, which act primarily on venous capacitance vessels and thus reduce preload, and hydralazine, which acts on arteriolar resistance vessels and thus reduces afterload, proved to be most effective on hemodynamic parameters (Fig. 5.1).

In 1986, Cohn and colleagues published the results of the first five-year, multicenter Veterans Administration Cooperative Vasodilator-Heart Failure Trial (V-HeFT I). It demonstrated that, compared with placebo, the

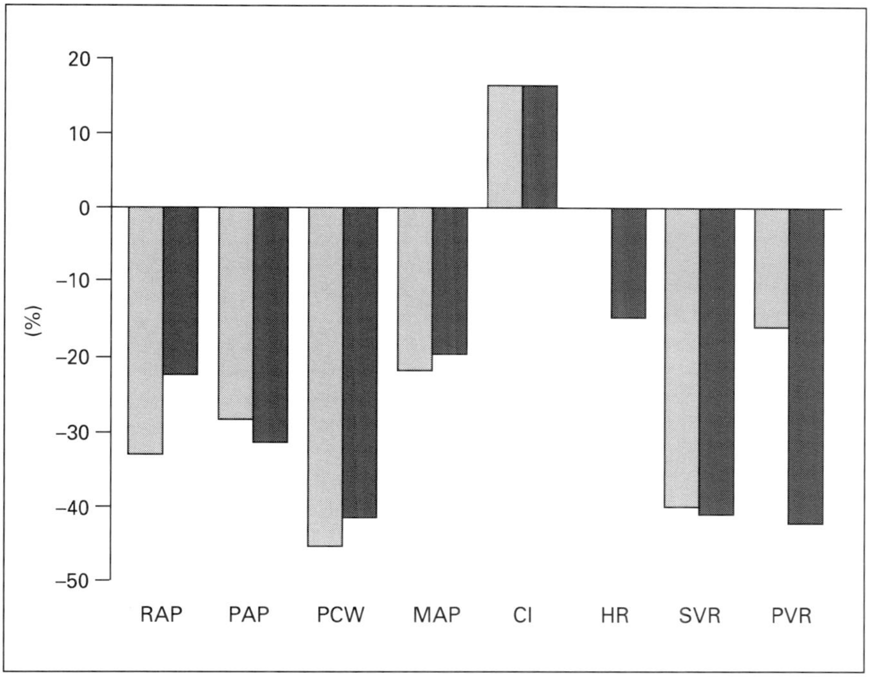

Figure 5.3
Effects of captopril on central hemodynamics (% change from baseline) in patients with congestive heart failure 1 hour after administration of 50 mg (hatched bars) and after 10 weeks of therapy with three 50 mg doses (closed bars). RAP = Right atrial pressure; PAP = Pulmonary artery pressure; PCW = Pulmonary capillary wedge pressure; MAP = Mean arterial pressure; CI = Cardiac index; HR = Heart rate; SVR = Systemic vascular resistance; PVR = Pulmonary vascular resistance.

combination of the vasodilators hydralazine and isosorbide dinitrate reduced mortality in patients with mild to moderate heart failure who were treated with digoxin and diuretics.[100] The reduction in mortality with this therapy was 38% after one year, 25% after two years and 28% over the entire follow-up period (mean 2.3 years).

Due to the various actions of the renin-angiotensin-aldosterone system on different target organs, ACE inhibitors attracted enormous interest for the treatment of CHF after their introduction (Fig. 5.2). Since 1977, various orally active ACE-inhibitors have been designed and synthesized according to a biochemical model of the enzyme site based on X-ray crystal structure of a related enzyme, carboxypeptidase A.[101,102]

In 1978, Gavras and coworkers reported on significant hemodynamic improvement in eight patients with CHF after acute intravenous administration of teprotide. They found a significant decline in arterial blood pressure, left ventricular filling pressure and systemic vascular resistance, along with an increase of cardiac index and stroke volume index.[103] One year later, Davis et al. and Turini et al. reported simultaneously on beneficial acute hemodynamic effects of captopril in patients with CHF.[104,105] Enalapril was introduced as the second orally active ACE inhibitor in 1980.[106] Numerous studies have been performed meanwhile, demonstrating beneficial effects on hemodynamics, exercise intolerance, neurohormonal systems and mortality.

Central and regional hemodynamic effects of ACE inhibitors

The acute hemodynamic effects of ACE inhibitors have been extensively studied and demonstrate wide interindividual variations. Our own results in 18 patients with moderate CHF are shown in Fig. 5.3. They are in accordance with results reported by other groups for either captopril or enalapril.[103,107–112] The initial hemodynamic response seems to be dependent on plasma renin activity, which shows an inverse correlation with serum sodium concentration; however these findings are not consistently reported.[113–116] Although angiotensin II is not thought to cause direct venoconstriction, the hemodynamic profile of ACE inhibitors suggest both arterial and venous dilation. The latter might be due to the interaction of angiotensin II with the sympathetic nervous system, which causes venoconstriction in the limb, or it might be due to increases of bradykinin or vasodilator prostaglandins.[117,118]

In 1983 Packer and coworkers reported on serial hemodynamic evaluation of 51 patients with CHF NYHA IV, who were refractory to digitalis and diuretics, over a period of up to eight weeks.[119] They received captopril at a target dose of 100–150 mg tid, which was reduced if renal impairment was present. There were four hemodynamic patterns of response:

1 minimal initial response (9 of 51) with delayed improvement (3 of 9)
2 sustained significant benefit (28 of 51)
3 triphasic response, i.e. initial response, that was attenuated after 48 hours and accompanied by spontaneous restoration after 2–8 weeks (7 of 51)
4 drug tolerance, i.e. initial response with attenuation without restoration (7 of 51).

Altogether 38 out of 51 patients (75%) demonstrated long-term hemodynamic benefit with oral ACE inhibition. The response to the initial dose of captopril reflected the long-term effects in 80% of these NYHA IV heart failure patients, although long-term response did not depend on the initial plasma renin activity. There is little data on hemodynamics over longer periods (one year), however their interpretation would be difficult because the spontaneous course of the disease might make a significant but unpredictable impact.

The increase in vascular resistance in patients with CHF is not uniform among the different vascular beds. Renal, cutaneous and splanchnic vascular resistance may be progressively increased, while cerebral and myocardial vascular resistance are nearly unchanged.

Only few data are available about effects of ACE inhibition on regional hemodynamics in the skin and splanchnic circulation.[120–122] In patients with CHF, captopril may reduce cerebral vascular resistance in order to preserve cerebral blood flow, despite a fall in blood pressure; even a slight increase of blood flow has been reported.[123,124]

The effects of ACE inhibitors on renal function are very complex. Depending on the initial fall of arterial pressure, a transient decrease of renal blood flow may occur. Thereafter a pronounced increase is seen, which may be further enhanced by coadministration of diuretics.[125] The acute effects of ACE inhibition and diuretics on renal function seem to be dose-dependent. Recently, Motwani and coworkers demonstrated that in patients with moderate CHF, 1 mg captopril did not significantly alter serum angiotensin II levels, mean arterial blood pressure, renal vascular resistance, glomerular filtration rate, filtration fraction and urinary sodium excretion. Coadministration of furosemide (median dose 80 mg) did not change glomerular filtration

Vascular bed	Regional blood flow		Regional vascular resistance	
	Status in CHF	Effect of ACE Inhibitors	Status in CHF	Effect of ACE Inhibitors
Myocardial	—(↑)	—	(Coronary reserve reduced in CAD)	↓
Cerebral	—(↘)	—(↑)	—	↓
Skeletal muscle				
- resting	↘	—	↗	↓
- exercising	↘↘	↑↑	↗↗	↓↓
Renal	↘	↑↑	↗	↓↓
Splanchnic	↘	—(↓)	↗	(↓)
Cutaneous	↘	?(—)	↗	?(↓)

CHF: Congestive heart failure
CAD: Coronary artery disease
—: no change
↘: progressive decrease
↓: decrease
ACE: Angiotensin converting enzyme
( ): small effect possible
↗: progressive increase
↑: increase
?: no data in humans available

Table 5.2
Regional hemodynamics in congestive heart failure.

rate but a significant increase of the natriuretic response to furosemide (about three-fold) was observed. In contrast, 25 mg captopril significantly reduced serum angiotensin II levels, mean arterial pressure, renal vascular resistance, glomerular filtration rate, and filtration fraction. The natriuretic response to furosemide was totally blunted. The authors propose, that in the face of furosemide-induced postglomerular vasodilation in CHF, captopril at a starting dose of 1 mg (but not 25 mg) preserves enough circulating angiotensin II to maintain efferent arteriolar tone and thus glomerular filtration rate, while offsetting the antinatriuretic renal tubular effects of angiotensin II.[126] Available long-term studies clearly indicate that during coadministration of ACE inhibitors and diuretics, there is a slight decline of glomerular filtration rate, which is a physiological response to the fall in angiotensin II and not a toxic effect of ACE inhibitors. However, renal sodium excretion is reversed despite vasodilation, which is in contrast (or superior) to direct vasodilators.

Myocardial blood flow is usually well preserved in CHF or may even be higher in dilated cardiomyopathy.[122] Recently, Kiowski and coworkers showed that cilazapril may reduce resting coronary vascular resistance and leave coronary blood flow unchanged along with an improved overall myocardial aerobic metabolism.[127] However, the changes in coronary vascular resistance were smaller for any given degree of systemic vasodilation.

In patients with CHF, resting skeletal muscle blood flow is initially not significantly reduced, but may decrease with progression of cardiac dysfunction.[128,129] Even in mild CHF, exercise-induced vasodilation is markedly impaired in addition to a reduction of blood flow to the non-exercising limb. Short-term administration of ACE inhibitors causes no change in resting limb blood flow, but long-term treatment may do so.[130] The direct effects of ACE inhibition on exercise-induced skeletal muscle blood flow are quite difficult to measure, however the generally observed increase of exercise tolerance may be either the result of or the reason for increased blood flow.[131.132]

A survey of regional hemodynamics in CHF is presented in Table 5.2.

Effects of ACE inhibitors on exercise tolerance

In mild to moderate CHF patients are limited by breathlessness during exertion at high workload or by fatigue in cases of prolonged low-level exertion.[133,134] The mechanisms of these important symptoms are not yet completely understood. CHF means that cardiac output on exertion is reduced and that filling pressures are elevated, but exercise performance relates poorly to filling pressure.[134] Circulating lactate rises at lower workload in patients with CHF compared to normal subjects.[135,136] Metabolic acidosis is prevented by an increase of ventilation.[137] The sensation of breathlessness might be due to an altered relationship of exercise workload and ventilation, however in normal subjects workload, ventilation and blood lactate levels are far higher.[138] It is possible that an increased dead space due to a ventilation/perfusion mismatch may play an additional role.[138] Disturbances of pulmonary function may also contribute to breathlessness. Extravasation of fluid due to elevated filling pressures, altered bronchial arterial and venous hemodynamics and reduced lymphatic clearance possibly induce thickening of the tissue barrier between the alveoli and the capillary blood resulting in the observed fall in membrane conductance.[139] As has been recently shown by Cabanes and coworkers, an increase in airways resistance might be related to exertional dyspnea in CHF, however, Eichacker did not confirm these findings.[140,141] Alterations of skeletal muscle blood flow and deconditioning (CHF-patients are unfit) have already been mentioned.

Breathlessness is the symptom most consistently improved by ACE inhibitors. The first 'well-designed' trial was reported in 1983. The Captopril Multicenter Research Group investigated patients with CHF NYHA II, III (and IV), refractory to digitalis and diuretics. Patients received additional captopril at a target dose of 100 three times daily (70%) or placebo. There was a significant (24%) mean increase in exercise tolerance with captopril (495 ± 22 to 614 ± 27 seconds) compared with placebo (480 ± 28 to 483 ± 43 seconds) ($P < 0.01$); the captopril group had an increase in the ejection fraction from a mean baseline value of 0.19 ± 0.02 to 0.22 ± 0.02 seconds, compared with 0.19 ± 0.02 to 0.18 ± 0.02 seconds ($P < 0.05$) in the placebo group.[142] This study emphasized very clearly the necessity of a placebo group, because a real treatment effect could be observed only after 12 weeks. Meanwhile, numerous studies have confirmed increases in exercise tolerance with all available ACE inhibitors.

The question of a dose-dependency, however, has been addressed by only one published study until now. In 1991, Riegger for the Quinapril Research Group reported a dose-dependency of exercise improvement. Patients with CHF NYHA II and III were studied in

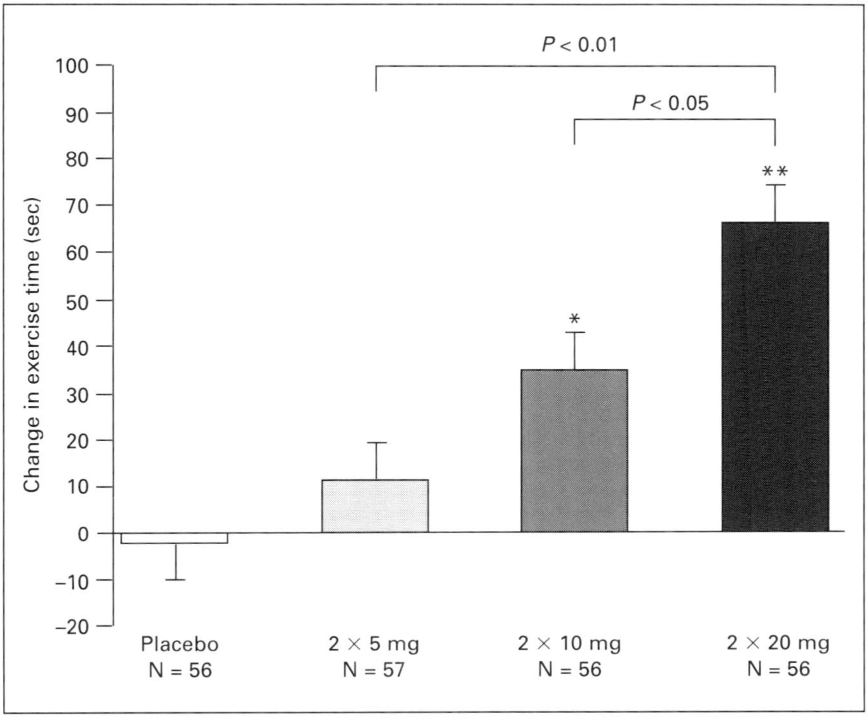

Figure 5.4 *Mean changes in exercise time from baseline after 12 weeks of treatment with placebo or increasing doses of quinapril for all randomized patients with congestive heart failure NYHA II and III (intention-to-treat analysis). Adapted from Riegger.*[143]
**: $P < 0.05$ versus placebo. **: $P < 0.01$ versus placebo.*

four parallel treatment groups receiving either placebo or quinapril in a dose of 5, 10 or 20 mg bid. Concomitant therapy with digitalis and diuretics was allowed. The overall clinical response rate (an increase in exercise time to more than 20% of baseline) was about 55% (out of the 225 patients studied). There was a striking dose-dependent increase of exercise time related to dose of quinapril (intention-to-treat analysis, Fig. 5.4). Because serum angiotensin converting enzyme activity was already completely suppressed 12 hours after ingestion of the lowest quinapril dose, the observed dose-relationship was attributed to possible effects on local renin angiotensin systems, for example in the blood vessels.[143]

Several studies have addressed the question of whether one ACE inhibitor is superior to others for the treatment of CHF. Giles for the Multicenter-Lisinopril-Captopril Congestive Heart Failure Study Group reported no difference of the increase in mean exercise tolerance after 12 weeks of treatment with the short-acting captopril (n = 95; +120 seconds) versus the long-acting lisinopril (n = 97; +137 seconds; P = not significant) on an intention-to-treat analysis.[144] In that study, left ventricular ejection fraction rose by 4.1% in the lisinopril subgroup versus 0.8% in the captopril subgroup (treatment effect: $P < 0.05$). Using appropriate doses, at the moment no major advantages have been documented favoring a specific ACE inhibitor for patients with CHF.[145] This contrasts with recent findings in men with arterial hypertension, where captopril and enalapril were indistinguishable according to clinical assessments of efficacy and safety, but had different effects on quality-of-life. In men with low quality-of-life scores at baseline, both enalapril and captopril were beneficial. However, in men with the highest quality-of-life scores at baseline, negative effects of treatment on General Perceived Health scores could be demonstrated for enalapril but not for captopril.[146] A possible explanation for the observed positive effects of captopril — which is more lipophilic — might be its better access to the central nervous system, with influences on peptide degradation or on cholinergic mechanisms.

The old controversy about the usefulness of digitalis in patients with CHF was addressed in the Captopril-Digoxin Multicenter Trial, published in 1988.[147] Patients with mild to moderate CHF (82% NYHA II), who were stable on adjustable doses of diuretics (initially 84%), were included. However, 164 out of 464 patients who entered the washout-stabilization period (more than 10 days) had to be excluded from final randomization, 20% due to worsening of CHF after withdrawal of inotropic or vasodilatory drugs, most of whom had been on digitalis. This was regarded to be a limiting bias against digitalis in this study. Patients were randomized to three treatment arms: placebo, digoxin (serum target level: 0.7–2.5 ng/ml) or captopril (target dose 3×50 mg). After six months, mean exercise tolerance as the primary objective increased from 552 to 587 seconds on placebo, from 564 to 618 seconds on digoxin and from 572 to 653 seconds on captopril. Captopril was significantly superior to placebo but not to digoxin (treatment effect). Left ventricular ejection fraction increased by 4.4% on digoxin, by 1.8% on captopril and by 0.9% on placebo (significant treatment effect of digoxin versus placebo and versus captopril). Interestingly, this study focused for the first time on hospitalization rate and emergency department visits as important endpoints. Both captopril and digoxin were significantly superior to placebo. It was concluded that 'captopril is an effective alternative to digoxin treatment in patients with mild to moderate heart failure...'.

A comparison of ACE inhibition with direct-acting vasodilators was performed in the Veterans Administration Cooperative Vasodilator-Heart Failure Trial II (V-HeFT II). Although the primary endpoint was mortality, there were several subsidiary endpoints including exercise tolerance, left ventricular function, neurohormones and arrhythmias.[7] Patients with CHF NYHA (I), II and III on digoxin and diuretics were randomized to enalapril (n = 403) or the combinationof hydralazine-isosorbide dinitrate (n = 401). Exercise tolerance was assessed by measuring maximal oxygen uptake during treadmill exercise. Oxygen consumption was increased significantly by hydralazine-isosorbide dinitrate but not by enalapril (*P* values for treatment differences: 0.01 after 13 weeks, 0.02 after 6 months and 0.02 after 2 years). However, whether a difference of peak oxygen consumption in the range of about 5% of its baseline value marks a relevant endpoint, might be questioned. Secondly, after two years mortality was significantly higher in the hydralazine-isosorbide dinitrate group (162 out of 401) compared to enalapril (141 out of 403), possibly indicating a positive selection of patients subjected to repeated exercise tests.

Recently, ACE-inhibition by enalapril was compared with the direct-acting vasodilator flosequinan in a double-blind, parallel-group design in 61 patients with CHF NYHA III, who were on constant doses of furosemide.[148] Exercise tolerance was repeatedly assessed over a period of 12 weeks and there was no significant difference between enalapril and flosequinan on an intention-to-treat analysis.

These data clearly indicate that ACE inhibitors substantially improve exercise tolerance in patients with CHF. The amount of improvement seems to relate inversely to the severity of CHF, and is probably dose-dependent. The time course of improvement is strikingly uncoupled from changes in central hemodynamics, indicating a complex response of the cardiovascular system, where systemic effects are not the major determinants. This approach is further supported by recent data from Aldigier and coworkers, who found, that the exercise-induced increase of plasma angiotensin II levels is not significantly reduced by captopril.[149] Finally, as may be inferred from the V-HeFT II trial, an improvement in exercise tolerance does not necessarily translate into improved survival.

Effects of ACE inhibitors on survival

In 1987, the Consensus I trial was published, demonstrating for the first time that ACE inhibition prolongs life in patients with severe CHF.[8] The trial was stopped prematurely (14.12.1986) by the Ethical Review Committee due to the observed efficacy of enalapril. At that time, 253 patients with CHF NYHA IV, who were refractory to digitalis, diuretics (mean furosemide dose 210 mg/day; mean spironolactone dose 80 mg/day) and vasodilator drugs other than ACE inhibitors (49%) were randomized to placebo (n = 126) or enalapril (n = 127; target dose: 40 mg/day; mean final dose 18.4 mg/day). Mean follow-up was 188 days. As soon as one month after initiation of treatment there was a striking reduction of mortality: –40% at six months and –31% at one year with an assumed prolongation of life of over one year by enalapril.[150] The entire treatment effect was due to a reduction of mortality from the progression of heart failure (–50%). Mortality in the placebo group was highest in those patients with marked neuroendocrine activation and reduction of mortality by enalapril was consistently largest in this group.[151] Beneficial effects of enalapril were confined to patients with coronary artery disease, however this was the cause of CHF in nearly 75% of the whole study population, therefore the remaining number of patients seems to be too small to exclude effects of ACE inhibition on survival.[152]

The V-HeFT II trial, which has already been mentioned, compared direct vasodilatation by hydralazine-isosorbide dinitrate with ACE inhibition by enalapril on mortality in 804 patients with mild to moderate CHF.[7] Exercise tolerance and left ventricular ejection fraction increased more with the combination of hydralazine-isosorbide dinitrate, however enalapril significantly attenuated the rise in plasma noradrenaline (norepinephrine) and reduced the extent of ventricular ectopy. After two years, mortality was reduced by 28% (P = 0.016) in the enalapril arm, predominantly due to a reduction of sudden death with or without premonitory worsening. These findings are in sharp contrast to other mortality studies, where sudden death is not significantly reduced by ACE inhibition. They might be attributed to the possible increase of plasma noradrenaline (norepinephrine) in the hydralazine-isosorbide dinitrate arm, possibly with deleterious effects on ventricular arrhythmias.

In the early and mid-1980s the SOLVD trial (Studies Of Left Ventricular Dysfunction) was designed at the Clinical Trials Branch of the National Institutions of Health to examine the effects of therapy on mortality in mild to moderate CHF.[153] Patients with a left ventricular ejection fraction of less than or equal to 35%, measured by radionuclide techniques (68%), contrast angiography (11%) or two-dimensional echocardiography (21%), were enrolled into the treatment arm if they had symptoms of CHF (n = 2569) or into the prevention arm if they were asymptomatic (n = 4228). Francis and coworkers reported on progressive neuroendocrine activation in a subgroup analysis.[44] In summary, they found increases in plasma noradrenaline (norepinephrine), plasma renin activity, plasma arginine-vasopressin and plasma atrial natriuretic peptide. Activation was more pronounced if symptoms of CHF were present. However, in patients with asymptomatic left ventricular dysfunction, a significant rise in plasma renin activity was seen only if they were on diuretic therapy (median values: 0.6 ng/ml/h in 56 control persons; 0.7 ng/ml/h in 121 patients without diuretics and 1.0 ng/ml/h in 30 patients on diuretics).

In the treatment arm, mean ejection fraction was 25%, 98% of the patients were in NYHA classes I–III, and coronary artery disease was the cause of CHF in 70%. Mean age was 61 years and 80% were males. Two-thirds were on digitalis, 85% on diuretics, 30% were taking calcium blockers and one-third antiplatelet drugs.[6] 1284 patients were randomized to placebo and 1285 patients to enalapril (target dose: 20 mg/day; mean dose [intention-to-treat]: 11.2 mg/day; mean dose [efficacy sample]: 16.6 mg/day; target dose achieved: 49.3%). Follow-up ranged from 22 to 55 months (average 41.4). By the end of the study 32.5% of patients in the enalapril group and 41.4% in the placebo group had stopped taking blinded medication. Treatment with enalapril reduced overall mortality by 16% ($P < 0.0036$) and cardiovascular mortality by 18%, which was due to a reduction of mortality from progressive heart failure (risk reduction 22%, $P < 0.0045$). There were no effects on deaths classified as due to arrhythmia without worsening of CHF. The number of hospitalizations was also consistently reduced

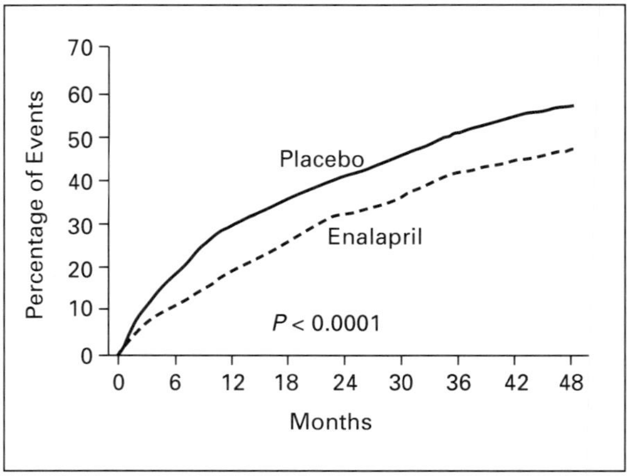

Figure 5.5
Percentage of events, defined as death or hospitalization for congestive heart failure, occurring in the placebo and enalapril groups of the SOLVD treatment trial. (Reproduced with permission from SOLVD Investigators 1991, N Engl J Med ***325:****293–302.)*

(–36%) by ACE inhibition. The reduction of mortality and of hospitalizations for CHF was observed soon after randomization and increased for as much as 24 months (Fig. 5.5). Subgroup analysis revealed that the benefit with respect to the combined endpoint of death or hospitalization appeared to be significantly smaller among the patients in the highest percentile for ejection fraction; however, even in these patients there was a trend toward fewer deaths or hospitalizations in the enalapril group (risk reduction 12%).

In the prevention arm, mean ejection fraction was 28%, 66% of the patients were in NYHA class I, and coronary artery disease was the cause of CHF in 83%. Mean age was 59 years and 89% were males. Seventy-five per cent of patients took neither digitalis nor diuretics, 35% were taking calcium blockers and over

50% anti-platelet drugs.[83] 2117 patients were randomized to placebo and 2111 patients to enalapril (target dose: 20 mg/day; mean dose [intention-to-treat]: 12.7 mg/day; mean dose [efficacy sample]: 16.7 mg/day; target dose achieved: 56.1%). Follow-up ranged from 14.6 to 62 months (average 37.4). By the end of the study 24% of patients in the enalapril group and 27% in the placebo group had stopped taking blinded medication. Treatment with enalapril reduced overall mortality by 8% (P = 0.3) and cardiovascular mortality by 12% (P = 0.12), which was due to a reduction of mortality from progressive heart failure (risk reduction 21%, P = 0.1). There were no effects on deaths classified as due to arrhythmia without worsening of CHF. Morbidity, however, was significantly reduced by ACE inhibition in patients with asymptomatic left ventricular dysfunction: the risk reduction for development of CHF was –37%; risk reduction for first hospitalization was –36%, and for multiple hospitalizations, –44%. Combining mortality and morbidity endpoints, enalapril was strikingly effective, as shown in Fig. 5.6. The median length of time to the first hospitalization for CHF was 13.2 months in the placebo group, the time before there were a similar number of hospitalizations in the enalapril group was 27.8 months. The median length of time to the development of heart failure was 8.3 months in the placebo group, the time to development of a similar number of events in the enalapril group was 22.3 months. The benefit of enalapril in preventing the development of CHF was evident as early as six weeks after randomization, and the difference between the two groups continued to increase until the end of the study. Among patients who did not have CHF, mortality rates were 12.1% in the placebo group and 11.5% in the enalapril group, contrasting an excess mortality rate of 33% for patients who were hospitalized for CHF. The benefits of ACE inhibition on morbidity were greatest in patients with the lowest ejection fractions. Comparing effects on mortality and morbidity, the large number of patients who developed CHF and were treated — in an open fashion — with ACE inhibitors may be responsible for an underestimation of the effects of enalapril on mortality in this trial.

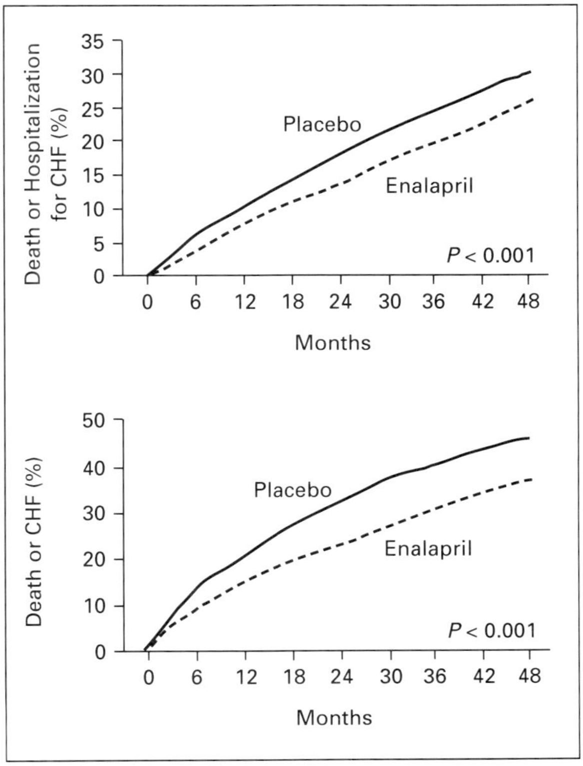

Figure 5.6
Death or hospitalization for congestive heart failure (upper panel) or death or development of congestive heart failure (lower panel) in the SOLVD prevention trial. CHF = congestive heart failure. (Reproduced with permission from SOLVD Investigators 1992, N Engl J Med ***327****:685–91.)*

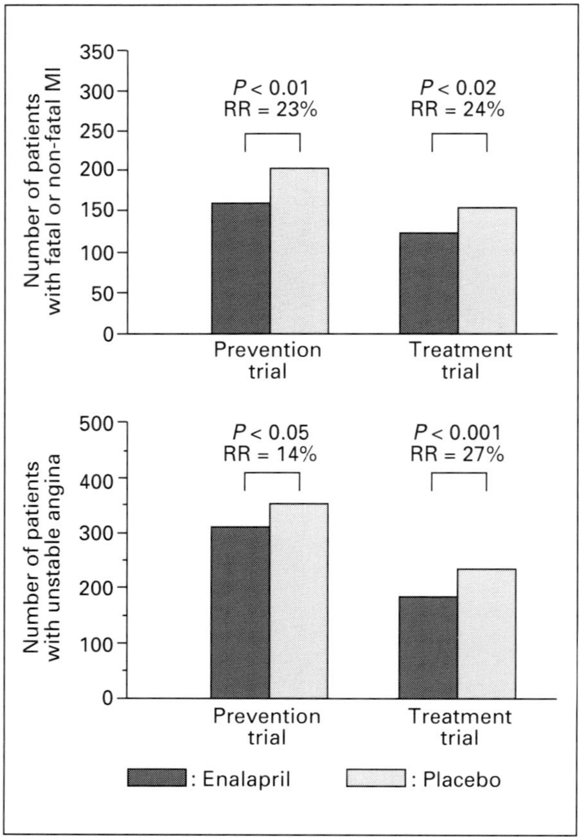

Figure 5.7

Effect of enalapril on myocardial infarction (upper panel) and unstable angina (lower panel) in patient with low ejection fractions. Modified from Yusuf et al.[54]

MI = myocardial infarction; RR = risk reduction.

A recent combined analysis of the treatment and prevention trial demonstrated that enalapril significantly reduced coronary events, that is fatal and non-fatal recurrent myocardial infarctions and the occurrence of unstable angina (Fig. 5.7).[154] Because these effects were not apparent until at least six months, they seem to be the result of different effects of ACE inhibition: reduction of blood pressure, restoration of endothelial function, coronary vasodilatation, prevention of myocardial hypertrophy and possibly prevention of atherosclerosis progression.

As coronary artery disease has become the leading cause of CHF in the western world, much effort has been made to investigate the pathogenesis of postinfarction CHF.[155,156] In the early phase after myocardial infarction, ventricular volumes increase due to infarct expansion and/or a rise in filling pressure. The increase in wall stress promotes remodeling of the residual myocardium, for example left ventricular volume and mass both increase, allowing filling pressure and stroke volume to be restored. However, if the infarct is sufficiently large in size, the increase in volume is out of proportion to mass, resulting in continuous elevations of filling pressure and wall stress. Over time, this leads to an unrelenting stimulus to progressive dilatation, finally resulting in the development of CHF.[156,157] As has been shown, systemic neuroendocrine activation is present in the acute phase of myocardial infarction, which turns to normal after about ten days in uncomplicated cases.[158] Messenger RNA levels for angiotensinogen and ACE may be elevated in the residual myocardium in experimental infarction in the rat, emphasizing the possible involvement of local renin-angiotensin systems.[45,159] In patients with myocardial infarction, left ventricular volumes are powerful predictors of mortality after myocardial infarction.[160,161]

In 1985, Pfeffer and coworkers reported beneficial effects of captopril on infarcted ventricles and survival in rats.[162,163] In 1988, the same group demonstrated that ACE inhibition by captopril could attenuate progressive ventricular dilatation in patients with anterior myocardial infarction.[164] The impact of these findings on survival was addressed by the

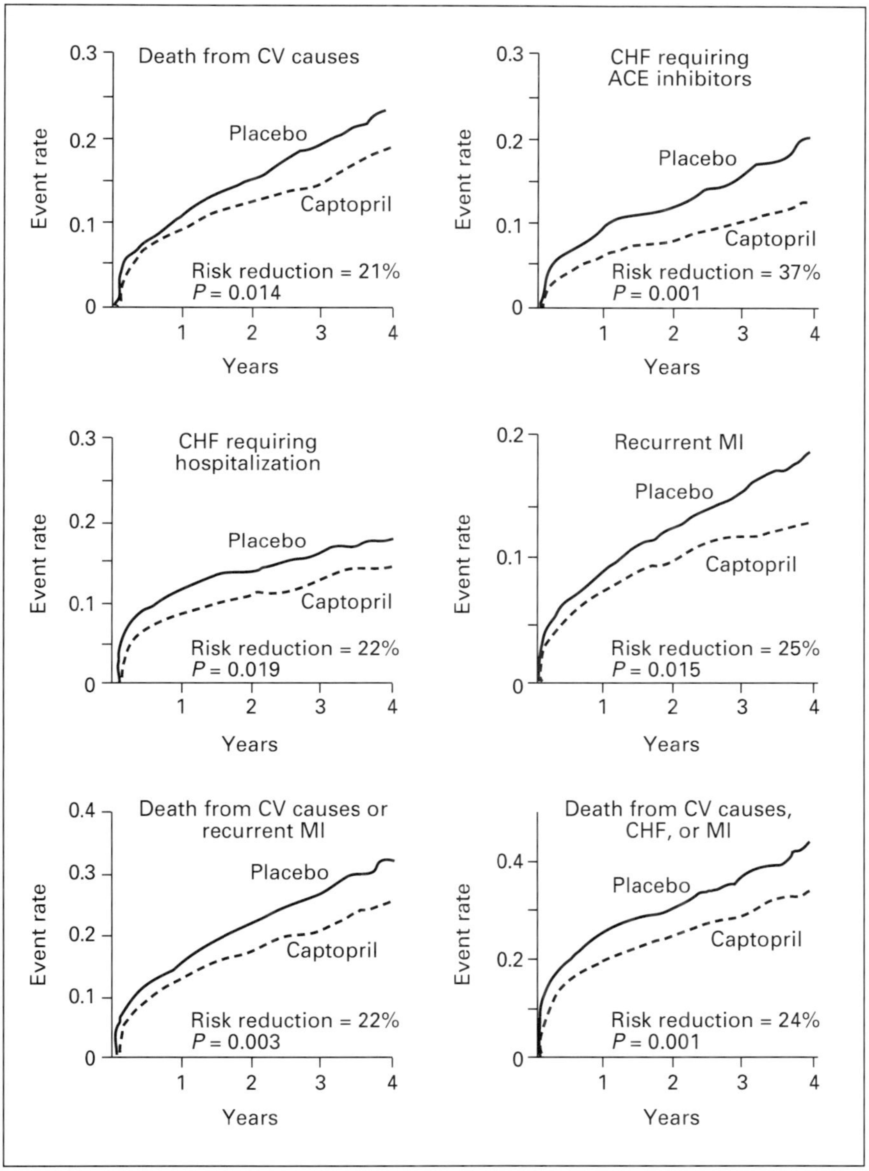

Figure 5.8 *Life tables for cumulative fatal and non-fatal cardiovascular events in the SAVE trial. CV = Cardiovascular. CHF = Congestive heart failure. ACE = Angiotensin converting enzyme. MI = Myocardial infarction. (Reproduced with permission from Pfeffer MA et al.* N Engl J Med *1992;* ***327****:669–77.)*

Survival And Ventricular Enlargement Trial (SAVE). Patients who survived the first three days after a myocardial infarction and had a left ventricular ejection fraction of up to 40% (radionuclide ventriculography) were included if they had no overt CHF.[165] Within 3 to 16 days after infarction (mean 11 days) they were randomized to placebo (n = 1116) or captopril (n = 1115; target dose 150 mg/day; target dose achieved: 79%). Mean ejection fraction was 31%, mean age was 59 years, 83% were males; 26% were on digitalis, 35% on diuretics, over

50% on nitrates and 59% on aspirin. Follow-up ranged from 24 to 60 months (average 42). At the last study visit, 73% of the surviving patients in the placebo group and 70% in the captopril group were still taking the study drug. Comedication was not significantly different. Treatment with captopril had striking beneficial effects on total mortality (risk reduction: 19%) and cardiovascular morbidity (risk reduction: 22%). As in the SOLVD trial, sudden, unexpected death was not significantly affected, although risk reduction was 19%.[166] The overall effects were independent from thrombolytic therapy, patency of the infarct-related artery and coadministration of β-blockers or aspirin.[167] The risk reduction for cardiovascular death and morbidity was more pronounced in patients with a more depressed ejection fraction. In accordance with the SOLVD trial, the development of CHF requiring open-label treatment with an ACE inhibitor was associated with a dramatically increased risk for subsequent death (4.5-fold). Recurrent non-fatal myocardial infarction, which also was associated with a consistently increased risk of death (seven-fold), was reduced by captopril treatment by 25%, together with a reduction of death after recurrent myocardial infarction by 32%.[168]

In conclusion, this trial demonstrated beneficial effects of ACE inhibition by captopril on cardiovascular mortality and particularly morbidity after myocardial infarction (Fig. 5.8). Treatment was started an average of 11 days after myocardial infarction (note that symptomatic hypotension after the first captopril 'challenge' with 6.25 mg occurred in 16 out of 2250 eligible patients) and proved to be safe and highly effective.

The second important mega-trial on ACE inhibition after myocardial infarction, the AIRE study, demonstrated a 27% reduction in mortality in ramipril-treated patients with clinical evidence of transient or on-going heart failure following myocardial infarction.[169] Oral treatment was started between the second and ninth days after myocardial infarction (mean 5.4 days; 170). The results of both the SAVE and the AIRE studies are in contrast to the findings of the CONSENSUS II trial, where ACE inhibition by enalapril/enalaprilat was started within the first 24 hours after onset of chest pain (1 mg enalaprilat intravenously within 2–3 hours). Follow-up was six months. In this study (3046 patients on placebo, 3044 patients on enalapril, target dose 20 mg/day) overall mortality was not improved by enalapril (relative risk 1.1, P = 0.26). These findings were attributed to an increased incidence of profound early hypotension (12% versus 3%) after early ACE inhibition, possibly inducing infarct expansion instead of infarct size reduction. However, a change of treatment due to development of overt CHF — as a secondary endpoint — was significantly reduced by enalapril.[170]

Conclusion

With respect to the above mentioned data, the role of ACE inhibitors in the treatment of CHF is well established. Using appropriate doses, for example captopril 50 mg tid or enalapril 10 mg bid, they improve hemodynamics, exercise tolerance and mortality in patients with severe CHF as well as in patients with mild to moderate CHF. Mortality and morbidity in patients with asymptomatic left ventricular dysfunction is reduced, possibly due to reduction of coronary events and/or prevention/delay of the development of postinfarction CHF. There are numerous specific pharmacological effects of ACE inhibitors, which may contribute to their overall clinical benefit, but their favourable interference with systemic and/or local neurohormonal activation in CHF seems to be of crucial importance. However, a number of important questions are still unresolved:

Do ACE inhibitors reduce mortality from sudden cardiac death?

Do ACE inhibitors prevent progression of atherosclerosis?

Does the combination of ACE inhibitors with aldosterone antagonists offer additional advantages?

Does the combination of ACE inhibitors with β-blockers offer additional advantages?

Does the combination of ACE inhibitors with neutral endopeptidase inhibitors offer additional advantages?

Are ACE inhibitors effective and safe in older patients (over 70 years) suffering from CHF?

Are lower doses of ACE inhibitors also effective?

Are pharmacological differences of ACE inhibitors really important?

Is the response to ACE inhibitors genetically determined?

Which patient should be treated with an ACE inhibitor and when should we start its administration?

References

1 Braunwald E. Pathophysiology of heart failure. In: Braundwald E, ed. Heart disease: a textbook of cardiovascular medicine, 4 Ed. Philadelphia, London: W.B. Saunders, 1992: 393–418.

2 Cohn JN. Current therapy of the failing heart. *Circulation* 1988; **78**:1099–1107.

3 Lutz JE. A XII century description of congestive heart failure. *Am J Cardiol* 1988; **61**:494–95.

4 Yusuf S, Thom T, Abbott RD. Changes in hypertension treatment and in congestive heart failure mortality in the United States. *Hypertension* 1989; **13**:74–79.

5 McKee PA, Castelli WP, McNamara PM, Kannel B. The natural history of congestive heart failure: The Framingham Study. *N Engl J Med* 1971; **285**:1441–45.

6 The SOLVD Investigators. Effect of enalapril on survival in patients with reduced left ventricular ejection fractions and congestive heart failure. *N Engl J Med* 1991; **325**:293–302.

7 Cohn JN, Johnson G, Ziesche S et al. A comparison of enalapril with hydralazin-isosorbide dinitrate in the treatment of chronic congestive heart failure. *N Engl J Med* 1991; **293**:303–10.

8 The CONSENSUS Trial Study Group. Effects of enalapril on mortality in severe congestive heart failure: results of the Cooperative North Scandinavian Enalapril Survival Study (CONSENSUS). *N Engl J Med* 1987; **316**:1429–35.

9 Franciosa JA, Wilen M, Ziesche S, Cohn JN. Survival in men with severe chronic left ventricular failure due to either coronary artery disease or idiopathic dilated cardiomyopathy. *Am J Cardiol* 1983; **51**:831–36.

10 White HD, Norris RM, Brown MA, Brandt PWT, Whitlock RMC, Wild CJ. Left ventricular endsystolic volume as the major determinant of survival after recovery from myocardial infarction. *Circulation* 1987; **76**:44–51.

11 Meinertz T, Hofmann T, Kasper W et al. Significance of ventricular arrhythmias in idiopathic dilated cardiomyopathy. *Am J Cardiol* 1984; **53**:902–7.

12 Swedberg K, Eneroth P, Kjekshus J, Wilhelmsen L for the CONSENSUS Trial Study Group. Hormones regulating cardiovascular function in patients with severe congestive heart failure and their relation to mortality. *Circulation* 1990; **82**:1730–36.

13 Gottlieb S, Kukin ML, Ahern D, Packer M. Prognostic importance of atrial natiuretic peptide in patients with chronic congestive heart failure. *J Am Coll Cardiol* 1989; **13**:1534–39.

14 Cohn JN, Levine TB, Olivari MT et al. Plasma norepinephrine as a guide to prognosis in patients with chronic congestive heart failure. *N Engl J Med* 1984; **311**:819–23.

15 Ross J. Afterload mismatch and preload reserve: A conceptual framework for the analysis of ventricular function. *Prog Cardiovasc Dis* 1976; **18**:255–64.

16 Chidsey CA, Braunwald E, Morrow AG, Mason DT. Myocardial norepinephrine concentration in man: Effects of reserpine and of congestive heart failure. *N Engl J Med* 1963; **269**:653–59.

17 Bristow MR, Ginsberg R, Minobe W et al. Decreased catecholamine sensitivity and beta-adrenergic receptor density in failing human hearts. *N Engl J Med* 1982; **307**:205–11.

18 Kromer EP, Riegger AJG. Effects of long-term angiotensin converting enzyme inhibition on myocardial hypertrophy in experimental aortic stenosis in the rat. *Am J Cardiol* 1988; **62**:161–63.

19 Brilla CG, Janicki JS, Weber KT. Cardioreparative effects of lisinopril in rats with genetic hypertension and left ventricular hypertrophy. *Circulation* 1991; **83**:1771–79.

20 Eckberg DL, Drabinsky M, Braunwald E. Defective parasympathetic control in patients with heart disease. *N Engl J Med* 1971; **285**:877–83.

21 Higgins CB, Vastner SF, Eckberg DL, Braunwald E. Alterations in the baroreceptor reflex in conscious dogs with heart failure. *J Clin Invest* 1972; **51**:715–24.
22 Zucker IH, Gilmore JP. Aspects of cardiovascular reflexes in pathologic state. *Fed Proc* 1985; **44**:2400–7.
23 Leimbach WN, Wallin G, Victor RG, Aylward E, Sundlof G, Marl AL. Direct evidence from intraneuronal recordings for increased sympathetic outflow in patients with heart failure. *Circulation* 1986; **73**:913–19.
24 Haskins GE, Esler MD, Jennings GL, Burton D, Johnson JA, Korner PI. Norepinephrine spillover to plasma in patients with heart failure. *Circulation* 1986; **73**:615–21.
25 Chidsey CA, Harrison DC, Braunwald E. Augmentation of the plasma norepinephrine response to exercise in patients with congestive heart failure. *N Engl J Med* 1962; **267**:650–54.
26 Vogel JHK, Jacobowitz D, Chidsey CA. Distribution of norepinephrine in the failing heart: Correlation of chemical analysis and fluorescence microscopy. *Circ Res* 1969; **24**:71–84.
27 Sole MJ, Helke CJ, Jacobowitz DM. Increased dopamine in the failing hamster heart: Transvesicular transport of dopamine limits the rate of norepinephrine synthesis. *Am J Cardiol* 1982; **49**:1682–90.
28 Ungerer M, Böhm M, Elce JS, Erdmann E, Lohse MJ. Altered expression of β-adrenergic receptor-kinase and β_1-adrenergic receptors in the failing human heart. *Circulation* 1993; **87**:454–63.
29 Ungerer M, Böhm M, Erdmann E, Lohse MJ. Veränderte Expression von β-adrenerger Rezeptor-kinase-1 und β_1-adrenergen Rezeptoren bei Herzinsuffizienz. *Zeitschr Kardiol* 1993; **82**:162A.
30 Bristow MR, Hershberger RE, Port JD et al. β-Adrenergic pathways in nonfailing and failing human ventricular myocardium. *Circulation* 1990; **82[suppl 1]**:I-12–I-25.
31 Böhm M, Gierschik P, Jakobs K-H et al. Increase of G_i in human hearts with dilated but not ischemic cardiomyopathy. *Circulation* 1990; **82**:1249–65.
32 Lohse MJ, Benovic JL, Codina J, Caron MG, Lefkowitz RJ. β-Arrestin: A protein that regulates β-adrenergic function. *Science* 1990; **248**:1547–50.
33 Craeger MA, Quigg RJ, Ren CJ, Roddy M-A, Colucci WS. Limb vascular responsiveness to β-adrenergic stimulation in patients with congestive heart failure. *Circulation* 1991; **83**:1873–79.
34 Limas CJ, Goldenberg IF, Limas C. Autoantibodies against β-adrenoceptors in human idiopathic dilated cardiomyopathy. *Circ Res* 1989; **64**:97–103.
35 Bristow MR, Minobe W, Rasmussen R. Differential regulation of alpha and beta-adrenergic receptor in the failing human heart. *Circulation* 1985; **72[suppl III]**:III-329.
36 Colucci WS, Craeger MA. Vascular adrenergic responsiveness in congestive heart failure. *Heart Failure* 1991; **7**:71–78.
37 Goldsmith SR, Francis GS, Cohn JN. Norepinephrine infusions in congestive heart failure. *Am J Cardiol* 1985; **56**:802–4.
38 Raab W. Adrenocortical compounds in the blood: Relation of their quantity to arterial hypertension, renal insufficiency and congestive heart failure. *Arch Int Med* 1941; **68**:713–39.
39 Merril AJ, Morrison L, Brannon ES. Concentration of renin in renal venous blood in patients with chronic heart failure. *Am J Med* 1946; **1**:468–72.
40 Holmer SR, Riegger AJG, Notheis W, Kromer EP, Kochsiek K. Hemodynamic changes and renal plasma flow in early heart failure: Implications for renin, aldosterone, norepinephrine, atrial natriuretic peptide and prostacyclin. *Basic Res Cardiol* 1987; **82**:101–8.
41 Mancia G, Romero JC, Shepherd JT. Continuous inhibition of renin release in dogs by vagally innervated receptors in the cardiopulmonary region. *Circ Res* 1975; **36**:529–35.
42 Elsner D, Kromer EP, Riegger AJG. Effect of vagal blockade on neurohormonal systems in conscious dogs with heart failure. *J Cardiovasc Pharmacol* 1990; **15**:586–91.

43 Francis GS, Siegel RM, Goldsmith SR et al. Acute vasoconstrictor response to intravenous furosemide in patients with chronic congestive heart failure: Activation of the neurohumoral axis. *Ann Int Med* 1985; **103**:1–6.
44 Francis GS, Benedict C, Johnstone DE et al. Comparison of neuroendocrine activation in patients with left ventricular dysfunction with and without congestive heart failure. *Circulation* 1990; **82**:1724–29.
45 Lindpaintner K, Lu W, Niedermajer N, et al. Selective activation of cardiac angiotensinogen gene expression in post-infarction ventricular remodelling in the rat. *J. Mol Cell Cardiol* 1993; **25**:133–43.
46 Hirsch AT, Talsness CE, Schunkert H, Paul M, Dzau VJ. Tissue-specific activation of cardiac angiotensin converting enzyme in experimental heart failure. *Circ Res* 1991; **69**:475–82.
47 Lee WH, Packer M. Prognostic importance of serum sodium concentration and its modification by converting enzyme inhibition in patients with severe chronic heart failure. *Circulation* 1986; **73**:257–67.
48 Urata H, Healy B, Stewart RW, Bumpus FM, Husain A. Angiotensin II receptors in normal and failing human hearts. *J Clin Endocrinol Metab* 1989; **69**:54–66.
49 Riegger AJG, Liebau G, Kochsiek K. Antidiuretic hormone in congestive heart failure. *Am J Med* 1982; **72**:48–52.
50 Share L. Interactions between vasopressin and the renin angiotensin system. *Fed Proc* 1979; **38**:2267–71.
51 Riegger AJG, Kochsiek K. Vasopressin, renin, and norepinephrine in congestive heart failure from idiopathic dilated cardiomyopathy before and after captopril. *Am J Cardiol* 1986; **75**:300–303.
52 Stewart DJ, Cernacek P, Costello KB, Rouleau JL. Elevated endothelin-1 in heart failure and loss of normal response to postural change. *Circulation* 1992; **85**:510–17.
53 Lerman A, Hildebrand FL, Aarhus LL, Burnett JC. Endothelin has biological actions at pathophysiological concentrations. *Circulation* 1991; **83**:1808–14.
54 Edwards BS, Ackermann MD, Lee ME, Reeder GS, Wold LE, Burnett JC Identification of atrial natriuretic factor within ventricular tissue in hamsters and humans with congestive heart failure. *J Clin Invest* 1988; **81**:82–86.
55 Richards AM, Cleland JFG, Tonolo G et al. Plasma alpha natriuretic peptide in cardiac impairment. *BMJ* 1986; **293**:409–12.
56 Lerman A, Gibbons RJ, Rodeheffer RJ, Bailey KR, McKinley LJ, Heublein DM, Burnett JC Jr. Circulating N-terminal atrial natriuretic peptide as a marker for symptomless left ventricular dysfunction. *Lancet* 1993; **341**:1105–9.
57 Riegger AJG. The role of atrial natriuretic factor in congestive heart failure. *Prog Pharmacol Clin Pharmacol* 1990; **7**:101–15.
58 Mukoyama M, Nakao K, Hosoda K et al. Brain natriuretic peptide as a novel cardiac hormone in humans. *J Clin Invest* 1991; **87**:1402–12.
59 Motwani JG, McAlpine H, Kennedy N, Struthers AD. Plasma brain natriuretic peptide as an indicator for angiotensin converting enzyme inhibition after myocardial infarction. *Lancet* 1993; **341**:1109–13.
60 Dzau VJ, Packer M, Lilly LS et al. Prostaglandins in severe congestive heart failure. *N Engl J Med* 1984; **310**:347–52.
61 Riegger AJG, Elsner D, Kromer EP. Circulatory and renal control by prostaglandins and renin in low cardiac output in dogs. *Am J Physiol* 1989; **256**:H1079–H1086.
62 Hall D, Zeitler H, Rudolph W. Counteraction of the vasodilator effects of enalapril by aspirin in severe heart failure. *J Am Coll Cardiol* 1992; **20**:1549–55.
63 Aiken JW, Vane JR. Intrarenal prostaglandin release attenuates the renal vasoconstrictor activity of angiotensin. *J Pharmacol Exp Therap* 1973; **184**:678–87.
64 Malik KU, McGiff JC. Modulation by prostaglandins of adrenergic transmission in the isolated perfused rabbit and rat kidney. *Circ Res* 1975; **36**:599–609.
65 Oliver JA, Sciacca RR, LeGren P, Cannon PJ. Modulation by prostaglandins of the renal vascular action of arginine vasopressin. *Prostaglandins* 1982; **24**:641–56.

66 Jabs K, Zeidel ML, Silva P. Prostaglandine E_2 inhibits Na-K-ATPase activity in the inner medullary collecting duct. *Am J Physiol* 1989; **257**:F424–F430.

67 Furchgott RF, Zawadzki JV. The obligatory role of endothelial cells in the relaxation of arterial smooth muscle by acetylcholine. *Nature* 1980; **288**:373–74.

68 Elsner D, Müntze A, Kromer EP, Riegger AJG. The systemic vasoconstriction induced by inhibition of nitric oxide synthesis is attenuated in conscious dogs with heart failure. *Cardiovasc Res* 1991; **25**:438–40.

69 Zimlichmann R, Levinson PD, Kelly G, Stull R, Keiser HR, Goldstein D. Derivation of urinary dopamine from plasma dopa. *Clin Sci* 1988; **75**:515–20.

70 Casagrande C. Dopamine and the kidney in heart failure. *Herz* 1991; **16**:102–15.

71 Missale C, Liberini P, Memo M, Carruba MO, Spano PF. Characterization of dopamine receptors associated with aldosterone secretion in rat adrenal glomerulosa. *Endocrinology* 1986; **119**:2227–32.

72 Felder CC, Campbell T, Albrecht F, Jose PA. Dopamine inhibits Na^+-H^+ exchanger activity in renal brush border membrane vesicles. *Am J Physiol* 1990; **259**:F297–F303.

73 Goldberg LI, McDonald RH, Zimmermann AM. Sodium diuresis produced by dopamine in patients with congestive heart failure. *N Engl J Med* 1963; **269**:1060–64.

74 Sullivan MJ, Green HJ, Cobb FR. Skeletal muscle biochemistry and histology in ambulatory patients with long-term heart failure. *Circulation* 1990; **81**:518–27.

75 Sinoway L, Prophet S, Gorman I et al. Muscle acidosis during static exercise is associated with calf vasoconstriction. *J Appl Physiol* 1989; **66**:429–36.

76 Zelis R, Delea CS, Coleman HN, Mason DT. Arterial sodium content in experimental congestive heart failure. *Circulation* 1970; **41**:312–20.

77 Blaes N, Boissel JP. Growth-stimulating effect of catecholamines on rat aortic smooth muscle cells in culture. *J Cell Physiol* 1983; **116**:167–72.

78 Campbell-Boswell M, Robertson AL. Effect of angiotensin II and vasopressin on human smooth muscle cells in vitro. *Exp Mol Pathol* 1981; **35**:266–76.

79 Geisterfer AAT, Owens GK. Arginine vasopressin-induced hypertrophy of cultured rat aortic smooth muscle cells. *Hypertension* 1989; **14**:413–20.

80 Kromer EP, Elsner D, Holmer SR, Müntze A, Riegger AJG. Aortic input impedance and neurohormonal activation in patients with mild to moderate chronic congestive heart failure. *Cardiovasc Res* 1992; **26**:265–72.

81 McMichael J. Cardiac venous congestion — its causes and consequences. *Am J Med* 1949; **6**:651–55.

82 Meredith IT, Broughton A, Jennings GL, Esler MD. Evidence of a selective increase in cardiac sympathetic activity in patients with sustained ventricular arrhythmias. *N Engl J Med* 1991; **325**:618–24.

83 The SOLVD-Investigators. Effect of enalapril on mortality and the development of heart failure in asymptomatic patients with reduced left ventricular ejection fractions. *N Engl J Med* 1992; **327**:685–91.

84 Howarth S, McMichael J, Sharpey-Schafer EP. The circulatory effect of theophylline ethylene diamine. *Clin Sci* 1946; **6**:125–29.

85 Steinberg FU, Hensen J. Effect of theophylline aminobutanol on circulation in congestive heart failure. *J Lab Clin Med* 1946; **31**:857–62.

86 Hayward GW. Tetraethylammonium bromide in hypertension and hypertensive heart failure. *Lancet* 1948; **251**:18–22.

87 Relman AS, Epstein FH. Effect of tetraethylammonium on venous and arterial pressure in congestive heart failure. *Proc Soc Exp Biol Med* 1949; **70**:11–15.

88 Freis ED, Stanton JR, Colbertson JW, Litter J, Halperin M, Burnett CH, Wilkins RW. The hemodynamic effects of hypotensive drugs in man. I. Veratrum viride. *J Clin Invest* 1949; **28**:353–58.

89 Braun J, Fryd CH. Effect of priscoline on the peripheral venous pressure. *Brit Heart J* 1951; **13**:294–98.

90 Kelley RT, Freis ED, Higgins TF. The effects of hexamethonium on certain manifestations of congestive heart failure. *Circulation* 1953; **7**:169–74.

91 Judson WE, Hollander W, Wilkins RW. The effects of intravenous Apresoline (hydralazine) on cardiovascular and renal function in patients with and without heart failure. *Circulation* 1956; **13**:664–74.

92 Johnson JB, Gross JF, Hale E. Effects of sublingual nitroglycerine on pulmonary artery pressure in patients with failure of the left ventricle. *N Engl J Med* 1957; **257**:1114–17.

93 Majid PA, Sharma B, Taylor SH. Phentolamine and cardiovascular performance. *Lancet* 1971; **298**:719–24.

94 Guyatt GH. Treatment of congestive heart failure. *Drugs* 1986; **32**:538–68.

95 Conradsen T-B, Ryden L, Ahlmark G, Saetre H, Persson S, Nyquist O, Wernersson B. Clinical efficacy of hydralazine in chronic heart failure: One-year double-blind placebo-controlled study. *Am Heart J* 1984; **108**:1001–6.

96 Franciosa JA, Weber KT, Levine TB, Kinasewitz GT, Janicki JS, West J, Henis MMJ, Cohn JN. Hydralazine in the long-term treatment of chronic heart failure: Lack of difference from placebo. *Am Heart J* 1982; **104**:587–94.

97 Packer M, Meller J, Gorlin R, Herman MV. Clinical and hemodynamic tachyphylaxis to prazosin mediated afterload reduction in severe congestive heart failure. *Circulation* 1979; **59**:531–39.

98 Weber KT, Kinasewitz GT, West JS, Janicki JS, Reichek N, Fishman AP. Long-term vasodilator therapy with trimazosin in chronic cardiac failure. *N Engl J Med* 1980; **303**:242–50.

99 Massie B, Chatterjee K, Werner J, Greenberg B, Hart R, Parmley WW. Hemodynamic advantage of combined administration of hydralazine orally and nitrates nonparenterally in the vasodilator therapy of chronic heart failure. *Am J Cardiol* 1977; **40**:794–801.

100 Cohn JN, Archibald DG, Ziesche S et al. Effect of vasodilator therapy on mortality in chronic congestive heart failure. Results of a Veterans Administration Cooperative Study. *N Engl J Med* 1986; **314**:1547–52.

101 Ondetti MA, Rubin B, Cushmann DW. Design of specific inhibitors of angiotensin-converting enzyme: New class of orally active antihypertensive agents. *Science* 1977; **196**:441–44.

102 Cushmann DW, Cheung HS, Sabo EF, Ondetti MA. Design of potent competitive inhibitors of angiotensin-converting enzyme. Carboxyalkanoyl and mercaptoalkanoyl amino acids. *Biochemistry* 1977; **16**:5484–91.

103 Gavras H, Faxon DP, Berkoben J, Brunner HR, Ryan TJ. Angiotensin converting enzyme inhibition in patients with congestive heart failure. *Circulation* 1978; **58**:770–6.

104 Davis R, Ribner HS, Sonnenblick EH, LeJemetel TH. Treatment of chronic congestive heart failure with captopril, an oral inhibitor of angiotensin converting enzyme. *N Engl J Med* 1979; **301**:117–21.

105 Turini GA, Brunner HR, Gribic M, Waeber B, Gavras H. Improvement of chronic congestive heart failure by oral captopril. *Lancet* 1979; **323**:1213–15.

106 Patchett AA, Harris E, Tristam EW et al. A new class of angiotensin converting enzyme inhibitors. *Nature* 1980; **288**:280–3.

107 Ader R, Chatterjee K, Ports T, Hiramatsu B, Parmley WW. Immediate and sustained hemodynamic and clinical improvement in chronic heart failure by oral angiotensin-converting enzyme inhibition. *Circulation* 1980; **61**:931–37.

108 Awan NA, Evenson MK, Needham KE, Win A, Mason DT. Efficacy of oral angiotensin converting enzyme inhibition with captopril therapy in severe chronic normotensive congestive heart failure. *Am Heart J* 1981; **101**:22–31.

109 Cleland JGF, Semple P, Hodsman P, Ball S, Ford I, Dargie H. Angiotensin II levels, hemodynamics and sympathoadrenal function after low dose captopril in heart failure. *Am J Med* 1984; **77**:880–6.

110 DiCarlo L, Chatterjee K, Parmley WW, Swedberg K, Atherton B, Curren D, Cucci M. Enalapril: A new converting enzyme inhibitor in chronic heart failure: acute and chronic hemodynamic evaluation. *J Am Coll Cardiol* 1983; **2**:865–71

111 Sharpe ND, Murphy J, Coxon R, Hannan SF. Enalapril in patients with chronic heart failure: a placebo-controlled, randomized, double-blind study. *Circulation* 1984; **70**:271–78.

112 Craeger MA, Massie BM, Faxon DP, Friedman SD, Kramer BL, Weiner DA, Ryan TJ, Topic N, Melidossian CD. Acute and long-term effects of enalapril on the cardiovascular response to exercise and exercise tolerance in patients with congestive heart failure. *J Am Coll Cardiol* 1985; **6**:163–70.

113 Dzau V, Packer M, Lilly LS, Swartz SL, Hollenbereg NK, Williams GH. Prostaglandins in severe congestive heart failure. *N Engl J Med* 1984; **310**:347–52.

114 Packer M, Medina N, Yushak M, Lee WH. Usefulness of plasma renin activity in predicting haemodynamic and clinical responses and survival during long-term converting enzyme inhibition in severe chronic heart failure. *Br Heart J* 1985; **54**:298–304.

115 Levine TB, Cohn JN. Determinants of acute and long-term response to converting enzyme inhibitors in congestive heart failure. *Am Heart J* 1982; **104**:1159–64.

116 Rouleau LJ, Kortas C, Bichet D, de Champlain J. Neurohumoral and hemodynamic changes in congestive heart failure: lack of correlation and evidence of compensatory mechanisms. *Am Heart J* 1988; **116**:746–51.

117 Benjamin N, Collier JG, Webb D. Angiotensin II augments sympathetically-induced venoconstriction in man. *Clin Sci* 1988; **75**:337–40.

118 Packer M. Interaction of prostaglandins and angiotensin II in the modulation of renal function in congestive heart failure. *Circulation* 1988; **77[Suppl I]**:1–64.

119 Packer M, Medina N, Yushak M, Meller J. Hemodynamic patterns of response during long-term therapy for severe heart failure. *Circulation* 1983; **68**:803–12.

120 Zelis R, Mason DT, Braunwald E. Partition of blood flow to the cutaneous and muscular beds of the forearm at rest and during leg exercise in normal subjects and in patients with heart failure. *Circ Res* 1969; **24**:799–802.

121 Levine TB, Olivari MT, Cohn JN. Hemodynamic and regional blood flow response to captopril in congestive heart failure. *Am J Med* 1984; **76[Suppl 5b]**:38–42.

122 Faxon DP, Craeger MA, Halperin JL, Bernard DB, Ryan TJ. Redistribution of regional blood flow following angiotensin-converting enzyme inhibition: comparison of normal subjects and patients with heart failure. *Am J Med* 1984; **76**:104–10.

123 Paulson OB, Jarden JO, Godtfredsen J, Vorstrup S. Cerebral blood flow in patients with congestive heart failure treated with captopril. *Am J Med* 1984; **76**:91–95.

124 Rajagopalan B, Raine AEG, Cooper R, Ledingham JGG. Changes in cerebral blood flow in patients with severe congestive cardiac failure before and after captopril treatment. *Am J Med* 1984; **76**:86–90.

125 Cleland JGF, Gillen G, Dargie HJ. The effects of frusemide and angiotensin-converting enzyme inhibitors and their combination on cardiac and renal haemodynamics in heart failure. *Eur Heart J* 1988; **9**:132–41.

126 Motwani JG, Fenwick MK, Morton JJ, Struthers AD. Furosemide-induced natriuresis is augmented by ultra-low-dose captopril but not by standard doses of captopril in chronic heart failure. *Circulation* 1992; **86**:439–45.

127 Kiowski W, Zuber M, Elsasser S, Erne P, Pfisterer M, Burkart F. Coronary vasodilation and improved myocardial lactate metabolism after angiotensin converting enzyme inhibition with cilazapril in patients with congestive heart failure. *Am Heart J* 1991; **122**:1382–88.

128 Zelis R, Longhurst J, Capone R, Lee G. Peripheral circulatory control mechanisms in congestive heart failure. *Am J Cardiol* 1973; **32**:481–88.

129 Wilson JR, Martin JR, Schwartz D, Ferraro N. Exercise intolerance in patients with chronic heart failure: role of impaired nutritive flow to skeletal muscle. *Circulation* 1984; **69**:1079–87.

130 Mancini DM, Davis L, Wexler JP, Chadwick B, Lejemetel TH. Dependence of enhanced maximal exercise performance on increased skeletal muscle perfusion during long-term captopril therapy in heart failure. *J Am Coll Cardiol* 1987; **10**:845–50.

131 Drexler H, Banhardt U, Meinertz T, Wollschläger H, Lehmann M, Just H. Contrasting peripheral short-term and long-term effects of converting enzyme inhibition in patients with congestive heart failure: a double-blind, placebo-controlled trial. *Circulation* 1989; **79**:491–502.

132 Timmis AD, Bojanowski LMR, Najm YC, Nelson DJ, Gosling RG. Captopril versus placebo in congestive heart failure: effects of oxygen delivery to exercising skeletal muscle. *Eur Heart J* 1987; **8**:1295–1304.

133 Cleland JGF, Stirling KW, Henderson E, Dargie HJ. Symptom-limited exercise and respiratory gas exchange in heart failure. *Br Heart J* 1986; **55**:519–25.

134 Lipkin DP, Canepa-Anson R, Stephens MR, Poole-Wilson PA. Factors determining symptoms in heart failure: comparison of fast and slow exercise tests. *Br Heart J* 1986; **55**:439–45.

135 Weber KT, Janicki JS. Lactate production during maximal and submaximal exercise in patients with chronic heart failure. *J Am Coll Cardiol* 1985; **6**:717–24.

136 Reddy KH, Weber KT, Janicki JS, McElroy PA. Hemodynamic, ventilatory and metabolic effects of light isometric exercise in patients with chronic heart failure. *J Am Coll Cardiol* 1988; **12**:353–58.

137 Wilson JR, Ferraro N, Weber KT. Respiratory gas analysis during exercise as a noninvasive measure of lactate concentration in chronic congestive heart failure. *Am J Cardiol* 1983; **51**:1639–43.

138 Sullivan MJ, Higginbotham MB, Cobb FR. Increased exercise ventilation in patients with chronic heart failure: intact ventilatory control despite hemodynamic and pulmonary abnormalities. *Circulation* 1988; **77**:552–59.

139 Weston A, Moore RP, Hughes JMB et al. Increased contribution of the alveolar-capillary membrane to resistance to gas exchange in chronic congestive heart failure. *Br Heart J* (in press).

140 Cabanes LR, Weber SN, Matran R et al. Bronchial hyperresponsiveness to metacholine in patients with impaired left ventricular function. *N Engl J Med* 1989; **320**:1317–22.

141 Eichacker PQ, Seidelman MJ, Rothstein MS, LeJemetel TH. Metacholine bronchial reactivity testing in patients with chronic congestive heart failure. *Chest* 1988; **93**:336–38.

142 Cannon PJ, Powers ER, Reison DS et al. A placebo-controlled trial of captopril in refractory chronic congestive heart failure. *J Am Coll Cardiol* 1983; **2**:755–63.

143 Riegger GAJ. Effects of quinapril on exercise tolerance in patients with mild to moderate heart failure. *Eur Heart J* 1991; **12**:705–11.

144 Giles TD, Katz R, Sullivan JM et al. Short- and long-term angiotensin-converting enzyme inhibitors: A randomized trial of lisinopril versus captopril in the treatment of congestive heart failure. *J Am Coll Cardiol* 1989; **13**:1240–47.

145 Packer M, Lee WH, Yushak M, Medina N. Comparison of captopril and enalapril in patients with severe chronic heart failure. *N Engl J Med* 1986; **315**:847–53.

146 Testa MA, Andersen RB, Nackley JA, Hollenberg NK, and the Quality-of-Life Study Group. Quality-of-life and antihypertensive therapy in men. *N Engl J Med* 1993; **328**:907–13.

147 The Captopril-Digoxin Multicenter Research Group. Comparative effects of therapy with captopril and digoxin in patients with mild to moderate heart failure. *JAMA* 1988; **259**:539–44.

148 Silke B, Fischer-Hansen J, Keller N, Salminen K. A double-blind, parallel-group comparison of flosequinan and enalapril in the treatment of chronic heart failure. *Eur Heart J* 1992; **13**:1092–1100.

149 Aldigier JC, Dalmay F, Lartigue M, Baussant T, Chassain AP, Leroux-Robert C, Galen FX. Angiotensin converting enzyme inhibition does not suppress plasma increase during exercise in humans. *J Cardiovasc Pharmacol* 1993; **21**:289–95.

150 Kjekshus J, Swedberg K, Snapinn S, for the CONSENSUS Trial Group. Effects of enalapril on long-term mortality in severe congestive heart failure. *Am J Cardiol* 1992; **69**:103–7.

151 Swedberg K, Eneroth P, Kjekshus J, Snapinn S, for the CONSENSUS Study Group. Effects of enalapril and neuroendocrine activation on prognosis in severe congestive heart failure. *Am J Cardiol* 1990; **66**:40D–45D.

152 Swedberg K, Kjekshus J for the CONSENSUS Study Group. Effects of enalapril on mortality in severe congestive heart failure: results of the CONSENSUS. *Am J Cardiol* 1988; **62**:60A–66A.

153 The SOLVD Investigators. Studies of left ventricular dysfunction (SOLVD): rationale, design and methods: two trials to evaluate the effects of enalapril in patients with reduced ejection fraction. *Am J Cardiol* 1990; **66**:315–22.
154 Yusuf S, Pepine CJ, Garces C et al. Effect of enalapril on myocardial infarction and unstable angina in patients with low ejection fractions. *Lancet* 1993; **3409**:1173–78.
155 Pfeffer MA, Braunwald E. Ventricular remodeling after myocardial infarction. *Circulation* 1990; **81**:1161–72.
156 Pfeffer JM, Pfeffer MA, Fletcher PJ, Braunwald E. Progressive ventricular remodeling in rats with myocardial infarction. *Am J Physiol* 1991; **29**:H1406–H1414.
157 Gaudron P, Eilles C, Kugler I, Ertl G. Progressive left ventricular dysfunction and remodeling after myocardial infarction: potential mechanisms and early predictors. *Circulation* 1993; **87**:755–63.
158 McAlpine HM, Morton JJ, Leckie B, Rumley A, Gillen G, Dargie HJ. Neuroendocrine activation after acute myocardial infarction. *Br Heart J* 1988; **60**:117–24.
159 Hirsch AT, Talsness C, Lage A, Dzau FJ. The effect of experimental myocardial infarction and chronic captopril treatment on plasma and tissue angiotensin converting enzyme activity. *Clin Res* 1989; **37**:266A.
160 Hammermeister KE, DeRouen TA, Dodge HT. Variables predictive of survival in patients with coronary disease: selection by univariate and multivariate analyses from the clinical, exercise, arteriographic and quantitative angiographic evaluations. *Circulation* 1979; **59**:421–30.
161 White HD, Norris RM, Brown MA, Bradt PWT, Whitlock RML, Wild CJ. Left ventricular end-systolic volume as the major determinant of survival after recovery from myocardial infarction. *Circulation* 1987; **76**:44–51.
162 Pfeffer JM, Pfeffer MA, Braunwald E. Influence of chronic captopril therapy on the infarcted left ventricle in the rat. *Circ Res* 1985; **57**:84–95.
163 Pfeffer MA, Pfeffer JM, Steinberg C, Finn P. Survival after an experimental myocardial infarction: beneficial effects of long-term therapy with captopril. *Circulation* 1985; **72**:406–12.
164 Pfeffer MA, Lames GA, Vaughan DE, Parisi AF, Braunwald E. Effect of captopril on progressive ventricular dilatation after anterior myocardial infarction. *N Engl J Med* 1988; **319**:80–6.
165 Pfeffer MA, Braunwald E, Moye LA et al. on behalf of the SAVE Investigators. Effect of captopril on mortality in patients with left ventricular dysfunction after myocardial infarction. *N Engl J Med* 1992; **327**:669–77.
166 Packer M, Rouleau JL, Moye LA et al. Effect of captopril on ventricular arrhythmias and sudden death in patients with left ventricular dysfunction after myocardial infarction. *J Am Coll Cardiol* 1993; **21**:130A.
167 Lamas G, Flaker G, Mitchell G et al. Effect of captopril therapy on post MI outcome in patients with and without a patent infarct related artery. *J Am Coll Cardiol* 1993; **21**:44A.
168 Rutherford JD, Pfeffer MA, Flakjer GC, et al. The salutary effects of captopril on recurrent myocardial infarction: The SAVE trial experience. *J Am Coll Cardiol* 1993; **21**:44A.
169 The Acute Infarction Ramipril Efficacy Study Investigators. Effect of ramipril on mortality and morbidity of survivors of acute myocardial infarction with clinical evidence of heart failure. *Lancet* 1993; **342**:821–8.
170 Swedberg K, Held P, Kjekshus J et al. Effects of the early administration of enalapril on mortality in patients with acute myocardial infarction. *N Engl J Med* 1992; **327**:678–84.

6

The use of ACE inhibitors in blacks and whites

W Dallas Hall and Zafar H Israili

Introduction

Angiotensin converting enzyme (ACE) inhibitors are used increasingly in the management of hypertension and congestive heart failure. Like the combination of nitrates and hydralazine, ACE inhibitors improve life expectancy in chronic heart failure.[1–5] Although ACE inhibitors are effective in a high percentage of patients, and are extremely well tolerated,[6–20] even by the elderly,[21–24] the efficacy of these drugs as monotherapeutic agents in black patients with hypertension is less than in white patients.

The prevalence and incidence of hypertension in the black population is higher than that in whites,[25–30] and the biochemical and hormonal differences observed in the two races may be, in part, responsible for differences in the occurrence of hypertension and response to ACE inhibitors. This chapter summarizes treatment of hypertension with ACE inhibitors in both blacks and whites, with special emphasis on black hypertensives.

Racial differences in the prevalence of hypertension

There are about 50 million people in the USA who have a single screening blood pressure reading of >140/>90 mmHg (or are taking antihypertensive medication),[29] and some 35 million have definite high blood pressure by the World Health Organization criteria, defined as a one-time blood pressure reading of >160/>95 mmHg. Of the latter, 5 million are black,[31] representing one in four black Americans compared to one in six white Americans. As a matter of fact, the frequency of hypertension in African Americans (American Blacks) is among the highest in the world,[27,29] and hypertension is the major health problem in adult African Americans.

More black hypertensives have uncontrolled hypertension compared to non-black hypertensives. Black men have the highest prevalence and the poorest control of hypertension.[27] The prevalence of hypertension increases more steeply with age in blacks compared to non-blacks.[27,29] and isolated systolic hypertension in the elderly is also more prevalent in blacks.[26–28,30–32]

Although the racial difference in blood pressure does not exist at birth, in the neonatal period,[33] or in young children (aged 3–4 years),[34] blacks are more likely to develop hypertension.[26,27,35,36] They often develop more severe hypertension earlier[37] and/or remain undiagnosed, or inadequately controlled for longer periods of time. As much as 30% of all deaths in hypertensive black men and 20% in hypertensive black women are attributable to high blood pressure.[38]

Racial differences in target organ damage

Several epidemiological studies show disproportionate morbidity and mortality associated with the higher incidence of high blood pressure in blacks.[26,27,30,39] The morbidity and mortality from hypertension-related diseases is two to five times greater in blacks than in whites in the USA,[40] and the UK.[41]

Kidney damage secondary to high blood pressure is greater in blacks than in whites.[31,32,42,43] Renal function declines in black hypertensives at about twice the rate of white hypertensives.[42,43] Adequate blood pressure control does not always stabilize or improve renal function in black hypertensives,[44] and treated black hypertensives are twice as likely to have elevation in serum creatinine than white patients.[43] The prevalence of hypertension-related renal disease is six times greater in blacks than in whites,[45] and hypertension is the leading cause of renal failure in this subgroup of the population.[45,46] The combination of diabetes and hypertension is twice as common in the black population,[47] and blacks have a four-fold higher incidence of diabetes-related renal disease when compared with whites.[45] These factors account, at least in part, for the increased susceptibility of blacks to develop end-stage renal disease.[45–47]

Cardiac involvement is more prevalent and severe in the hypertensive black population,[48] and the mortality of hypertensive heart disease is two- to three-fold higher in blacks than in whites.[48] Electro- and echocardiographic evidence of left ventricular hypertrophy, an independent cardiovascular risk factor, occurs up to four times more often in blacks than in whites.[49–52] Since diabetes mellitus accentuates the development of left ventricular hypertrophy in patients with essential hypertension, independent of arterial pressure, black hypertensive patients with diabetes are at a higher risk for cardiovascular morbidity and mortality.[53]

American blacks do not always exhibit the expected nocturnal decline in blood pressure.[54,55] Since elevated nighttime blood pressure has been related to the development of left ventricular hypertrophy,[55–57] a lack of nocturnal decline in blood pressure might contribute to higher cardiovascular morbidity in blacks.

Stroke is a common complication of hypertension in the black community.[58–60] Blacks are two- to four-fold more likely to have strokes than are whites,[31] and the death rate from stroke among blacks is two- to three-fold higher in blacks than in whites in the USA[31,58–60] and the UK.[41]

Racial differences in hormonal and physiological factors that may influence antihypertensive therapy

There is no evidence for a racial difference in the basic mechanism(s) of hypertension, although familial aggregation of essential hypertension has been documented,[61–63] and a T235 mutation of the angiotensin gene (associated with hypertension) is found more frequently in blacks than in whites.[64] Despite reports of association of darker skin color with higher levels of blood pressure,[65] it is unlikely that skin color, independent of socioeconomic status, is correlated with elevated blood pressure.[66–68] Further, it is difficult to define or identify a homogeneous black population in the Americas and Europe since many ethnic groups are heterogeneous, with origins from many regions of the world and admixture of several races

(African, Caucasian, Native American, etc.). Nonetheless, a number of clinical, biochemical, pathophysiological and epidemiological differences have been observed when comparison is made between 'black' and 'white' hypertensives as two distinct subgroups. These differences may be important enough to provide insight into the causes of hypertension and therapeutic intervention. Table 6.1 provides a summary of the data on racial differences in the hormonal and physiological factors which may influence therapy of hypertension.

Blacks, in general, have low plasma renin activity (PRA),[37,69–76,78] and a high prevalence of low-renin hypertension.[27,73,75,76] Stimulation of PRA by sodium restriction is also less in blacks.[71,73] A characteristic low renin profile could result from primary volume expansion due to factors that promote renal sodium and water retention, such as a disruption in the normal homeostatic response to renin release as a result of elevation of intracellular calcium, increased Na^+-H^+ antiport activity, inefficient conversion of prorenin to renin, or alteration in renal blood flow.[76–78]

Serum levels of angiotensin converting enzyme do not differ between blacks and whites,[79] but serum levels of atrial natriuretic peptide (ANP) may be higher in hypertensive blacks.[80] The defective natriuresis in blacks, despite higher levels of ANP, suggests the possibility of target organ resistance to ANP.[80] Both plasma levels and urinary excretion of aldosterone are similar in black and white adults.[69,71,74,75,78,79] The maintenance of normal aldosterone levels despite a low renin profile suggests that there could be increased adrenal sensitivity to angiotensin II in black individuals with low PRA.

Blacks excrete lower amounts of kallikrein than whites.[74,81,82] Deficiency in the natriuretic and vasodilatory renal kallikrein-kinin system might relate to excessive vasoconstriction in blacks.[81] The increase in kallikrein excretion with salt restriction is also less in blacks than in whites.[71,74] Similarly, lower levels of vasodilatory prostaglandins in the serum and urine[69,74] of blacks could contribute to increased vascular resistance and hypertension.

Black individuals with low-renin hypertension have decreased β-adrenergic drive, suggesting that unopposed α-adrenergic activity could be a pathogenic factor.[83] Increased vascular reactivity to α-adrenergic stimulation is comparable to a chronic stressful lifestyle.[83] Several studies have reported higher cardiovascular reactivity to mental and physical stressors in blacks,[83–85] but other studies show no racial difference.[86]

In general, blacks compared to whites have decreased renal blood flow and a higher renal vascular resistance.[87] They also have a higher capacity for renal absorption of oral or intravenous sodium loads. The increased sodium retention results in volume expansion,[87–91] which may relate to the excellent response to diuretics.[76,88,89,91–94] However, some studies report no difference in plasma volume between blacks and whites when patients were matched for age, sex, body surface area, obesity and level of blood pressure.[51,95] Both normotensive and hypertensive blacks tend to be salt-sensitive,[71,89,91,96,97] that is, their mean arterial blood pressure rises with increased salt intake. Other features of salt sensitivity include renal sodium retention, increased plasma volume, low PRA, and non-modulator status.[71,78,88,89,91,96–98] The latter implies lack of the normal 10–20% increase in renal blood flow following high salt intake, and impaired aldosterone release in response to angiotensin II on a low salt intake.[89,96,98] Individuals who are predisposed to develop salt-sensitive hypertension may not adequately mobilize dopamine on salt loading.[99]

Hormonal and Physiological Parameters	Blacks versus Whites
Plasma renin activity (PRA)	
PRA	Lower in normotensive and hypertensive black adults and children[27,37,69–76,78]
PRA stimulation by sodium restriction	Less in blacks[71,73]
Aldosterone	
Plasma levels in adults	Similar in normotensive and hypertensive black and white adults.[69,71,74,75,78,79]
Urinary aldosterone	Similar in hypertensive and normotensive blacks[69]
Angiotensin converting enzyme (serum)	Similar[79]
Urinary kallikrein excretion (normotensives and hypertensives)	Lower in blacks[74,81,82]
Atrial natriuretic peptide	Higher in hypertensive blacks[80]
Prostaglandins	
Serum levels of PGE_2, PGF_2-alpha and 6-keto-PGF-1-alpha	Lower in normotensive black males[74]
Urinary excretion ratio of TXB_2/6-keto-PGF-1-alpha	Higher ratio in hypertensive versus normotensive black males[69]
Sympathetic nervous system indices:	
Adrenergic drive	Lower in low PRA hypertensive blacks[83]
Cardiovascular reactivity to mental and physical stressors (via increase in peripheral vascular resistance)	Higher in blacks[83–85]
Reactivity to cold pressor test	Higher in blacks[83,85]
Hemodynamic factors	
Renal reabsorption of oral or intravenous sodium loads	Higher in blacks[71,78,88]
Plasma volume	Expanded in blacks[78,88–91]

Blacks excrete less potassium than whites, reflecting a lower dietary potassium intake.[71–73] Since 24-hour sodium excretion is similar in the two races, the urinary sodium/potassium ratio is higher in most black hypertensives. Calcium and magnesium excretion are also lower in blacks than in whites.[100]

A number of studies have identified differences in ion regulation between blacks and whites (see Table 6.1). Intracellular Na^+ is higher in blacks compared to whites,[101–106] and erythrocyte Na^+-K^+-ATPase activity, Na^+-K^+ cotransport or Na^+-Li^+ countertransport are lower in blacks.[77,103–107] Accelerated Na^+ turnover in blacks may be related to increased activity of Na^+-H^+ antiport.[77,108] Na^+-K^+-ATPase activity is lower in hypertensive versus normotensive blacks.[77,105] Platelet and transmembrane Na^+/H^+ exchange and erythrocyte Na^+ transport are higher in hypertensive than in normotensive black females,[109] but erythrocyte intracellular K^+ is similar.[109] Serum levels of ionized calcium are also lower in normotensive blacks and in low-renin hypertensives.[101,110,111]

Cellular transport mechanisms	
Erythrocyte Na^+-K^+-ATPase activity, Na^+-K^+ cotransport or Na^+-Li^+ countertransport	Lower in blacks[77,104–109]
Na^+ turnover rate (related to increased activity of Na^+-H^+ antiport)	Accelerated in blacks[77,108]
Platelet Na^+/H^+ exchange	Higher in hypertensive than in normotensive black females[109]
Erythrocyte intracellular Na^+	Higher in blacks[102,109]
Erythrocyte Na^+ transport	Higher in hypertensive than in normotensive black females[109]
Erythrocyte intracellular K^+	Similar in hypertensive and normotensive black females[109]
Serum ionized Ca^{2+}	Decreased in low renin hypertensive blacks;[102,110] lower in normotensive blacks[111]
Platelet cytosolic ionized Ca^{2+}	Lower in normotensive blacks[101,102,111]
Dietary sodium, potassium, magnesium and calcium intake in normotensives (based on urinary excretion)	
Potassium	Lower in blacks[71–73,82,88]
Sodium	Similar[73]
Calcium	Lower in blacks[100,111]
Magnesium	Lower in blacks[100,102]
Salt-sensitivity	Higher in normotensive and hypertensive blacks[71,89,91,96,97]

Table 6.1
Hormonal and physiological differences in blacks and whites

Management of hypertension

Lifestyle modifications

Lifestyle modifications are the first step in the management of patients with the mildest levels of hypertension. The following measures are recommended:

- weight reduction to levels < 15% above normal;
- moderate sodium restriction (< 2300 mg/day);
- reduction of alcohol intake to less than two drinks or one ounce of ethanol per day;
- cessation of smoking;
- a regular (aerobic) exercise program;
- and stress management.[27,112]

Aggressive treatment of hypertension in blacks is indicated because of the worse cardiovascular, renal and cerebrovascular complications. The 60 000 additional deaths annually of American blacks compared with whites can be accounted for in large part by complications of undetected, untreated, or uncontrolled hypertension.[31,40]

Pharmacologic therapy

The goal of antihypertensive drug therapy is to control blood pressure with the fewest drugs at their lowest dosage levels. A hypertension management plan that emphasizes patient education, an adequate therapeutic trial, and appropriate dose titration can maximize the efficacy of antihypertensive agents. Patient counseling can improve understanding of the goals, the antihypertensive regimen, and the importance of controlling mild degrees of hypertension.

Racial differences in pharmacologic therapy of hypertension

In general, hypertensive blacks do not respond as well as hypertensive whites to monotherapy with β-blockers or angiotensin converting enzyme (ACE) inhibitors. Monotherapy with both groups of drugs shows less of a blood pressure lowering effect when compared with diuretics or calcium channel blockers in hypertensive blacks in the USA,[17–20,27,29,30,76,91,92,112–142] Africa[143–151] and Brazil.[152] However, combination therapy using a diuretic with either a β-blocker or an ACE inhibitor is equally effective in blacks and whites.[27,91,112,113,121,134,138,142,143]

Beta-blockers

As a group, β-blockers show less of a hypotensive effect when used as monotherapeutic agents in black hypertensives compared to white hypertensives. This racial difference in the therapeutic efficacy has been demonstrated for acebutolol,[124] atenolol,[115,118,119,135,147,150,152] bisoprolol,[135] betaxolol,[113] metoprolol,[127,129] nadolol,[142] oxprenolol,[143] pindolol,[116] propranolol,[92,93,114,117,141,144,145] and timolol.[146] Some reports suggest good antihypertensive response to acebutolol,[151] and penbutolol[153] in black hypertensives. In some studies, labetalol was found to be equally effective in black and white hypertensives,[114,131] while in others, both atenolol and labetalol were less effective in black compared to white hypertensives.[115] In general, the low response of β-blockers in black hypertensives can be explained, in part, by the low renin status of many black hypertensives.

Diuretics

Black hypertensives are very responsive to diuretic therapy.[27,30,76,89,91–94,112,113,116,119,121,124,125,131,133,138,143,145–147,154,155] The excellent response to diuretics is related to the low renin profile, expanded plasma volume, salt sensitivity, and higher intracellular levels of sodium related to decreased activity of erythrocyte Na^+-K^+-ATPase and Na^+-K^+ cotransport.[76,77] Current emphasis is on low doses of diuretics (for example, 12.5–25 mg of hydrochlorothiazide or chlorthalidone), which accomplish about 80% of the blood pressure lowering efficacy of the higher doses, but with a markedly lower risk of adverse metabolic effects such as hypokalemia, glucose intolerance and hyperuricemia.[27,76,112] Any adverse initial lipid changes with diuretic therapy often revert to baseline levels following therapy for 12 months or more. In two different reports, use of relatively low doses of a diuretic with adequate K^+ supplementation in black hypertensives was not associated with worsening of high grade ventricular ectopy or other arrhythmias.[156,157] Nevertheless, in one recent study, hydrochlorothiazide used alone in doses of less than 25 mg/day did not effectively control blood pressure in African blacks.[149]

Calcium antagonists

The effect of monotherapy with calcium antagonists in black hypertensives is significantly greater than with ACE inhibitors or β-blockers, and compares favorably with the response to diuretics.[27,30,76,112,117–119,124,129,130,136,144,148,150,152,154,155,158–161] The response to calcium antagonists may be enhanced by a low renin profile, lower extracellular ionized calcium, higher intracellular sodium,[150] and a high dietary salt intake.[162,163] There does not appear to be a racial difference in the antihypertensive response to calcium antagonists.[112,154,158]

ACE inhibitors

In general, monotherapy with ACE inhibitors is less effective in controlling blood pressure in black than in white hypertensive patients[17–20,27,29,30,76,112,118–128,130,132,133,136,139,140,146] (Table 6.2). Hypertensive blacks compared to whites usually have less reduction in blood pressure by a given dose of an ACE inhibitor and a lower per cent of black patients achieve blood control at a given dose, or they require a higher dose to produce the same reduction in blood pressure as that in white patients. For example, Table 6.2 illustrates that the average decrease in systolic and diastolic blood pressures was much less in 848 blacks (–7.2/–6.8 mmHg) than in 1,453 whites (–15.3/–11.0 mmHg). This racial difference was also indicated by the per cent of patients whose blood pressure was controlled with ACE inhibitors alone. Studies have rarely suggested absence of racial differences in the antihypertensive response to certain ACE inhibitors.[137,158] Nevertheless, addition of a diuretic (usually in low doses) to ACE inhibitors abolishes any racial differences in the antihypertensive response to these agents.[27,91,112,121,134,138,143,164] Such a combination is advantageous because of relatively low adverse effect profile, favorable cardiac and renal effects, and reduction in risk for cardiovascular morbidity and mortality.

Alpha-adrenergic blockers and other antihypertensive agents

Alpha-adrenergic blockers appear to be equally effective in black and white hypertensives,[27,76,112,119,124,165] although studies are less extensive than with the previously discussed therapies. Other groups of antihypertensive drugs such as central adrenergic agonists and vasodilators, including hydralazine and minoxidil, also appear to be equally effective in black and white hypertensive patients.[27,76,112,119]

Advantages of ACE inhibitors in the treatment of hypertension

ACE inhibitors appear to be well tolerated with relatively few side-effects compared to the many other classes of antihypertensive agents.[6–24,27,76,112,118,122,123,137] These drugs can be used in patients with diabetes mellitus, peripheral vascular disease, cerebrovascular disease, chronic obstructive pulmonary disease, and congestive heart failure. In hypertensive patients with congestive heart failure, treatment with ACE inhibitors has the added benefit of reducing the risk of cardiovascular morbidity and mortality.[1–5]

In the general population, ACE inhibitors are very effective monotherapeutic antihypertensive agents. In an unselected population of hypertensives, ACE inhibitors will reduce

Table 6.2
Race and high blood pressure monotherapy with ACE inhibitors

Drug	Dose (mg/day)	N	Change in BP SBP/DBP (mmHg)
Hypertensive whites			
Captopril[119]	(25–100)	94	−11.0/−11.0
Captopril[122]	(37.5–150)	162	−14.7/−10.7
Captopril[125]	(150–450)	129	−22.0/−15.3
Captopril[125]	(150–450)	154	−22.0/−15.6
Captopril[138]	(450)	18	/−5.0
Enalapril[120]	(5)	6	−16.0/−11.0
Enalapril[123]	(10–40)		−32.0/−17.0
Enalapril[124]	(5)	145	−6.9/−3.9
Enalapril[139]	(10–40)	15	/−11.8
Lisinopril[127]	(40–80)	80	/−16.8
Ramipril[18]	(2.5–10)	641	−7.8/−8.7
Spirapril[120]	(5–10)	9	−5.0/−5.0
All ACE inhibitors		1 453	−15.3/−11.0
Hypertensive blacks			
Captopril[118]	(25–100)	92	−9.6/−8.2
Captopril[119]	(25–100)	92	−8.0/−8.0
Captopril[121]	(75)	32	−0.7/−4.1
Captopril[122]	(37.5–150)	190	−9.1/−8.0
Captopril[125]	(150–450)	20	−9.2/−10.7
Captopril[125]	(150–450)	23	−12.8/−12.0
Enalapril[120]	(5)	16	−6.0/−3.0
Enalapril[124]	(5)	35	−2.7/−1.4
Enalapril[126]	(10–40)		+2.6/−12.3
Enalapril[128]	(20)	28	−10.8/−4.0
Enalapril[130]	(20)	41	/−4.6
Enalapril[136]	(10)	25	/−5.2
Enalapril[139]	(10–40)	10	/−4.6
Lisinopril[127]	(10–80)	35	/−10.4
Lisinopril[132]	(10)	21	−15.0/−8.0
Lisinopril[133]	(20)	41	−8.8/−7.2
Lisinopril[137]	(20–80)	23	/−11.0
Ramipril[18]	(2.5–10)	117	−2.9/−3.0
Spirapril[120]	(5–10)	7	−8.0/−4.0
All ACE inhibitors		848	−7.2/−6.8

blood pressure significantly in more than 60% of patients. Addition of a diuretic increases the response rate to more than 80%. In hypertensive blacks, the combination of ACE inhibitor with low-dose thiazide is very effective. In addition, ACE inhibitors partially counteract some of the adverse metabolic effects of thiazides, such as the elevation in renin and secondary aldosteronism that is characteristic of long-term diuretic treatment.[27,91,112,121,164] They also help protect the patient from excessive loss of potassium.

Additional advantages of ACE inhibitors include:

- Favorable cardiac hemodynamic effects, which include increase in cardiac output by increasing stroke volume and reducing afterload with little or no change in heart rate (Table 6.3), and reduction in right atrial and left ventricular filling pressures. ACE inhibitors also regress left ventricular hypertrophy[166,167] and improve arterial compliance.[168,169]
- Decrease in peripheral resistance producing both arteriolar and venous dilatation with reduction of afterload and preload.
- Reduction in supine and standing blood pressures without excess orthostasis.
- Favorable renal hemodynamic effects, including an increase in renal blood flow, usually with no impairment of glomerular filtration rate (Table 6.3).
- Reduction in aldosterone secretion with no retention of sodium and water (Table 6.3).
- Some ACE inhibitors produce moderate reduction in blood pressure during sleep,[170,171] which may be important for the prevention of left ventricular hypertrophy,[55–57] especially in black patients.
- Redistribution of regional blood flow with preservation of cerebral, cardiac, and renal blood flow despite a decreased systemic blood pressure (Table 6.3). In some patients, ACE inhibitors have improved cerebral, cardiac or renal blood flow.[172–175]
- Abrupt cessation of ACE inhibitor therapy does not result in a rebound rise in blood pressure.
- Prolonged treatment with ACE inhibitors attenuates vascular hyperreactivity, which may play a role in preventing hypertensive episodes in response to vasoconstrictor stimuli in hypertensive subjects.[176]
- ACE inhibitors usually improve quality of life.[177–179] A part of this improvement may result from a central nervous system effect (mood elevation, anxiolytic action and improvement in cognitive function[180]), an improvement in cardiac function, or increased alertness.[181]

ACE inhibitors in other special populations

ACE inhibitors are safe and effective in treating hypertensive diabetics, since these agents do not adversely affect glucose regulation[182,183] or lipid profiles.[183–187] ACE inhibitors have been reported to improve insulin resistance[183] and delay the onset of diabetic glomerulonephropathy.[188–191]

ACE inhibitors are effective in elderly hypertensives,[21–25,27,29,30,76] and in patients with a low renin profile.[23,27,76] These drugs are also efficacious in the treatment of isolated systolic hypertension in the elderly.[192,193] Since ACE inhibitors improve exercise tolerance and do not worsen bronchospasm, they should be considered as antihypertensive therapy in elderly individuals with chronic obstructive pulmonary disease.

Captopril and enalapril have been administered to children with hypertension, congestive heart failure and diabetes with proteinuria.[194–197] Since ACE inhibitors (and their active metabolites) cross the placenta and are secreted in milk, these drugs should be avoided in pregnant women and in nursing mothers.

Hemodynamic		Biochemical	
Blood pressure	↓	Plasma angiotensin I	↑
Total peripheral resistance	↓	Plasma angiotensin II	↓
Heart rate	↔	Plasma ACE	↓
Stroke volume	↑*/↔	Plasma aldosterone	↓/↔
Cardiac output	↑*/↔	Plasma renin activity	↑
Ejection fraction	↑*/↔	Blood kinins	↑/↔
Pulmonary capillary wedge pressure	↓*/↔	Plasma noradrenaline (norepinephrine)	↔
		Serum glucose	↔
Right atrial pressure	↔	Glucose tolerance	↔
Capacitance and distensibility of large arteries	↑	Serum potassium	↑
		Serum uric acid	↔
		Serum lipids	↔
Coronary blood flow	↔	Serum creatinine	↔
Forearm blood flow	↑	Creatinine clearance	↔
Renal vascular resistance	↓	Urinary kinins	↔/↑
Renal blood flow	↑	Urinary kallikrein	↔/↑
Glomerular filtration rate	↑/↔	Urinary prostaglandins	↔/↑
Cerebral blood flow	↑	Urinary proteins	↓
Hepatic blood flow	↓/↔		

*In patients with congestive heart failure

Table 6.3
Hemodynamic and biochemical effects of ACE inhibitors

Adverse effects and drug–drug interactions of ACE inhibitors

The main adverse effects of ACE inhibitors include first dose acute hypotension, dizziness, hyperkalemia, cough,[198] gastrointestinal distress, dysgeusia and angioneurotic edema. In patients with bilateral renal artery stenosis or dehydration, ACE inhibitors can induce reversible, acute renal failure.[11,12,199] In patients with autoimmune diseases, ACE inhibitors can cause neutropenia.

All ACE inhibitors have the potential to interact adversely with antacids (decreased bioavailability), thiazide diuretics (excessive hypotensive effect), potassium-sparing diuretics and potassium supplements (hyperkalemia due to increased potassium retention), digoxin (decreased digoxin clearance), anesthetics (increased hypotensive effects), immunosuppressants (increased risk of neutropenia), and NSAIDs (hyperkalemia and interference with reduction in blood pressure).

Conclusions

The prevalence of essential hypertension is strikingly higher in blacks than in whites in the USA and the Caribbean. Hypertensive blacks also have disproportionately more hypertension-related mortality and target organ damage than whites. Basic pathophysiologic differences exist in the mechanism(s) of hypertension between these two racial groups. Blacks, in general, have lower plasma renin activity, lower urinary potassium and kallikrein, and higher prevalence of salt sensitivity. Other factors including genetic, socioeconomic, dietary and psychologic, have been suggested as possible explanations for the higher prevalence and severity of hypertension in blacks. Black hypertensives need aggressive treatment because of the worse cardiovascular, renal and cerebrovascular complications.

The strategy for the treatment of hypertension has shifted from a stepped-care approach to more individualized adjustment to the needs of the patient, based upon age, race, and concurrent medical problems and cardiovascular risk factors. The aim of therapy is not only to decrease blood pressure but also to reduce risk factors associated with cardiovascular and renal morbidity and mortality. The choice of initial antihypertensive therapy in hypertensive patients should be based on not only the usual response pattern to the main groups of antihypertensive agents in a particular subgroup of patients, but also on their effects on lipids, potassium, glucose, uric acid, renal function, and left ventricular anatomy and function. Associated medical conditions such as asthma, diabetes mellitus, congestive heart failure, renal failure, gout, coronary artery disease, hyperlipidemia and peripheral vascular disease are determinants of choice of antihypertensive medications.

ACE inhibitors are efficacious antihypertensive agents. They are well tolerated and have a relatively low incidence of adverse effects. ACE inhibitors preserve renal function and decrease the progression of renal disease in diabetics. Since they improve life expectancy in chronic heart failure, ACE inhibitors are the drugs of choice in hypertensive patients with congestive heart failure.

ACE inhibitors are more effective in young and white hypertensive patients. Although they are not as effective as monotherapeutic agents in black hypertensives as some other classes of antihypertensive drugs, coadministration with low doses of a diuretic is efficacious in all subgroups of hypertensives, including blacks.

References

1 The CONSENSUS Trial Study Group. Effect of enalapril on mortality in severe congestive heart failure: results of the Cooperative Scandinavian Enalapril Survival Study (CONSENSUS). *N Engl J Med* 1987; **316**:1429–35.
2 Newman TJ, Maskin CS, Dennick LG et al. Effects of captopril on survival in patients with congestive heart failure. *Am J Med* 1988; **84**:140–4.
3 Cohn JN, Johnson G, Ziesche S et al. A comparison of enalapril with hydralazine-isosorbide dinitrate in the treatment of chronic congestive heart failure. *N Engl J Med* 1991; **325**:303–10.
4 The SOLVD Investigators. Effect of enalapril on survival in patients with reduced left ventricular ejection fractions and congestive heart failure. *N Engl J Med* 1991; **325**:293–302.
5 Kjekshus J, Swedberg K, Snapinn S. Effect of enalapril on long-term mortality in severe congestive heart failure. CONSENSUS Trial Group. *Am J Cardiol* 1992; **69**:103–7.
6 Davies RO, Irvin JD, Kramsch DK et al. Enalapril worldwide experience. *Am J Med* 1984; **77**:23–35.
7 Armayor GM, Lopez LM. Lisinopril: a new angiotensin-converting enzyme inhibitor. *Drug Intell Clin Pharm* 1988; **22**:365–72.
8 Brogden RN, Todd PA, Sorkin EM. Captopril. An update of its pharmacodynamic and pharmacokinetic properties, and therapeutic use in hypertension and congestive heart failure. *Drugs* 1988; **36**:540–600.
9 Inman WH, Rawson NSB, Wilton LV et al. Postmarketing surveillance of enalapril. I: Results of prescription-event monitoring. *Br Med J* 1988; **297**:826–29.
10 Weber MA. Fosinopril: a new generation of angiotensin-converting enzyme inhibitors. *J Cardiovasc Pharmacol* 1990; **20[Suppl 10]**:S7–S12.
11 Weber MA. Safety issues during antihypertensive treatment with angiotensin converting enzyme inhibitors. *Am J Med* 1988; **84[Suppl 4A]**:16–23.
12 Warner NJ, Rush JE. Safety profiles of the angiotensin-converting enzyme inhibitors. *Drugs* 1988; **35[Suppl 5]**:89–97.
13 Knapp LE, Frank GJ, McLain R et al. The safety and tolerability of quinapril. *J Cardiovasc Pharmacol* 1990; **15[Suppl 2]**:S47–S55.
14 Balfour JA, Goa KL. Benazepril: A review of its pharmacodynamic and pharmacokinetic properties, and therapeutic efficacy in hypertension and congestive heart failure. *Drugs* 1991; **42[Suppl]**:S11–S39.
15 Todd PA, Benfield P. Ramipril. A review of its pharmacologic properties and its efficacy in cardiovascular disorders. *Drugs* 1990; **39**:110–35.
16 Jenkins AC, Knill JR, Dreslinski GR. Captopril in the treatment of the elderly hypertensive patient. *Arch Intern Med* 1985; **145**:2029–35.
17 Napolitano LV. Monopril (Fosinopril). In: *Physicians desk reference*. Montvale: Medical Economics Co, 1993: 1457–60.
18 Napolitano LV. Altace (Ramipril). In: *Physicians desk reference*. Montvale: Medical Economics Co, 1993; 1112–14.
19 Napolitano LV. Lotensin (Benazepril). In: *Physicians desk reference*. Montvale: Medical Economics Co, 1993: 902–4.
20 Accupril (Quinapril). In: *Physicians desk reference*. Montvale: Medical Economics Co, 1993: 1741–43.
21 Laher MS, Nadin D, Rao SK et al. Lisinopril in elderly patients with hypertension. *J Cardiovasc Pharmacol* 1987; **9[Suppl 3]**:S69–S71.
22 Reid JL, Macdonald NJ, Nees KR, Elliott HL. Angiotensin-converting enzyme inhibitors in the elderly. *Am Heart J* 1989; **117**:751–55.
23 Cox JP, Duggan J, O'Boyle CA et al. A double-blind evaluation of captopril in elderly hypertensives. *J Hypertens* 1989; **7**:299–303.

24 Schnaper HW. Use of quinapril in the elderly patient. *Am J Hypertens* 1990; **3**:278S–282S.

25 Stamler J, Stamler R, Riedlinger WF et al. Hypertension screening of 1 million Americans: Community Hypertension Evaluation Clinic (CHEC) Program, 1973 through 1975. *JAMA* 1976; **235**:2299–2306.

26 National Center for Health Statistics. *National health nutrition examination survey No. 2 (NHANES-II), 1976–1980*. Bethesda: National Institutes of Health, 1981.

27 Saunders E. Hypertension in African-Americans. *Circulation* 1991; **83**:1465–67.

28 Cornoni-Huntley J, LaCroix AZ, Havlik RJ. Race and sex differentials in the impact of hypertension in the United States. The National Health-Nutrition Examination Survey. I. Epidemiologic follow-up study. *Arch Intern Med* 1989; **149**:780–8.

29 Joint National Committee (1993). The Fifth Report of the Joint National Committee on Detection, Evaluation, and Treatment of High Blood Pressure (JNC V). *Arch Intern Med* 1993; **153**:154–83.

30 Joint National Committee (1988). The 1988 report of the Joint National Committee on Detection, Evaluation, and Treatment of High Blood Pressure. *Arch Intern Med* 1988; **148**:1023–38.

31 US Department of Health and Human Services. Secretary's Task Force on Black and Minority Health. *US Report of the Secretary's Task Force on Black and Minority Health*. DHHS Publ no. 186-620-638: 40716. Washington, D.C.: US Government Printing Office, 1986.

32 Horan MJ, Lenfant CJM. Hypertension in blacks: future research directions. *Ethnicity Dis* 1992; **2**:115–19.

33 Edwards R, Falkner B, Hulman S et al. Blood pressure in new born infants: comparison by race. Presented at the Sixth International Interdisciplinary Conference on Hypertension in Blacks, International Society on Hypertension in Blacks. Long Beach, Calif, 1990: 19 (Abstract Monograph).

34 Greaves K, Puhl J, Gruben D, Baranowski T. Patterns in blood pressures of black-, anglo-, and mexican-Americans 3–4 year-old children and their parents [Abstract]. *J Hum Hypertens* 1988; **1**(**5**):280.

35 Brandon LJ, Fillingham L. Body composition and blood pressure in children based on age, race and sex. *Am J Prev Med* 1993; **9**:34–38.

36 Alpert BS, Fox ME. Racial aspects of blood pressure in children and adolescents. *Ped Clinic North Am* 1993; **40**:13–22.

37 Gillum RF. Pathophysiology of hypertension in blacks and whites; a review of the basis of racial blood pressure differences. *Hypertension* 1979; **1**:468–75.

38 Tyroler HA. Twenty-year mortality in black residents of Evans County, Georgia. *J Clin Hypertens* 1987; **3[Suppl]**:9S–15S.

39 Aubert RE, Blount SB. Variability of hypertension-related mortality in the United States, 1968–1986. *J Hum Hypertens* 1990; **4**:97–99.

40 Otten MW Jr, Teutsch SM, Williamson DF, Marks JS. The effect of known risk factors on the excess mortality of black adults in the United States. *JAMA* 1990; **263**:845–50.

41 Chaturvedi N, McKeigue PM, Marmot MG. Hypertension in blacks and whites in the UK. Presented at the Sixth International Interdisciplinary Conference on Hypertension in Blacks, International Society on Hypertension in Blacks. Bahia, Brazil, 1991:9 (Abstract Monograph).

42 Rostand SG, Kirk KA, Rutsky EA, Pate BA. Racial differences in the incidence, and treatment for end-stage renal disease. *N Engl J Med* 1982; **306**:1276–79.

43 Rostand SG, Brown G, Kirk KA et al. Renal insufficiency in treated essential hypertension. *N Engl J Med* 1989; **320**:684–88.

44 Walker WG, Neaton JD, Cutler JA et al. Renal function change in hypertensive members of the Multiple Risk Factor Interventional Trial. Racial and treatment effects. *JAMA* 1992; **268**:3085–91.

45 United States Data System 1989 Annual Report. *Am J Kidney Dis* 1990; **16[Suppl 2]**:22–27.

46 Shulman NB. Renal disease in hypertensive blacks. In: Hall WD, Saunders E, Shulman NB, eds. *Hypertension in blacks: epidemiology, pathophysiology and treatment*. Chicago: Year Book Medical Publishers, 1985: 106–112.

47 Bild D, Teutsch SM. The control of hypertension in persons with diabetes: A public health approach. *Public Health Rep* 1987; **102**:522–29.

48 Gillum RF. Cardiovascular disease in the United States: an epidemiologic overview. In: Saunders E, ed. *Cardiovascular diseases in blacks*. Philadelphia: F.A. Davis Co, 1991: 3–16.

49 Devereux RB. The importance of left ventricular mass as a predictor of cardiovascular morbidity in hypertension. *Am J Hypertens* 1989; **2**:650–4.

50 Levy D, Garrison RJ, Savage DD et al. Prognostic implications of echocardiographically determined left ventricular mass in the Framingham Heart Study. *N Engl J Med* 1990; **322**:1561–66.

51 Dunn FG, Oigman W, Sungaard-Riise K et al. Racial differences in cardiac adaptation to essential hypertension determined by echocardiographic indexes. *J Am Coll Cardiol* 1983; **1**:1348–51.

52 Savage DD. Echocardiographic left ventricular hypertrophy in blacks. *J Clin Hypertens* 1987; **3[Suppl 3]**:61S–68S.

53 Grossman E, Shemesh J, Shamiss A et al. Left ventricular mass in diabetes-hypertension. *Arch Intern Med* 1992; **152**:1001–4.

54 Harshfield GA, Hwang C, Grim CE. Circadian variation of blood pressure in blacks: influence of age, gender and activity. *J Hum Hypertens* 1990; **4**:43–47.

55 Murphy MB, Lang RL, Nelson KS et al. Diurnal blood pressure differences are associated with inter-racial differences in cardiac hypertrophy. Presented at the Fifth International Interdisciplinary Conference on Hypertension in Blacks, International Society on Hypertension in Blacks. Nairobi, Kenya, 1989; **38**: (Abstract monograph).

56 Palatini P, Penzo M, Racioppa A et al. Clinical relevance of nighttime blood pressure and of daytime blood pressure variability. *Arch Intern Med* 1992; **152**:1855–60.

57 Verdeccia P, Schillaci G, Gueniere M et al. Circadian blood pressure changes and left ventricular hypertrophy in essential hypertension. *Circulation* 1990; **81**:528–36.

58 Hypertension Detection and Follow-up Program Cooperative Group. Five-year findings of the Hypertension Detection and Follow-up Program: III. Reduction in stroke incidence among persons with high blood pressure. *JAMA* 1982; **247**:633–38.

59 Loue S, Molgaard CA, Eisenman P et al. Stroke in four race-ethnic populations in San Diego County, 1983–1985. Presented at the Fifth International Interdisciplinary Conference on Hypertension in Blacks, International Society on Hypertension in Blacks, Long Beach, Calif, 1990: 43 (Abstract monograph).

60 Cooper ES, Cerebrovascular diseases in hypertensive blacks. In: Saunders E, ed. *Cardiovascular diseases in blacks*. Philadelphia: F.A. Davis Co, 1991: 145–155.

61 Schachter J, Kuller LH, Perfetti C. Blood pressure during the first five years of life: Relation to ethnic group (black or white) and to parental hypertension. *Am J Epidemiol* 1984; **119**:541–53.

62 Williams RR, Hunt SC, Hasstedt SJ et al. Multigenic human hypertension: evidence for subtypes and hope for haplotypes. *J Hypertens* 1990; **8[Suppl 7]**:S39–S46.

63 Ward R. Familial aggregation and genetic epidemiology of blood pressure. In: Laragh JH, Brenner BM, eds. *Hypertension: pathophysiology, diagnosis and management*. New York: Raven Press, 1990: 81–100.

64 Lifton RD. New observations on the genetics and molecular biology of human hypertension. Presented at the Eighth Scientific Meeting of the American Society of Hypertension. New York, NY, 1993.

65 Klag MJ, Whelton PK, Coresh J et al. The association of skin color with blood pressure in US blacks with low socio-economic status. *JAMA* 1991; **265**:599–602.

66 Moorman PG, Hames CG, Tyroler HA. Socioeconomic status and morbidity and mortality in hypertensive blacks. In: Saunders E, ed. *Cardiovascular diseases in blacks*. Philadelphia: F.A. Davis Co, 1991: 179–194.

67 Brandao AP, Brandao AA, Oliveira RC et al. Skin color and blood pressure in children and adolescents — The Rio de Janeiro Study. Presented at the Sixth International Interdisciplinary Conference on Hypertension in Blacks, International Society on Hypertension in Blacks. Salvadore, Brazil, 1991; **33**: (Abstract monograph).

68 Dressler WW. Social class, skin color, and arterial blood pressure in two societies. *Ethnicity Dis* 1991; **1**:60–77.

69 Somova LI, Mufunda JJ. Renin-angiotensin-aldosterone system and thromboxane A2/prostacyclin in normotensive and hypertensive black Zimbabweans. *Ethnicity Dis* 1992; **2**:27–34.

70 Poulter NR, Watkins H, Thompson AV, Sever PS. Plasma renin activity in black hypertensives, black normotensives and white normotensives in Kenya. [Abstract]. *J Hum Hypertens* 1988; **1**(5):290.

71 Israili ZH, Hall WD, Douglas F et al. Effect of dietary sodium on weight, blood pressure, serum aldosterone, plasma renin activity and urinary excretion of potassium and eicosanoids in black and white hypertensive and normotensive subjects. Presented at the Sixth International Conference on Hypertension in Blacks, International Society on Hypertension in Blacks. Salvadore, Brazil, 1991; **10**:(Abstract monograph).

72 Hohn AR, Riopel DA, Keil JE. Childhood familial and racial differences in physiology and biochemical factors related to hypertension. *Hypertension* 1983; **5**:56–70.

73 Grim CE, Luft FC, Miller JZ et al. Racial differences in blood pressure in Evans County, Georgia: relationship to sodium and potassium intake and plasma renin activity. *J Chronic Dis* 1980; **33**:87–94.

74 Lijnen P, M'Buyama-Kabangu JR, Fagard R et al. Racial differences in renin, aldosterone, prostaglandins and kallikrein of normal male subjects. *Ann Soc Belg Med Trop* 1985; **65**:293–301.

75 Cohen SL, Jhetam D, DaSilva J et al. Sodium and potassium status, plasma renin and aldosterone profiles in normotensive and hypertensive Johannesburg blacks. *S Afr Med J* 1982; **62**:941–44.

76 Hall WD. Pharmacologic therapy of hypertension in blacks. In: Hall WD, Saunders E, Shulman NB, eds. *Hypertension in blacks: epidemiology, pathophysiology and treatment.* Chicago: Year Book Medical Publishers, 1985: 182–208.

77 Aviv A, Gardner J. Racial differences in ion regulation and their possible links to hypertension in blacks. *Hypertension* 1989; **14**:584–89.

78 Savage DD, Watkins LD, Grim CE, Kumanyika SK. Hypertension in black populations. In: Laragh JH, Brenner BM eds. *Hypertension: pathophysiology, diagnosis and management.* New York: Raven Press Ltd, 1990: 1837–52.

79 Israili ZH, Hall WD. Serum levels of angiotensin converting enzyme in hypertensive blacks. [Abstract]. *J Hum Hypertens* 1990; **4**:184.

80 Brown RD, Berg S. Evidence for end organ resistance to atrial natriuretic peptide in black hypertension. Presented at the Fifth International Interdisciplinary Conference on Hypertension in Blacks, International Society on Hypertension in Blacks, Long Beach, Calif 1990; **13**: (Abstract monograph).

81 Warren SE, O'Connor DT. Does a renal vasodilator system mediate racial differences in essential hypertension? *Am J Med* 1980; **69**:425–29.

82 Israili ZH, Olutade B, Arepally A, Hall WD. Blood pressure response to captopril and urinary kallikrein excretion. [Abstract]. *Clin Pharmacol Ther* 1991; **49**:146.

83 Anderson NB, McNeilly M. Autonomic reactivity and hypertension in blacks: towards a contextual model. In: Fray JCS, Douglas JG, eds. *Pathophysiology of hypertension in blacks.* New York: Oxford University Press, 1993: 107–39.

84 Light KC, Obrist PA, Sherwood A et al. Effect of race and marginally elevated blood pressure on responses to stress. *Hypertension* 1987; **10**:555–63.

85 Thomas J, Semenya K, Thomas CB et al. Precursors of hypertension in black compared to white medical students. *J Chronic Dis* 1987; **40**:721–27.

86 Johnson EH, Nazzaro P, Gilbert DC et al. Similarities in cardiovascular reactivity to behavioral stressors in African-American and white males. *Ethnicity Dis* 1992; **2**:232–45.

87 Frohlich ED. Hemodynamic differences between black patients and white patients with essential hypertension. State of the art lecture. *Hypertension* 1990; **15**:675–80.

88 Luft FC, Grim CE, Fineberg N, Weinberger MC. Effects of volume expansion and contraction in normotensive whites, blacks, and subjects of different ages. *Circulation* 1979; **59**:643–50.

89 Hall WD. Pathophysiology of hypertension in blacks. *Am J Hypertens* 1990; **3**:366S–371S.

90 Chrysant SG, Danisa K, Kem DC, Dillard BL, Smith WJ, Frohlich ED. Racial differences in pressure, volume and renin interrelationships in essential hypertension. *Hypertension* 1979; **1**:136–41.

91 Hall WD. Hypertension in blacks: rationale for pharmacologic therapy. *J Natl Med Assoc* 1988; **80[Suppl]**:17–22.

92 Veterans Administration Cooperative Study Group on Antihypertensive Agents: Comparison of propranolol and hydrochlorothiazide for the initial treatment of hypertension. I. Results of short-term titration with emphasis on racial differences in response. *JAMA* 1982; **248**:1996–2003.

93 Veterans Administration Cooperative Study Group on Antihypertensive Agents: Comparison of propranolol and hydrochlorothiazide for the initial treatment of hypertension. II. Results of long-term therapy. *JAMA* 1982; **248**:2004–11.

94 Hawkins DW, Dieckmann MR, Horner RD. Diuretics and hypertension in black adults. *Arch Intern Med* 1988; **148**:803–5.

95 Messerli FH, DeCarvalho JGR, Christie B, Frohlich ED. Essential hypertension in black and white subjects. Hemodynamic findings and fluid volume state. *Am J Med* 1979; **67**:27–31.

96 Sowers JR, Zemel MB, Zemel P et al. Salt sensitivity in blacks: salt intake and natriuretic substances. *Hypertension* 1988; **12**:485–90.

97 Mufunda J, Sparkes HV Jr. Salt sensitivity and hypertension in African blacks. In: Fray JCS, Douglas JG, eds. *Pathophysiology of hypertension in blacks*. New York: Oxford University Press, 1993: 143–65.

98 Dluhy RG, Smith K, Taylor T et al. Prolonged converting enzyme inhibition in non-modulating hypertension. *Hypertension* 1989; **13**:371–77.

99 Critchley JAJH, Sriwatanakul K, Balali-Mood M et al. Ethnic differences in the renal dopamine response to an oral salt load. Presented at the Fourth International Interdisciplinary Conference on Hypertension in Blacks, International Society on Hypertension in Blacks. Nairobi, Kenya, 1989: 17 (Abstract monograph).

100 Bell NH, Greene A, Epstein S et al. Evidence for alterations of vitamin D-endocrine system in blacks. *J Clin Invest* 1985; **76**:470–3.

101 Touyz RM, Milne FJ, Reinach SG. Intracellular Mg^{2+}, Ca^{2+}, Na^+ and K^+ in platelets and erythrocytes of essential hypertension patients: relation to blood pressure. *Clin Exp Hypertens* 1992; **14(A)**:1189–1209.

102 Cooper RC, Aina O, Chaco L et al. Cell cations and blood pressure in US whites, US blacks, and west African blacks. *J Hum Hypertens* 1990; **4**:477–83.

103 Cooper RS, Borke JL. Intracellular ions in hypertension in blacks. In: Fray JCS, Douglas JG eds. *Pathophysiology of hypertension in blacks*. New York: Oxford University Press, 1993: 181–213.

104 Lasker N, Hopp L, Grossman S et al. Race and sex differences in erythrocyte Na^+, K^+, and Na^+-K^+-adenosine triphosphatase. *J Clin Invest* 1985; **75**:1813–20.

105 Tuck ML, Gross C, Maxwell MH et al. Erythrocyte Na^+, K^+ cotransport and Na^+, K^+ pump in black and caucasian hypertensive subjects. *Hypertension* 1984; **6**:536–44.

106 Aviv A, Aladjem M. Essential hypertension in blacks: epidemiology, characteristics, and possible roles of racial differences in sodium, potassium, and calcium regulation. *Cardiovasc Drug Ther* 1990; **4**:335–42.

107 Beutler E, Kuhl W, Sacks P. Sodium-potassium-ATPase activity is influenced by ethnic origin and not by obesity. *N Engl J Med* 1983; **309**:756–60.

108 Hatori N, Gardner JP, Tomonari H et al. Na^+-H^+ antiport activity in skin fibroblasts from blacks and whites. *Hypertension* 1990; **15**:140–45.

109 Langford HG, Zhang ZM, Hutchinson RG, Files JC. Platelet Na^+-H^+ exchange and erythrocyte Na^+ transport in essential hypertension in blacks. Presented at the Fifth International Interdisciplinary Conference on

Hypertension in Blacks, International Society on Hypertension in Blacks. Long Beach, Calif 1990: 41 (Abstract monograph).

110 Resnick LM, Muller FB, Laragh JH. Calcium-regulating hormones in essential hypertension: relation to plasma renin activity and sodium metabolism. *Ann Intern Med* 1986; **105**:649–54.

111 Brickman AS, Nyby MD, Griffith RF et al. Racial differences in platelet cytosolic calcium and calcitropic hormones in normotensive subjects. *Am J Hypertens* 1993; **6**:46–51.

112 Hall WD, Kong W. Hypertension in blacks: Nonpharmacologic and pharmacologic therapy. In: Saunders E, ed. *Cardiovascular diseases in blacks*. Philadelphia: F.A. Davis Co, 1991: 157–69.

113 Burris JF, Davidov ME, Jenkins P et al. Comparison of the antihypertensive effects of betaxolol and chlorthalidone as monotherapy and in combination. *Arch Intern Med* 1989; **119**:2437–41.

114 Flamenbaum W, Weber MA, McMahon FG et al. Monotherapy with labetalol compared with propranolol. Differential effects by race. *J Clin Hypertens* 1985; **1**:56–69.

115 Wright JT Jr, DiPette DJ, Goodman RP et al. Renin profile, race, and anti-hypertensive efficacy with atenolol and labetalol. *J Hum Hypertens* 1991; **5**:193–98.

116 Moser M, Lunn J. Comparative effect of pindolol and hydrochlorothiazide in black hypertensive patients. *Angiology* 1981; **32**:561–68.

117 Cubbeddu LX, Aranda J, Singh B et al. A comparison of verapamil and propranolol for the initial treatment of hypertension. Racial differences in response. *JAMA* 1986; **256**:2214–21.

118 Saunders E, Weir MR, Kong BW et al. A comparison of the efficacy and safety of a beta-blocker, a calcium channel blocker and a converting enzyme inhibitor in hypertensive blacks. *Arch Intern Med* 1990; **150**:1707–13.

119 Materson BJ, Reda DJ, Cushman WC et al. Single-drug therapy for hypertension in men: a comparison of six antihypertensive agents with placebo. *N Engl J Med* 1993; **328**:914–21.

120 Ram CVS, Tjoa HI, Kaplan NM et al. Response to ACE inhibitors in black patients with hypertension — clinical implications. Presented at the Fifth International Interdisciplinary Conference on Hypertension in Blacks, International Society on Hypertension in Blacks. Long Beach, Calif 1990: 52 (Abstract monograph).

121 Weinberger MH. Blood pressure and metabolic responses to hydrochlorothiazide, captopril, and the combination in black and white, mild-to-moderate hypertensive patients. *J Cardiovasc Pharmacol* 1985; **7**[**Suppl 1**]:S52–S55.

122 Veterans Administration Cooperative Study Group on Antihypertensive Agents: Low dose captopril for the treatment of mild to moderate hypertension. I. Results of a 14-week trial. *Arch Intern Med* 1984; **144**:1947–53.

123 Gavras H, Biollaz J, Waeber B et al. Antihypertensive effect of a new oral angiotensin-converting enzyme inhibitor "MK 421". *Lancet* 1981; **2**:542–46.

124 Flack JM, Grimm RH, Lewis C et al. Comparison of response to low dose antihypertensive medication between black and white participants in the treatment of Mild Hypertension Study (TOMHS). Presented at the Seventh International Conference on Hypertension in Blacks, International Society on Hypertension in Blacks. Atlanta, Georgia, 1992: 20 (Abstract monograph).

125 Moser M, Lunn J. Response to captopril and hydrochlorothiazide in black patients with hypertension. *Clin Pharmacol Ther* 1982; **32**:307–12.

126 Wilkins LH, Dustan HP, Walker JF, Oparil S. Enalapril in low-renin essential hypertension. *Clin Pharmacol Ther* 1983; **34**:297–302.

127 Zachariah PK, Bonnet G, Chrysant SG et al. Evaluation of antihypertensive efficacy of lisinopril compared to metoprolol in moderate to severe hypertension. *J Cardiovasc Pharmacol* 1987; **9**[**Suppl 3**]:S53–S58.

128 Freier PA, Wollam GL, Hall WD et al. Blood pressure, plasma volume, and catecholamine levels during enalapril therapy in blacks with hypertension. *Clin Pharmacol Ther* 1984; **36**:731–37.

129 Cruickshank JK, Anderson NM, Wadsworth J et al. Treating hypertension in black compared with white non-insulin dependent diabetics: a double blind trial of verapamil and metoprolol. *BMJ* 1988; **297**:1155–59.
130 Eisner G, Johnson B, McMahon G et al. Isradipine and enalapril in treatment of black hypertensives. [Abstract]. *J Hum Hypertens* 1990;**4**:180.
131 Lucas C, Jenkins P, Mendels J et al. The effectiveness of labetalol compared to hydrochlorothiazide in hypertensive black patients. *J Natl Med Assoc* 1991; **83**:866–71.
132 Cummings DM, Amadio P Jr, Taylor EJ Jr et al. The antihypertensive response to lisinopril: the effect of age in a predominantly black population. *J Clin Pharmacol* 1989; **29**:25–32.
133 Davidson JL, Gomez HJ, Bolognese JA et al. Comparative efficacy of lisinopril and hydrochlorothiazide in black hypertensive patients. [Abstract]. *J Human Hypertens* 1988; **1**(5):276.
134 Veterans Administration Cooperative Study Group on Antihypertensive Agents: Racial differences in response to low-dose captopril are abolished by the addition of hydrochlorothiazide. *Br J Clin Pharmacol* 1982; **14[Suppl 2]**:97s–101s.
135 Neutal JM, Smith DH, Ram CV et al. Comparison of bisoprolol with atenolol for systemic hypertension in four population groups (young, old, black and nonblack) using ambulatory blood pressure monitoring. Bisoprolol Investigators Group. *Am J Cardiol* 1993; **72**:41–46.
136 Maharaj B, van der Byl K. A comparative study of isradipine SRO and enalapril in black patients with mild-to-moderate hypertension. *Am J Hypertens* 1993; **6[Suppl]**:80S–81S.
137 Weir MR, Kong BW, Jenkins P, Lavin PT. Comparison of the efficacy and tolerability of lisinopril and sustained-release verapamil in black patients with hypertension. *Clin Ther* 1991; **13**:409–16.
138 Holland OB, von Kuhnert L, Campbell WB, Anderson RJ. Synergistic effect of captopril with hydrochlorothiazide for the treatment of low-renin hypertensive black patients. *Hypertension* 1983; **5**:235–39.
139 Thind GS. Demography predicts blood pressure response to once-daily enalapril monotherapy of mild to moderate essential hypertensive patients. *Am J Hypertens* 1988; **1[Suppl]**:227S–229S.
140 Drayer JI, Weber MA. Monotherapy of essential hypertension with a converting-enzyme inhibitor. *Hypertension* 1983; **5**:III-108–III-113.
141 Massie B, MacCarthy EP, Ramanathan KB et al. Diltiazem and propranolol in mild to moderate essential hypertension as monotherapy or with hydrochlorothiazide. *Ann Intern Med* 1989; **107**:150–7.
142 Veterans Administration Cooperative Study Group on Antihypertensive Agents: Efficacy of nadolol alone and combined with bendroflumethiazide and hydralazine for systemic hypertension. *Am J Cardiol* 1983; **52**:1230–7.
143 Obel AO. Effects of chlorthalidone, oxprenolol, and their combination in hypertensive blacks: a randomized double-blind crossover study. *J Cardiovasc Pharmacol* 1989; **299**:456–57.
144 Addy JH. Differentiation of essential hypertensive types by therapeutic response to beta-blockers and/or calcium entry blockers. Presented at the Sixth International Interdisciplinary Conference on Hypertension in Blacks, International Society on Hypertension in Blacks. Salvadore, Brazil, 1991: 51 (Abstract monograph).
145 Yonga G, Ogola EN, Juma FD, Orinda DA. Propranolol versus hydroflumethiazide on cardiovascular risk factors in Kenyan hypertensives. Presented at the Fifth International Interdisciplinary Conference on Hypertension in Blacks, International Society on Hypertension in Blacks. Long Beach, Calif, 1990: 23 (Abstract monograph).
146 Habte B. The efficacy of hydrochlorothiazide, timolol and enalapril in Ethiopians with essential hypertension. *Ethiop Med J* 1992; **30**:163–67.
147 Seedat YK. Trial of atenolol and chlorthalidone for hypertension in black South Africans. *BMJ* 1980; **281**:1421–23.
148 M'Buyamba-Kabangu JR, Fagard R, Lijnen P et al. Calcium entry blockade or beta-block-

ade in long-term management of hypertension in blacks. *Clin Pharmacol Ther* 1987; **41**:45–54.
149 Stein CM, Neill P, Kusemamuriwo F. Antihypertensive effects of low doses of hydrochlorothiazide in hypertensive black Zimbabweans. *Int J Cardiol* 1992; **37**:231–35.
150 M'Buyamba-Kabangu JR, Lepira B, Lijnen P et al. Intracellular sodium and the response to nitrendipine or atenolol in African blacks. *Hypertension* 1988; **11**:100–5.
151 Abengowe CV. A double-blind comparison of acebutolol (Sectral) and propranolol (Inderal) in the treatment of hypertension in black Nigerian patients. *J Int Med Res* 1985; **13**:116–21.
152 Feitosa GS, Protasio MS. Efficacy of nitrendipine and atenolol in a black population. A comparison with white and mulato population. Presented at the Sixth International Interdisciplinary Conference on Hypertension in Blacks, International Society on Hypertension in Blacks. Salvadore, Brazil, 1991: 52 (Abstract monograph).
153 Venter CP, Joubert PH. Penbutolol in black hypertensive patients. *S Afr Med J* 1986; **69**:286–87.
154 Moser M, Lunn J, Materson BJ. Comparative effects of diltiazem and hydrochlorothiazide in blacks with systemic hypertension. *Am J Cardiol* 1985; **56**:101H–104H.
155 Olutade BO, Hall WD, Hildebrandt K et al. Efficacy of nifedipine GITS vs hydrochlorothiazide in the management of mild to moderate hypertension in the black hypertensive. [Abstract]. *J Hum Hypertens* 1988; **1**(5):196.
156 Olutade BO, Hall WD, Porter VD. Effect of calcium channel blockers on high grade ectopy not superior to diuretic with potassium supplement in hypertensive blacks. Presented at the Sixth International Interdisciplinary Conference on Hypertension in Blacks, International Society on Hypertension in Blacks. Salvadore, Brazil, 1991: 41 (Abstract monograph).
157 Kostis JB. The prognostic significance of ventricular ectopic activity [editorial comment]. *Am J Cardiol* 1992; **70**:807–8.
158 Weir MR, Levin PT. Comparison of the efficacy and tolerability of Prinivil and Procardia XL in black and white hypertensive patients. *Clin Ther* 1992; **14**:730–9.
159 Carr AA, Prisant LM. The new calcium antagonist isradipine. Effect on blood pressure and the left ventricle in black hypertensive patients. *Am J Hypertens* 1990; **3**:8–15.
160 Moser M, Lunn J, Nash DT et al. Nitrendipine in the treatment of mild to moderate hypertension. *J Cardiovasc Pharmacol* 1984; **6**[**Suppl** 7]:S1085–89.
161 Lore W, Anjichi GM. Studies on isradipine in the management of adult hypertensive patients at Moi University Teaching Hospital, Eldoret, Kenya. *East Afr Med J* 1992; **69**:69–73.
162 Erne P, Bolli P, Bertel O et al. Factors influencing the hypotensive effects of calcium antagonists. *Hypertension* 1983; **5**[**Suppl II**]:II-97–II-102.
163 Nicholson JP, Resnick LM, Laragh JH. The antihypertensive effect of verapamil at extremes of dietary sodium intake. *Ann Intern Med* 1987; **107**:309–34.
164 Weinberger MH. Angiotensin converting enzyme inhibitors enhance the antihypertensive efficacy of diuretics and blunt or prevent adverse metabolic effects. *J Cardiovasc Pharmacol* 1989; **13**[**Suppl** 3]:S1–S4.
165 Luther RR, Klepper MJ, Laddu AR et al. Efficacy of terazosin in the treatment of essential hypertension in black patients. Presented at the Fourth International Interdisciplinary Conference on Hypertension in Blacks, International Society on Hypertension in Blacks. Nairobi, Kenya, 1989; 33 (Abstract monograph).
166 Cruickshank JM, Lewis J, Moore V, Dodd C. Reversibility of left ventricular hypertrophy by differing types of antihypertensive therapy. *J Hum Hypertens* 1992; **6**:85–90.
167 Dahlof B, Pennert K, Hansson L. Reversal of left ventricular hypertrophy in hypertensive patients: A meta-analysis of 109 treatment studies. *Am J Hypertens* 1992; **95**:95–110.
168 De Luca N, Ricciardelli B, Rosiello G et al. Stable improvement in large artery compliance after long-term antihypertensive treatment with enalapril. *Am J Hypertens* 1988; **1**:181–83.

169 Simon ACh, Levenson JA, Bouthier JL et al. Captopril-induced changes in large arteries in essential hypertension. *Am J Med* 1984; **76(5B)**:71–75.

170 Yasky J, Ferrario CM. Nocturnal hypotension and ACE inhibitors. *J Hum Hypertens* 1993; **7**:299–304.

171 White WB, McCabe EJ, Hagar D, Schulman P. The effects of the long-acting angiotensin-converting enzyme inhibitor cilazapril on casual, exercise and ambulatory blood pressure. *Clin Pharmacol Ther* 1988; **44**:173–78.

172 Munger MA, Chance M, Nair R et al. Evaluation of quinapril on regional blood flow and cardiac function in patients with congestive heart failure. *J Clin Pharmacol* 1992; **32**:70–6.

173 Ventura HO, Frohlich ED, Messerli FH et al. Immediate regional blood flow distribution following angiotensin converting enzyme inhibition in patients with essential hypertension. *Am J Med* 1984; **76[Suppl 5B]**:58–61.

174 Rajagopalan B, Raine AEG, Cooper R, Ledingham JGG. Changes in cerebral blood flow in patients with severe congestive heart failure before and after captopril treatment. *Am J Med* 1984; **76[Suppl 5B]**:86–90.

175 Magrini F, Shimizu M, Roberts N et al. Converting enzyme inhibition and coronary blood flow. *Circulation* 1987; **75**:I-168–I-174.

176 Zimlichman R, Gasper N, Eidelman L, Gold B. Prolonged treatment with captopril attenuates vascular hyper-reactivity in patients with essential hypertension. [Abstract]. *Am J Hypertens* 1992; **5**:12A.

177 Beto JA, Bansal VK. Quality of life in treatment of hypertension: A metaanalysis of clinical trials. *Am J Hypertens* 1992; **5**:125–33.

178 Croog SH, Kong BW, Levine S et al. Hypertensive black male and female. Quality of life and effects of antihypertensive medications. Black Hypertension Quality of Life Multicenter Trial Group. *Arch Intern Med* 1990; **150**:1733–41.

179 Testa MA, Anderson RB, Nackley JF et al. Quality of life and antihypertensive therapy in men: a comparison of captopril with enalapril. *N Engl J Med* 1993; **328**:907–13.

180 Barnes JM, Barnes NM, Costall B et al. ACE inhibition and cognition. In: *Current advances in ACE inhibition.* Proceedings of an International Symposium, London, 1989. Edinburgh: Churchill Livingstone, 1989: 159–171.

181 Kostis JB. Angiotensin-converting enzyme inhibitors. Emerging differences and new compounds. *Am J Hypertens* 1989; **2**:57–64.

182 WGHDPP: The Working Group on Hypertension in Diabetes. Statement on hypertension in diabetes mellitus. *Arch Intern Med* 1987; **147**:830–42.

183 De Feo P, Torlone E, Perriello G et al. Short-term metabolic effects of the ACE-inhibitor benazepril in type 2 diabetes mellitus associated with arterial hypertension. *Diabetes Metab* 1992; **18**:283–88.

184 Monzato E, Capurso A, Crepaldi G. Modification of cardiovascular risk factors during antihypertensive treatment: a multicentre trial with quinapril. *J Int Med Res* 1993; **21**:15–25.

185 Youssef S, Osman L, Sabbour MS. Serum lipoprotein profile under different antihypertensive therapy. *Int J Clin Pharmacol Res* 1992; **12**:109–16.

186 Keilani T, Schlueter WA, Levin ML, Battle DC. Improvement of lipid abnormalities associated with proteinuria using fosinopril, an angiotensin-converting enzyme inhibitor. *Ann Intern Med* 1993; **118**:247–54.

187 Costa FV, Borghi C, Mussi A, Ambrosioni E. Hypolipidemic effects of long-term antihypertensive treatment with captopril. A prospective study. *Am J Med* 1988; **84(3A)**:159–61.

188 Mathiesen ER, Hommel E, Giese J, Parving HH. Efficacy of captopril in postponing nephropathy in normotensive insulin dependent diabetic patients with microalbuminuria. *Br Med J* 1991; **303**:81–87.

189 Kasiske BL, Kalil RSN, Ma JZ et al. Effect of antihypertensive therapy on the kidney in patients with diabetes: a meta-regression analysis. *Ann Intern Med* 1993; **118**:129–38.

190 Hunsicker L, Lewis E, Bain R, McGill J. Outcome of a multi-center randomized clinical trial of ACE-inhibition in diabetic nephropathy [Abstract]. Diabetes. 1993; **42[Suppl 1]**:98A.

191 Slataper R, Vicknair N, Sadler R, Bakris GL. Comparative effects of different antihypertensive treatments on progression of diabetic renal disease. *Arch Intern Med* 1993; **153:** 973–80.
192 Silagy CA, McNeil JJ, McGrath BP. Crossover comparison of atenolol, enalapril, hydrochlorothiazide and isradipine for isolated systolic systemic hypertension. *Am J Cardiol* 1992; **70**:1299–1305.
193 Mann SJ. Systolic hypertension in the elderly. Pathophysiology and management. *Arch Intern Med* 1992; **152:**1977–84.
194 Sinaiko AR. Pharmacologic management of childhood hypertension. *Ped Clinics North America* 1993; **40**:195–212.
195 Miller K, Atkin B, Rodel PV Jr, Walker JF. Enalapril: a well tolerated and efficacious agent for the pediatric hypertensive patient. *J Cardiovasc Pharmacol* 1987; **10[Suppl 7]**:S154–56.
196 Schneeweiss A. Cardiovascular drugs in children: angiotensin-converting enzyme inhibitors. *Ped Cardiol* 1988; **9:**109–15.
197 Callis L, Vila A, Castellano J, Gras X. Long-term treatment with captopril in paediatric patients with severe hypertension and chronic renal failure. *Postgrad Med J* 1986; **62[Suppl 1]**:106–7.
198 Israili ZH, Hall WD. Cough and angioneurotic edema associated with angiotensin-converting enzyme inhibitor therapy. *Ann Intern Med* 1992; **117**:234–242.
199 Toto RD, Mitchell HC, Lee H-C et al. Reversible renal insufficiency due to angiotensin converting enzyme inhibitors in hypertensive nephrosclerosis. *Ann Intern Med* 1991; **115**:513–19.

7

Adverse reactions to ACE inhibitors

Wilfred W Yeo and Alyn H Morice

Introduction

In this chapter we will examine the side-effect profile of the ACE inhibitors. From pre-existing knowledge of the pathophysiology of hypertension and atherosclerosis, certain side-effects such as first-dose hypotension can be predicted. However a series of unique and bizarre side-effects are now recognized with inhibition of angiotensin converting enzyme (ACE). The mechanism of these side-effects is probably inhibition of the metabolism of peptides other than ANG I, such as bradykinin and substance P (Fig. 7.1). Which peptides are responsible for which side-effects is currently the source of much debate. Within the side-effect profile there is clearly an endogenous predisposition and recent genetic analysis has revealed that individuals vary in their serum ACE levels according to an insertion/deletion polymorphism at intron 16 of the human ACE gene.[1] Of the known side-effects of ACE inhibitors, cough has provoked the most interest. It must be classed as one of the least predictable of all side-effects, which meant that there was a long delay between drug discovery and the realization that this was a common adverse reaction. In this chapter we will therefore start with the unpredicted side-effects of ACE inhibition.

Idiosyncratic adverse reactions

Cough

That an orally active drug should have cough as a major side-effect is perhaps surprising. Indeed it was many years after the introduction of captopril that the typical dry, irritating cough was recognized as a side-effect of ACE inhibitor therapy.[2] Today this side-effect is still a source of referral to respiratory clinics and the lack of recognition of the drug-related nature of the cough by both doctor and patient is typical. Perhaps one reason for the lack of association of the cough as a side-effect is the delayed onset following institution of therapy. Life-table analysis reveals a gradual onset of clinically recognized cough such that the true incidence is noted only at 19 months.[3] Given this information, it is not surprising that early studies into the incidence of ACE inhibitor cough failed to demonstrate the common nature of this side-effect. ACE inhibitor coughers were thought to be rare individuals with an atypical response, but it has become clear that up to one-fifth of the population is affected. Cough will therefore have a major impact on the prescribing of ACE inhibitors and, because of the chronic nature of the symptom, will affect adversely the quality of life of those patients requiring this medication.

Natural history of ACE inhibitor cough

The insidious nature of ACE inhibitor cough blends the symptom into the 'naturally occurring' cough seen in everyday practice. Certain features are, however, characteristic. The sensation is described as a tickle at the back of the throat which leads to paroxysms of coughing, sometimes so intense that vomiting follows. No sputum is produced and wheezing is only rarely associated with the cough. Patients and their relatives are frequently kept awake at night as the cough may be exacerbated when supine. Traditional cough remedies, whether over the counter or prescribed by general practitioners, are ineffective in relieving the symptom. Some slight improvement is obtained by sucking sweets or sipping water. In chest clinics, patients are seen with a history of a very similar syndrome that is not associated with the use of ACE inhibitors. It is possible that the underlying abnormality in subjects with ACE inhibitor cough may be present in the normal population.

Spontaneous reporting of persistent dry cough occurs in about 15% of patients started on ACE inhibitor treatment.[4–12] The majority of these patients will report their cough within the first month but in about 20–30% the onset is delayed for some months. The incidence of cough in women is twice that in men,[3,12,13] and appears to be more common in non-smokers,[13] perhaps because non-smokers are more ready to recognize a cough as unusual. Once present, the cough persists until treatment is withdrawn. Surprisingly, the cough does not appear to be related to bronchoconstriction although one study has identified a slight increase in bronchial hyperreactivity.[14] This observation has not been converted into an increased incidence of asthma in patients taking ACE inhibitors, or an increase in bronchoconstriction in patients with established asthma. However, a recent report suggests that bronchoconstriction can be a rare adverse reaction to ACE inhibitors.[15]

In subjects who have had ACE inhibitor cough proven by repeated rechallenge, the reoccurrence of cough is rapidly detected. The rate of onset and offset appears to be associated with pharmcokinetics, in that short-acting drugs produce a more rapid onset of cough. However this is not directly related to plasma levels; even long-acting ACE inhibitors disappear from the plasma such that the return of serum ACE activity can be detected after 24 hours. In contrast, ACE inhibitor cough in subjects who are aware of the side-effect takes several days to reappear, even with the most rapidly acting drugs. These clinical observations suggest that serum ACE activity is unlikely to be an important factor in the aetiology of this side-effect and that inhibition tissue ACE is more likely to reflect the important compartment which leads to ACE inhibitor cough.

Mechanism of ACE inhibitor cough

ACE inhibitor cough has been described with all clinically available drugs from this class.[12] Thus it appears that the side-effect is due to enzyme inhibition rather than residing in any unique molecular configuration of the antagonist. ACE or kininase II is a non-specific metallopeptidase with two putative active sites.[16] Angiotensin I, the substrate after which the enzyme is named, is without known physiological activity and it seems more likely that accumulation of another peptide substrate of ACE is responsible for the generation of ACE inhibitor cough.

ACE metabolizes several small peptides including bradykinin, the tachykinins and enkephalins. Which of these peptides accumulates in the perineuronal tissue within the lung to give rise to the afferent stimulus is unknown. Animal studies have demonstrated that cough may be generated by stimulation of rapidly adapting receptors located in the larynx and upper airways. These receptors respond to stimulation from a wide range of tussive mediators, and stimulate Aδ-fibres in the vagus. In addition, small unmyelinated C-fibres with a more peripheral distribution are also responsible for carrying tussive impulses and these substance-P-containing neurones discharge in response to the pungent extract of capsaicin.

Of the candidate peptides, bradykinin is the only substance which causes cough when given by direct inhalation in man. It is thought to stimulate C-fibre afferents via irritant J-receptors.[17] Inhibition of ACE has been demonstrated to increase the biological activity of bradykinin both in the wheal and flare reaction in human skin[18,19] and in the bronchoconstriction produced in the guinea pig. In asthmatic subjects, inhalation of bradykinin causes bronchoconstriction[20] while ACE inhibitor cough is characterized by a lack of associated wheeze and ACE inhibitors do not cause exacerbation of pre-existing asthma. We and others have been unable to demonstrate a greater sensitivity to inhaled bradykinin in patients taking ACE inhibitors.[17] Whilst it is possible that excess bradykinin may accumulate around C-fibres and not bronchial smooth muscle, it seems more likely that bradykinin is not an important mediator in the production of ACE inhibitor cough.

Substance-P and other tachykinins are present in sensory nerves and are released following stimulation with the tussive agent capsaicin. In the lung, substance-P is broken down by ACE and it has been suggested that perineuronal accumulation of substance-P may be responsible for ACE inhibitor cough.[21] ACE is not the only enzyme which metabolizes substance-P and in normal subjects, blockade of ACE does not lead to increased vasodilator responses in the forearm. In healthy volunteers up to 1 mg of substance-P may be inhaled without apparent effects, whereas infusion of very low concentrations of the peptide leads to vasodilation. Thus the degrading enzymes present in the lung epithelium are highly efficient. Whether ACE inhibition in those subjects who are susceptible to the cough reveals a previously unrecognized enzyme deficiency remains to be determined.

ACE gene polymorphism

The human endothelial ACE molecule consists of 1306 amino acid residues which show a high degree of internal homology, suggesting that it originated by gene duplication.[16] A DNA polymorphism within the human ACE gene, consisting of the insertion (I) or deletion (D) of a DNA fragment of approximately 250 base pairs was described by Rigat et al.[22] The two allele polymorphism was detected by restriction fragment length polymorphism (RFLP) analysis, and the allelic distribution correlated with concentrations of serum ACE measured by radioimmunoassay. We have recently investigated methods of improving the detection of this RFLP, and used the polymorphism as a marker genotype to investigate the relationship between genotype and serum ACE activity, and the incidence of side-effects of ACE inhibitors.

In a population of 50 normal subjects the frequencies of the alleles were 46% for the insertion (I) and 54% for the deletion (D). The distribution of the genotypes was 24% II, 44% ID and 32% DD, which is in agreement with that observed by Rigat et al.[22] Analysis of variance revealed that there was a highly significant relation between serum ACE activity and genotype. The mean ACE activity differed significantly for the three genotypes from 24 nmol/ml/min (II) to 38 nmol/ml/min (DD).[23] Thus the polymorphism accounted for 48% of the total phenotypic variance of serum ACE activity.

Preliminary investigation of the relationship between the ACE genotype and ACE-inhibitor-induced cough in 33 patients has failed to show a strong association.[24,25] However the ACE gene polymorphism will clearly be an important factor in the profile of ACE inhibitor therapy, since a recent report has demonstrated an association between the DD genotype and myocardial infarction which was independent of other risk factors.[26] The therapeutic efficacy of ACE inhibitors in the prevention of myocardial infarction and unstable angina may be reliant on an individual's genotype.

Assessment of patients with suspected ACE inhibitor cough

Patients taking ACE inhibitors who present with a dry cough should be investigated for other causes of this symptom. Perhaps the commonest cause in clinical practice would be lung carcinoma and other associated features such as haemoptysis may provide a clue. A chest radiograph will be abnormal in the majority of cases but a small number of proximal lesions will be missed if bronchoscopy is not considered. To avoid unnecessary further investigation formal withdrawal and rechallenge protocol is recommended to establish the diagnosis of ACE inhibitor cough. Patients are asked to stop the medication and record symptoms and peak flow on a diary record card. If the cough disappears, then rechallenge with the agent for a minimum of two weeks should establish the diagnosis. The rechallenge protocol confirms the diagnosis because patients are aware of the symptom complex and the delay in recognition that is so typical of the initial presentation does not occur.

Treatment of ACE inhibitor cough

Because of its efficacy in congestive cardiac failure, withdrawal of ACE inhibitor therapy may be undesirable. Therefore attempts have been made to minimize or abolish ACE inhibitor cough in such patients. In general, the results have proved disappointing. Modification of the drug regimen either by substitution of a different ACE inhibitor or alteration of dosage may occasionally lead to a slight improvement. In the authors' experience, a halving of the dose leads to a reduction in cough frequency and severity of about 25% when assessed by visual analogue scale (unpublished observations).

Other drugs may be used to inhibit cough and non-steroidal anti-inflammatory drugs, such as sulindac, were initially reported to abolish ACE inhibitor cough. Such therapy may rationally be considered, as bradykinin is thought to cause cough by production of prostaglandins. Formal studies[27,28] have been disappointing, with small shifts in evoked cough and little improvement in diary record card symptom scores. The effect of inhaled steroids on ACE inhibitor cough has been investigated in the hope that local steroid deposition would induce production of ACE and diminish the cough. In ten non-asthmatic subjects, reported cough was less whilst taking inhaled Becloforte but the effect was not statistically significant.[29] At present, no treatment can be recommended which diminishes the symptoms of ACE inhibitor cough.

Angioneurotic oedema

Angioneurotic oedema is a rare but distressing side-effect of ACE inhibition which may lead to life threatening laryngeal oedema. In three-quarters of affected patients, symptoms occur within the first three weeks of therapy and the reaction occurs in approximately 1 in 2200

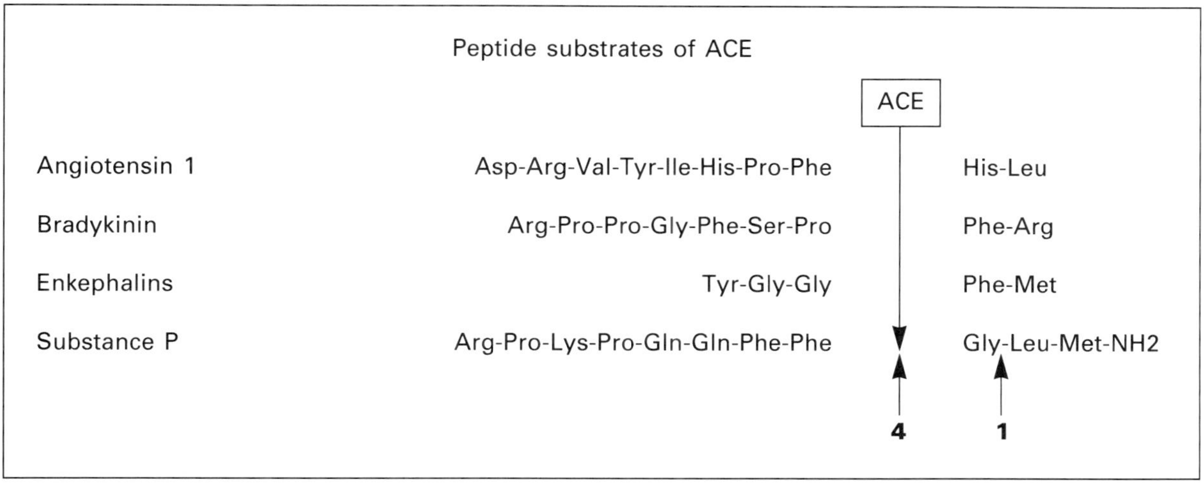

Figure 7.1
Peptide substrates of human angiotensin converting enzyme (ACE) showing the site of action of ACE. The activity of ACE at site Phe-Gly in substance P is four times that at the Gly-Leu site.

new prescriptions.[30,31] Angio-oedema has been described with all currently marketed drugs. There appears to be no recognizable pattern in predisposition and, unlike ACE inhibitor cough, patients occur equally within the sexes. The previous prescription of antiallergic medication does not protect against the development of angioneurotic oedema and patients who have had previous episodes of idiopathic angio-oedema may be at particular risk.

The clinical syndrome of angioneurotic oedema associated with ACE inhibition may run the full gamut from mild swelling of the skin to the full-blown features of anaphylaxis, with oedema of the mouth and mucous membranes, laryngeal oedema, bronchospasm and death.[30] Some patients appear to be sensitive to particular ACE inhibitors and reports have described development of angio-oedema de novo following swapping from one ACE inhibitor to another. Rechallenge with the original ACE inhibitor does not lead to recurrence of symptoms. Thus patients who have an absolute need for ACE inhibition, but have developed angio-oedema as a side-effect of one compound may be tried on an alternative drug.

The mechanism by which ACE inhibition leads to angio-oedema is unknown, but at present biochemical rather than immunological mechanisms seem likely. Bradykinin is known to alter vascular permeability and the wheal following intradermal injection of this vasodilator is potentiated by ACE inhibition. Infusion of bradykinin in normal volunteers produces flushing following ACE inhibition. Once again, a genetic predisposition in the form of a relative lack of other bradykinin degrading enzymes has been hypothesized as the basis of this side-effect, but this does not explain the ability of some patients to successfully swap therapy.

Table 7.1
Other idiosyncratic side-effects of ACE inhibitors

Idiosyncratic side-effects	Approximate incidence from placebo contrclled studies/reports
Throat discomfort[54]	10%
Voice changes[54]	10%
Taste disturbance*	2–4%
Sore mouth*	2–4%
Rash*	1–4%
Dizziness	1–4%
Diarrhoea (or constipation[55]	1–4%
Headache	1–3%
Fatigue	1–3%
Deafness [56]	reported
Anaemia	reported
Peripheral neuropathy*	reported

*higher incidence reported with sulphydryl group

The immunological basis for the angioneurotic reaction to ACE inhibition is supported by the demonstration of tissue-specific auto-antibodies.[30] Antibodies to conjugates of ACE inhibitors and protein have also been demonstrated; this is particularly evident with captopril. A factor against this hypothesis is that antibody reactions are present only with chronic therapy and angio-oedema occurs early in treatment. A further hypothesis suggests that patients with hereditary angio-oedema have abnormalities of the complement system, usually manifest by a decreased activity of the complement 1 esterase inactivator leading to unopposed activation of complement.[30,31] Patients with ACE inhibitor induced angio-oedema do not, however, demonstrate abnormalities in the first component of complement and so if this mechanism is involved, other, as yet undetermined, abnormalities of the complement system must be responsible.

Other idiosyncratic reactions

Other subjective side-effects and adverse reactions to ACE inhibitors are summarized in Table 7.1. Of note among these are disturbance of taste, often described as metallic, and rash. These appear to be related to the sulphydryl group in some ACE inhibitors and are therefore observed somewhat more commonly with captopril than with other ACE inhibitors which do not have this group.[32]

Predictable adverse reactions

First-dose hypotension

Precipitous falls in blood pressure after a first dose of ACE inhibitor are more of a problem in heart failure than hypertension. However such falls have been observed in hypertensive patients treated with high doses of diuretics, and those with resistant hypertension or renovascular disease — that is, in those more likely to have high plasma renin before treatment. Elderly patients are probably more susceptible to first-dose hypotension, perhaps because of their higher initial blood pressures and poor reflex compensatory mechanisms. In patients at risk, falls in systolic blood pressure of more than 50 mmHg may occur and can cause syncope, or poor-perfusion stroke or myocardial infarction.[33]

When ACE inhibitors are started, it is wise to omit diuretic therapy for one or two days and advise the patient to take the first dose at night while recumbent. Patients at very high risk of first-dose hypotension, for example those on high doses of loop diuretic or with concomitant heart failure may need to be admitted to hospital to start the ACE inhibitor. In the Consensus II study,[34] acute administration of ACE inhibitors following myocardial infarction was associated with a 15% excess of clinically significant hypotension, and it was hypothesized that this may have been responsible for the lack of beneficial effect seen in this trial.

First-dose hypotension has generally been considered a class effect of ACE inhibitors. Its duration is shorter but may be of more rapid onset with drugs such as captopril, which have a short half-life, and this drug may be preferred for the first dose administered to high-risk patients. Of some interest is a recent study in elderly patients with heart failure showing qualitative differences in the first dose effects of enalapril and perindopril, although these drugs have similar structure, metabolism and half-life.[35] When first-dose hypotension occurs, simple measures such as elevating the foot of the bed may be all that is required to alleviate symptoms. Since hypotension is due to relative hypovolaemia, plasma expansion with a colloid may be required. Theoretically infusion of ANG II would counteract catastrophic hypotension.

ACE inhibitors and renovascular hypertension

ACE inhibitors can cause severe and progressive renal failure in patients with advanced renovascular disease, particularly those with tight stenosis of the artery supplying a solitary kidney, those with unilateral renal artery thrombosis and contralateral renal artery stenosis (Fig. 7.2), or bilateral renal artery stenosis. In such circumstances glomerular filtration is thought to be maintained by angiotensin II induced constriction of efferent arterioles. ACE inhibitors abolish this constriction, dilate the efferent arterioles and thus cause filtration pressure to fall to such low levels that renal failure develops. There are suggestions that ACE inhibitors may also precipitate renal artery thrombosis,[36] but such reports may reflect the natural history of renovascular disease rather than an adverse reaction. The transplanted kidney may be at particular risk. ACE inhibitors may also cause important deterioration of renal function in some patients with preexisting renal disease. The presumption is that these patients have renovascular insufficiency caused by intrarenal blood vessel changes.

The effect of ACE inhibitors on advanced renovascular disease is of considerable practical importance. Patients 'at risk' are relatively common among those with resistant hypertension,[37] the elderly, and those with peripheral vascular disease. Renovascular disease, even of such a severe degree, is often clinically 'silent'. When prescribing an ACE inhibitor it is essential to measure serum creatinine before starting treatment and remeasure it early in treatment. If renal function declines during treatment,

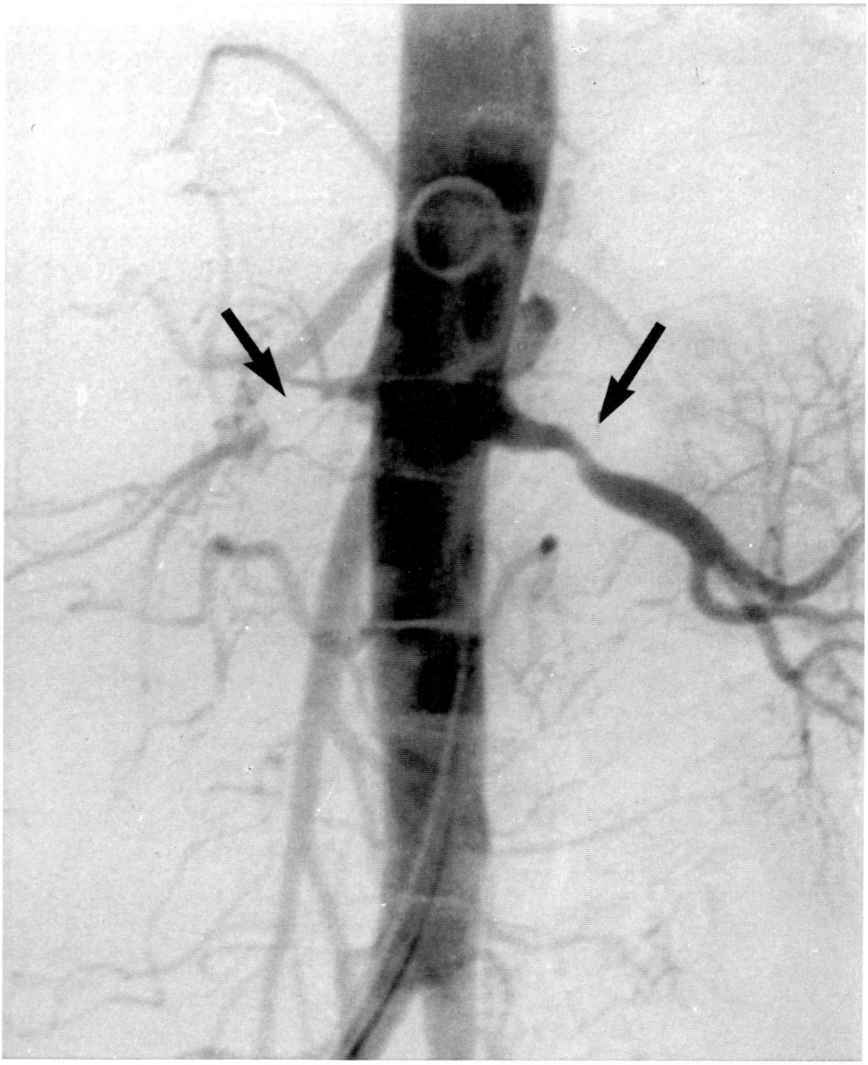

Figure 7.2
Renal arteriogram of a patient with right renal artery occlusion and left renal artery stenosis (arrows). The patient had normal renal function and, when treated with an ACE inhibitor for hypertension before diagnosis of the renovascular disease, developed reversible renal failure.

renovascular disease should always be suspected.

ACE inhibitors were reported to cause nephrotic syndrome and membranous glomerulonephritis, but this was at a time when captopril was used at very high doses, exceeding 300 mg daily. This adverse effect was attributed to the sulphydryl group in captopril, and ACE inhibitors without this chemical group have not been implicated in causing membranous glomerulonephritis. It does not appear to occur with the low doses of captopril in current use.

Hyperkalaemia

Angiotensin II is a potent stimulus for release of aldosterone. It is not surprising therefore that inhibition of ACE may be associated with elevation of serum potassium because of a decrease in aldosterone secretion. When ACE inhibitors are used as a single agent, this effect is trivial and clinically unimportant. However when ACE inhibitors are combined with potassium-sparing diuretics and/or potassium supplements, potentially hazardous hyperkalaemia may result.

Drug interactions and contraindications

ACE inhibitors are relatively free of contraindications, and can be used safely in patients with coexisting disease states such as diabetes, chronic obstructive airways disease, and heart failure. They should be avoided in pregnancy, and used with caution in patients with renovascular disease and those with impaired renal function, as discussed above. There are few important drug interactions involving ACE inhibitors. They should be prescribed together with potassium supplements or potassium-sparing diuretics only when essential and with meticulous biochemical monitoring, as dangerous hyperkalaemia may occur. ACE inhibitors may increase serum lithium concentration in patients taking lithium. However overall, these drugs are simple to prescribe, simple to take, and relatively free of limitations imposed by contraindications or drug interactions.

ACE inhibitors in pregnancy

ACE inhibitors are teratogenic in animals and possibly so in humans, and should therefore not be used in pregnancy.[38] A number of reports have described oligohydramnios in pregnant women taking captopril or enalapril,[38] and the majority of the fetuses have died either in utero[39] or soon after birth from anuria.[40] Oligohydramnios and neonatal anuria have been reported only when the mother continues treatment, and not in women who conceive while taking an ACE inhibitor but then stop the drug. When ACE inhibitors are stopped in the first trimester, the baby is likely to be born at or near term and to be of normal birth weight. Continued treatment carries a significant risk of premature delivery and low birth weight.[41] However pregnancies have gone successfully to term despite continued treatment with ACE inhibitors, and infants born prematurely with intrauterine growth retardation have generally done well if they survived the neonatal period.[41–43] The persistence of a ductus arteriosus has been reported in some fetuses[41,42] and may be attributed to prematurity, or perhaps to increased levels of bradykinin and prostaglandins with ACE inhibitor treatment. The manufacturers advise caution in use of ACE inhibitors when breast feeding.

Quality of life with ACE inhibitors

Because of this relative freedom from side-effects, there have been suggestions that ACE inhibitor treatment may be associated with a better 'quality of life' than other classes of antihypertensive drug. The evidence for this initially came from a single study which compared captopril to methyldopa and propranolol.[44] Not surprisingly, captopril proved superior to methyldopa in several respects, but of more note was the advantage of captopril over propranolol as regards quality of life. However propranolol is no longer generally used as the β-blocker of first choice. Furthermore, women were not included in this study and, as will be discussed below, ACE inhibitor cough is twice as common in women as in men. Since this study, the quality of life with ACE inhibitors in comparison to other antihypertensive drugs has been examined in several randomized double-blind studies. Captopril has been compared to atenolol,[45–49] propranolol,[46] verapamil,[47] and nifedipine;[49] and enalapril to atenolol[46,50] and propranolol.[46] These studies have not confirmed important benefits of ACE inhibitors on quality of life; in particular, they have not proved superior to β-blockers such as atenolol. Overall, the evidence available suggests that ACE inhibitors are as well tolerated as thiazides, β-blockers, and probably verapamil, and are better tolerated than the dihydropyridine calcium antagonists.[51–53]

Conclusions

ACE inhibitors are an important class of drugs which have a good safety profile. Care should be taken with certain groups of patients in whom side-effects will occur as a result of the pharmacological actions of these agents. Distressing but usually not dangerous side-effects may occur in an idiosyncratic manner, of which cough is the most common and perhaps most interesting. These latter adverse reactions have provided a stimulus for a large amount of basic research into the mechanisms of peptide activation and degradation.

References

1 Rigat G, Hubert C, Corvol P, Soubrier F. PCR detection of the insertion/deletion polymorphism of the human angiotensin converting enzyme gene (DCP1) (dipeptyl carboxypeptidase 1). *Nucl Acid Res* 1992; **20**:1433.

2 Sesoko S, Kaneko Y. Cough associated with the use of captopril. *Arch Int Med* 1985; **145**:1524.

3 Yeo WW, Ramsay LE. Persistent dry cough with enalapril: incidence depends on method used. *J Hum Hypertens* 1990; **4**:517–20.

4 Inman WHW, Rawson NSB, Wilton LV, Pearce GL, Speirs CJ. Postmarketing surveillance of enalapril. 1: results of prescription event-monitoring. *BMJ* 1988; **297**:286–89.

5 Coulter DM, Edwards IR. Cough associated with captopril and enalapril. *BMJ* 1987; **294**:1521–23.

6 Hood S, Nicholls MG, Gilchrist NL. Cough with angiotensin converting-enzyme inhibitors. *N Z Med J* 1987; **100**:6–7.

7 Town GI, Hallwright GP, Maling TJB, O'Donnell TV. Angiotensin converting enzyme inhibitors and cough. *N Z Med J* 1987; **100**:161–63.

8 Gibson GR. Enalapril-induced cough. *Arch Intern Med* 1989; **149**:2701–3.

9 Strocchi E, Valtancoli G, Ambrosioni E. The incidence of cough during treatment with angiotensin converting enzyme inhibitors. *J Hypertens* 1989; 7[**Suppl** 6]:S308–9.

10 Yeo WW, Maclean D, Richardson PJ, Ramsay LE. Cough and enalapril: assessment by spontaneous reporting and visual analogue scale under double-blind conditions. *Br J Clin Pharmacol* 1991; **31**:356–59.

11 Sebastian JL, McKinney WP, Kaufman J, Young MJ. Angiotensin converting enzyme inhibitors and cough. *Chest* 1991; **99**:36–39.

12 Just PM. The positive association of cough with angiotensin converting enzyme inhibitors. *Pharmacotherapy* 1989; **9**:82–87.

13 Os I, Bratland B, Dahloff B, Gisholt K, Syvertsen JO, Tretli S. Female sex as an important determinant of lisinopril-induced cough. *Lancet* 1992; **339**:372.

14 Bucknall CE, Neilly JB, Carter R, Stevenson RD, Semple PF. Bronchial hyperreactivity in patients who cough after receiving angiotensin converting enzyme inhibitors. *BMJ* 1988; **296**:86–88.

15 Lunde H, Hedner T, Samuelsson O, et al. Dyspnoea, asthma, and bronchospasm in relation to treatment with angiotensin converting inhibitors. *BMJ* 1994; **308**:18–21.

16 Soubrier F, Alhenc-Gelas F, Hubert C, Allegrini J, John M, Tregear G et al. Two putative active centers in human angiotensin 1-converting enzyme revealed by molecular cloning. *PNAS USA* 1988; **85**:9386–390.

17 Choudry NB, Fuller RW, Pride NB. Sensitivity of the human cough reflex: effect of inflammatory mediators prostaglandin E2, bradykinin and histamine. *Am Rev Respir Dis* 1989; **140**:137–41.

18 Ferner RE, Simpson JM, Rawlins MD. Effects of intradermal bradykinin after inhibition of angiotensin converting enzyme. *BMJ* 1987; **294**:1119–120.

19 Ferner RE, Wilson D, Paterson JR, Wilkinson R, Rawlins MD. The effects of intradermal bradykinin are potentiated by angiotensin converting enzyme inhibitors in hypertensive patients. *Br J Clin Pharmacol* 1989; **27**:337–42.

20 Fuller RW, Dixon CMS, Cuss FMC, Barnes PJ. Bradykinin-induced bronchoconstriction in humans. *Am Rev Respir Dis* 1987; **135**: 176–80.

21 Morice AH, Lowry R, Brown MJ, Higgenbottam T. Angiotensin converting enzyme and the cough reflex. *Lancet* 1987; **II**:1116–18.

22 Rigat B, Hubert C, Alhenc-Gelas F, Cambien F, Corvol P, Soubrier F. An insertion/deletion polymorphism in the angiotensin I-converting enzyme gene accounting for half the variance of serum enzyme levels. *J Clin Invest* 1990; **86**:1343–46.

23 Higgins KS, Goodeve AC, Yeo WW, Morice AH, Peake IR. Improved detection of the insertion/deletion polymorphism of the ACE gene, and relationship to serum ACE activity in a normotensive UK population. *Br J Clin Pharmacol* 1992; **34**:151P.
24 Yeo WW, Ramsay LE, Morice AH. ACE inhibitor cough: a genetic link? *Lancet* 1991; **337**:187.
25 Yeo WW, Higgins KS, Morice AH, Jackson PR, Peake IR, Ramsay LE. Investigation of relation between ACE gene polymorphism and ACE inhibitor cough. *Br J Clin Pharmacol* 1993; **35**:66P.
26 Cambien F, Poirier O, Lercef L, Evans A, Cambou J-P, Arveiler D et al. Deletion polymorphism in the gene for angiotensin-converting enzyme is a potent risk factor for myocardial infarction. *Nature* 1992; **359**:641–44.
27 Gilchrist NL, Richards AM, March R, Nicholls MG. Effect of sulindac on angiotensin converting enzyme inhibitor-induced cough: randomised placebo controlled double-blind cross-over study. *J Hum Hypertens* 1989; **3**:451–55.
28 Fuller RW, McEwan JR, Choudry NB. The abnormal cough reflex: role of prostaglandins. *Am Rev Resp Dis* 1989; **139**:A586.
29 Higgins K, Yeo WW, Ramsay LE, Morice AH. Effect of inhaled beclomethasone on ACE inhibitor induced cough. London: British Thoracic Society, 1992.
30 Hedner T, Samuelsson O, Lunde H, Lindholm L, Andren L, Wiholm B-E. Angio-edema in relation to angiotensin converting enzyme inhibitors. *BMJ* 1992; **304**:941–46.
31 Slater EE, Merrill DD, Guess HH et al. Clinical profile of angioedema associated with angiotensin converting-enzyme . inhibition. *JAMA* 1988; **260**:967–70.
32 Williams GH. Converting-enzyme inhibitors in the treatment of hypertension. *N Engl J Med* 1988; **319**:1517–25.
33 Furner RE. Adverse effects of angiotensin-converting-enzyme inhibitors. *Adverse Drug Reaction Bulletin* 1990; **141**:528–31.
34 Consensus II Study Group. Effects of early administration of enalapril on mortality in patients with acute myocardial infarction. Results of the Cooperative New Scandinavian Enalapril Survival Study II (CONSENSUS II). *N Engl J Med* 1992; **327**:678–84.
35 MacFayden RJ, Lees KR, Reid JL. Differences in first dose response to angiotensin converting enzyme inhibition in congestive heart failure: a placebo controlled study. *Br Heart J* 1991; **66**:206–11.
36 Hoefnagels WHL, Thien T. Renal artery occlusion in patients with renovascular hypertension treated with captopril. *Br Med J* 1986; **292**:24–25.
37 Cameron HA, Close CF, Yeo WW, Jackson PR, Ramsay LE. Role of the rapid sequence intravenous urogram in investigating hypertension: diagnostic yield and effect on outcome. *Lancet* 1992; **339**:658–61.
38 Anonymous. Are ACE inhibitors safe in pregnancy? *Lancet* 1989; **2**:482–83.
39 Knott PD, Thorpe SS, Lamont CAR. Congenital renal dysgenesis possibly due to captopril. *Lancet* 1989; **1**:451.
40 Scott AA, Purohit DM. Neonatal renal failure: a complication of maternal antihypertensive therapy. *Am J Obstet Gynecol* 1989; **160**:1223–24.
41 Huralt de Ligney B, Ryckelnynck J, Mintz P, Levy G, Muller G. Captopril therapy in pre-eclampsia. *Nephron* 1987; **46**:329–30.
42 Boutroy MJ, Vert P, Huralt de Ligney B, Milton A. Captopril administration in pregnancy impairs fetal angiotensin converting enzyme activity and neonatal adaptation. *Lancet* 1984; **2**:935–36.
43 Smith AM. Are ACE inhibitors safe in pregnancy? *Lancet* 1989; **2**:750–51.
44 Croog SH, Levine S, Testa MA et al. The effects of antihypertensive therapy on the quality of life. *N Engl J Med* 1986; **314**:1657–64.
45 Fletcher AE, Bulpitt CJ, Hawkins CM et al. Quality of life on antihypertensive therapy: a randomised double blind controlled trial of captopril and atenolol. *J Hypertens* 1990; **8**:463–66.
46 Steiner SS, Friedhoff AJ, Wilson BL, Wecker JR, Santo JP. Antihypertensive therapy and quality of life: a comparison of atenolol, captopril, enalapril and propranolol. *J Hum Hypertens* 1990; **4**:217–25.

47 Croog SH, Kong W, Levine S, Weir MR, Baume RM, Saunders E. Hypertensive Black men and women. Quality of life and effects of antihypertensive medications. *Arch Intern Med* 1990; **150**:1733–41.

48 Palmer AJ, Fletcher AE, Rudge P, Andrews C, Callaghan TS, Bulpitt CJ. Quality of life in hypertensives treated with atenolol or captopril: a double blind cross over trial. Presented at Second International Symposium on ACE Inhibition. London, 1991.

49 Fletcher AE, Bulpitt CJ, Chase D et al. Quality of life on three antihypertensive treatments: cilazapril, atenolol, nifedipine. Presented at Second International Symposium on ACE Inhibition. London, 1991.

50 Herick AL, Waller PC, Bern KE et al. Comparison of enalapril and atenolol in mild to moderate essential hypertension. *Am J Med* 1989; **86**:421–26.

51 Klein WW, Stuhlinger W, Mahr G. Cross-over comparison between captopril and nifedipine. *J Postgrad Med* 1986; **62[Suppl 1]**:108–10.

52 Morlin C, Baglivo H, Boeijinga JK et al. Comparative trial of lisinopril and nifedipine in mild to severe essential hypertension. *J Cardiovasc Pharmacol* 1987; **9[Suppl 3]**:s48–s52.

53 Maclean D, Ramsay LE, Richardson PJ. Enalapril and nifedipine in the treatment of mild to moderate essential hypertension: a six month comparison. *Br J Clin Pharmac* 1990; **30**:203–11.

54 Yeo WW, Foster G, Ramsay LE. Prevalence of cough in patients taking enalapril: controlled study versus nifedipine. *Q J Med* 1991; **293**:763–70.

55 Edwards IR, Coulter DM, Macintosh D. Intestinal effects of captopril. *BMJ* 1992; **304**:359–60.

56 Inman WHW. Enalapril induced deafness. *Lancet* 1987; **1**:872.

SECTION III

THERAPEUTIC PROSPECTS

8

ACE inhibitors, hypertension and cardiovascular structure

JFM Smits, HJMG Nelissen-Vrancken and HAJ Struijker Boudier

Introduction

Since the introduction of captopril in the late seventies, ACE inhibitors have become a major therapeutic tool in the treatment of hypertension. In fact, in many countries they are now regarded as part of the first-step approach to treating elevated blood pressure. In spite of their wide use in hypertension, the mode of antihypertensive action of ACE inhibitors remains largely unknown. It is now widely accepted that the inhibition of angiotensin I converting enzyme is a predominant molecular mechanism of action of captopril and related drugs. However, the inhibition of this enzyme does not only affect the conversion of angiotensin I to angiotensin II in the circulation. It also affects local, tissue-bound forms of angiotensin I conversion and may even influence the metabolism of other vasoactive peptides, such as bradykinin.

The molecular mode of action of ACE inhibitors is discussed extensively in other chapters in this book. The purpose of this chapter is to review how ACE inhibition affects organ and tissue function relevant to the control of blood pressure. We shall use a systems analysis approach, like that previously used for analyzing the mode of antihypertensive action of diuretics,[1] beta-blockers,[2] and calcium antagonists.[3] According to this approach the ultimate blood pressure lowering effect of a drug is the result of its primary pharmacological effect on different tissues involved in the control of blood pressure and the reactions of blood pressure control mechanisms to these initial perturbations. Since the structure of the heart and blood vessels may represent a very powerful and ACE-inhibitor-sensitive long-term cardiovascular control mechanism, we shall pay special attention to the role of cardiovascular structure.

This analysis is based upon literature published before August 1992. We have tried to focus as much as possible on data derived from studies in humans. However, research on structural control mechanisms has thus far been performed predominantly in experimental animals, or even in isolated tissues. Caution should be exercised in the interpretation of observations based on studies using one ACE inhibitor, sometimes administered in only one dose. From a chemical point of view, ACE inhibitors are a markedly heterogeneous group of substances. It is not inconceivable that this heterogeneity is extended to their effects on tissue and organ function.

ACE inhibitors: effects on vascular structure and function in hypertension

Hypertension is associated with several changes in structure and function of the vascular system. The nature of these changes depends on the observed level within the vascular tree, the model of hypertension investigated, and the phase of development of the hypertensive disease process.[4,5] Table 8.1 summarizes the most important structural alterations in blood vessels and their functional consequences in established hypertension.

Large artery changes in hypertension

Large arteries contribute little to resistance to flow; their major role is that they determine the conduit and compliance properties of the arterial system. The visco-elastic properties of the arterial wall determine the distensibility of the large arteries. These properties are usually investigated by studying the pressure–volume relationship. The change in volume (dV) divided by the change in pressure (dP) represents the arterial compliance. The pressure–volume relation is curvilinear, because of the non-linear behaviour of the major constituents of the arterial wall, primarily arterial smooth muscle cells, elastin and collagen fibres. Thus, arterial compliance can only be defined for a given pressure.

Compliance changes can be the mechanical consequence of the alteration in blood pressure or can reflect intrinsic changes in the structure of the arterial wall. It has now been established that in essential hypertension, compliance changes in the large arteries are at least partly due to intrinsic changes in arterial wall properties.[6,7] Non-invasive Doppler measurements show that both in young and older hypertensive patients, large artery diameters are not decreased, or are even slightly increased in comparison with normotensive age-matched controls. Still, the pressure–diameter curves are shifted towards higher pressure levels, usually with a decreased slope. This suggests a decreased arterial compliance on the basis of a change in intrinsic properties of the arterial vessel wall. A reduced compliance has been observed for the whole systemic circulation in hypertension as well as for individual large arteries, namely the aorta, the brachial artery, the common carotid artery, the femoral artery and the forearm arterial circulation.[6–11] Studies in experimental animals have provided more clarity about the mechanisms of the large artery compliance changes in hypertension. In particular, the elegant in situ studies by Levy and co-workers have defined the nature of the compliance changes in large arteries in various animal models of hypertension.[12,13] On the basis of this experimental work the following factors were deemed to be of major importance in the stiffer arterial wall in hypertension:

- increased smooth muscle mass;
- altered extracellular matrix, that is, the amount and geometrical arrangement of collagen and elastin content;
- and an increased vasomotor tone.

With respect to the latter, Levy et al. showed an important contribution of an ACE-sensitive endothelium-derived mechanism in hypertension. Since the endothelial lining is very thin it was, until recently, thought not to contribute to the mechanical properties of blood vessels. However, the endothelium may contribute importantly to the structure and function of arteries as a source of vasoactive and growth-affecting substances.[4]

Whatever its underlying mechanisms, decreased large artery compliance represents an

Site	Change	Consequence
Large arteries	VSMC hypertrophy	
		distensibility ↓
	ECM composition	
		compliance ↓
Small arteries	VSMC hyperplasia & hypertrophy	
		distensibility↓ compliance ↓
	ECM composition	
		resistance ↑
	decreased diameter	
Microcirculation	arteriolar constriction	
		resistance ↑
	arteriolar & capillary rarefaction decreased diameter	

VSCM = vascular smooth muscle cell
ECM = extracellular matrix

Table 8.1
Major structural alterations in the arterial system in hypertension and their functional consequences

important cardiovascular risk factor in hypertension.[14] A reduction in large artery compliance simultaneously increases systolic pressure and decreases diastolic pressure, thus causing an increase in pulse pressure. It has recently been suggested that increased pulse pressure influences cardiovascular mortality and morbidity independently of the mean arterial pressure level.[14–16]

ACE inhibitors and large arteries

The study of drug-induced changes in arterial compliance is complicated by the pressure dependence of arterial compliance. Thus, it has to be established to what degree drug-induced compliance changes are pressure-dependent or reflect intrinsic modifications of the arterial wall. Over the past decade, the non-invasive Doppler ultrasound approach, especially as used by Safar's group, has allowed the comparative investigation of the effects of different antihypertensive drugs on large artery mechanics in hypertensive patients.[14,15,17] These studies show that pressure changes alone do not explain the effects of antihypertensive drugs on arterial compliance. As a drug class, ACE inhibitors are remarkedly effective in increasing arterial compliance, both during acute and long-term administration. In healthy volunteers, increasing doses of the ACE inhibitor perindopril augmented the diameter of different large arteries, while at the same time increasing their compliance.[18,19] In hypertensive patients, similar effects were observed following acute and long-term administration of captopril,[17,20] enalapril,[21] and perindopril.[22,23] Four weeks after stopping long-term perindopril treatment, blood pressure and arterial compliance returned to pretreatment values, suggesting the reversibility of the ACE-inhibitor-induced large artery changes.[22]

Definitive proof of the pressure independence of ACE-inhibitor-induced compliance changes came from a study in spontaneously hypertensive rats (SHR) which showed that local incubation of in situ isolated carotid arteries with lisinopril or saralasin increased arterial compliance, independent of arterial pressure changes.[13,24] The mechanisms underlying ACE-inhibitor-induced large artery compliance increase have been studied in detail in animal models of hypertension. At present the following mechanisms are regarded as important contributors.

ACE inhibitors affect large artery smooth muscle tone

Angiotensin II is a potent vasoconstrictor of large artery smooth muscle, either by a direct effect or by facilitating the release of noradrenaline. Blockade of this effect by angiotensin II-receptor antagonism or angiotensin II formation may thus lead to a decrease of large artery smooth muscle tone. Experimental evidence for such an effect has been found in isolated precontracted rabbit aorta rings[25] and in situ carotid arteries of SHR,[13,24] although other authors could not show direct effects of ACE inhibitors on resting tone of large arteries.[26] Recent work indicates a potentially important role for the endothelium in ACE-inhibitor-induced decrease of large artery smooth muscle tone.[13,24,27,28] Lisinopril increased compliance of the in situ isolated carotid artery in the presence of an intact endothelium but did not alter it after endothelium removal in SHR.[13,24] In addition both perindoprilat[27] and cilazapril[29] improved endothelium-dependent arterial dilatation in SHR. In hypertensive patients captopril improved endothelium-dependent vasodilatation.[28] There is still uncertainty about the mechanism of this effect: specifically, it still has to be determined whether the endothelium-dependent effect is caused by an interference with a local renin-angiotensin system in the endothelium.

ACE inhibitors reduce vascular smooth muscle mass

Freslon and Giudicelli were the first to report a decrease in large artery hypertrophy during chronic captopril treatment in SHR.[30] Later studies have confirmed this observation for other ACE inhibitors in different models of hypertension.[31] ACE inhibitors can both attenuate large artery hypertrophy during the development of spontaneous or renal forms of hypertension and diminish existing hypertrophy in established forms of experimental hypertension.[12,30,32–36] This suggests that they interfere with a fundamental mechanism of vascular smooth muscle cell growth. In this respect effects on oncogene expression, mitogen synthesis or second messengers involved in cell growth and proliferation have been proposed.[37,38] Inhibition of angiotensin II-induced activation of intracellular phosphoinositol pathways could be the key event in ACE inhibitor associated inhibition of arterial hypertrophy.[38]

ACE inhibitors influence the composition and geometry of the extracellular matrix

Levy et al. showed that chronic perindopril treatment of SHR not only decreases blood pressure and aorta wall hypertrophy, but also reduces aorta collagen content.[2,36] A reduction in collagen content was not observed during perindopril treatment of 2-kidney–1-clip renal (2K–1C) hypertensive rats, but this may have been due to the short duration of treatment.[12] Recently, these authors showed that long-term treatment of SHR with non-hypotensive doses of quinapril reduces aorta collagen content (Levy, personal communication). More information on the in vivo effects of ACE inhibitors on the large artery extracellular matrix would be of interest in view of the potential effects of angiotensin II on extracellular matrix composition.[39]

Small artery changes in hypertension

The smaller arteries (diameter less than 250 µm) are major sites of resistance control. Together with the arterioles (diameter 5–125 µm) they determine basal vascular resistance and are primary sites of the resistance increase in hypertension.[4,5] The major mechanisms of resistance increase at the small arteriolar level are decreased internal diameter and increased wall/lumen ratio. Both mechanisms contribute to the development and maintenance of the resistance increase in hypertension, the relative roles depending on the model and tissue investigated and the stage of development of hypertension.[5] There are surprisingly few direct observations in vivo or in situ on small artery structure and function in hypertension. Studies in vitro using myograph-based mechanical analyses in combination with histomorphological measurements suggest an important role for lumen narrowing in early stages of most models of hypertension with an increasing importance of vessel wall hypertrophy during later stages.[5] The molecular basis of these changes is currently the subject of intense research. Primary defects in the cellular calcium-mediated control of vascular tone may play a fundamental role in the early diameter decrease.[5] Vascular smooth muscle cell calcium metabolism is controlled by several membrane-bound receptors, including the AT_1 receptor.[40] With respect to changes in wall/lumen ratio both mechanical — increased pressure and wall stress — and chemical mediators have been implicated.[4,41] The latter would act through mechanisms of vascular smooth muscle cell growth and proliferation. Angiotensin II is one of these mediators, as discussed above.

ACE inhibitors and small arteries

ACE inhibitors in general lack a direct effect on tone or diameter of small arteries.[42,43] Yet they lower blood pressure by reducing vascular resistance. At least two factors explain this apparent discrepancy:

- the acute vasodilator effect of ACE inhibitors is limited to specific vascular beds;
- and the long-term effects of ACE inhibitors are based on their actions on vessel wall structure rather than tone or diameter.

The evidence for a restricted acute vasocilator effect of ACE inhibitors will be discussed in detail later in this chapter. ACE inhibitors have a preferential acute vasodilator effect on the renal vasculature. This may be due to the relatively high ACE concentration in the kidney[44] and the high sensitivity of the renal vasculature to the constrictor effects of angiotensin II.[45] Angiotensin II has no or only little contractile activity on resistance arteries of several other vascular beds, for example bovine retinal arteries,[46] rat mesenteric arteries,[47] or rabbit cerebral arteries.[48] The mechanisms underlying regional heterogeneity of angiotensin II induced vasoconstriction are unknown, but may include differences in AT receptor density,[46] differential interference by angiotensin II with noradrenaline release from sympathetic nerve terminals,[49] or regional differences in angiotensin II mediated effects on the release of endothelium dependent vasoactive substances.[48,50]

The major long-term effect of ACE inhibitors on small arteries in hypertension is their action on the structure of these vessels. Chronic ACE inhibitor treatment of SHR reduces the cross sectional wall area of various types of small arteries, including small mesenteric arteries (using captopril,[30,33,51] perindopril,[32,51] and cilazapril[35]), cerebral arteries (using cilazapril[52,53]) and coronary arteries (using cilazapril[35]). Within vascular beds there may be heterogeneity in the response to ACE inhibitors: the effect of cilazapril was more marked in large coronary arteries and arterioles than in the smaller vessels.[35] Captopril prevented hypertrophy of SHR superior and large mesenteric arteries and reduced the number of smooth muscle cell layers to the same number as in WKY, wheras in SHR small mesenteric arteries, cross sectional wall area reduction was not associated with an effect on the number of smooth muscle cell layers.[33] Previous observations suggest that large artery hypertrophy in SHR is mainly due to enlargement of the medial smooth muscle cells, secondary to the development of hypertension, whereas in smaller arteries hyperplasia of these cells seems a primary, pressure-unrelated phenomenon.[41] In addition, large and small arteries may differ in their relative content of growth-prone and growth-insensitive vascular smooth muscle cells.[54] Altogether, these data suggest that part of the ACE-inhibitor-induced reduction of vascular hypertrophy is related to the arterial pressure reduction and another part is due to a local effect, for example inhibition of vascular angiotensin II formation or local sympatho-inhibition.

Mulvany and co-workers[5,55] have recently challenged the importance of regression of small artery hypertrophy as a mechanism of chronic blood pressure reduction. First, in studies on isolated small arteries of hypertensive patients who were treated effectively with different antihypertensive drugs they could not find a complete regression of the increased wall/lumen ratio.[56] Secondly, the changes in media/lumen ratio seem to follow rather than precede changes in blood pressure during the development or treatment of hypertension.[5,57] Finally, in SHR treated chronically with different types of antihypertensives, withdrawal of treatment induced a rise in blood pressure, the rate of which was determined more by the drug type than by the effects on vascular structure.[51] With respect to the latter phenomenon, ACE inhibitors may be exceptional drugs. Both captopril[30,51] and perindopril[32,34] have persistent effects on blood pressure after withdrawal in SHR. With perindopril, even a short treatment period of four weeks during the early phase of rapid pressure rise in young SHR caused a persistent blood pressure reduction up to 30 weeks after treatment had been stopped.[58] Interestingly, co-infusion of angiotensin II with perindopril treatment during the four-week period completely reversed the antihypertensive effect of the ACE inhibitor during and after treatment. This suggests that local trophic actions of angiotensin II, in addition to adaptive responses to blood pressure changes, determine small artery structure and resistance.

The microcirculation in hypertension

The microcirculation is an important site of resistance control in hypertension.[4,59,60] At the level of the small arterioles and capillaries three mechanisms determine vascular resistance: blood viscosity; the number of small vessels, and the arteriolar diameter. Recent reviews discuss the relative importance of these factors in the development and maintenance of hypertension.[59,60] Intravital microscopic observations indicate that in the early stages of human essential hypertension and genetic models of hypertension in rats, a decrease in the number of small vessels ('rarefaction') is an important mechanism of resistance increase. Rarefaction may be due to the actual disappearance of existing vessels[59] or to a decreased angiogenic potential of various microvascular beds in hypertensive species.[60] The mechanisms underlying these phenomena are still largely unknown. Microvascular rarefaction has been interpreted as an adaptation to increased tissue oxygen supply.[59,61] Alternatively, the growth of small vessels might be controlled by angiogenic substances, one of which is angiotensin II.[62,63]

Small arteriolar diameter decrease is not generally observed in early stages of human essential hypertension or genetic animal models of hypertension.[60] However, it is a key phenomenon in later stages of these forms of hypertension. Moreover, renal and endocrine animal models of hypertension are usually characterized by early constrictions of the small arterioles.[59,60] It is not yet certain to what degree angiotensin II contributes to small arteriolar vasoconstriction in hypertension. A study by Vicaut et al.[64] in normotensive rat cremaster muscle shows that angiotensin II is able to constrict small arterioles.

ACE inhibitors and the microcirculation

In view of the factors discussed above, an analysis of ACE inhibitor associated microvascular changes should focus on blood viscosity, arteriolar diameter and the number of small vessels. The first of these factors has not yet been the subject of systematic investigation. With respect to arteriolar diameter changes, ACE inhibitors show a surprising lack of effect in the tissues investigated thus far. Striated muscle small arterioles of conscious SHR are not dilated by acute administration of captopril or enalaprilat, as opposed to the strong dilatations exerted by various calcium antagonists, α-blockers or potassium channel openers.[65–68] Similarly, long-term captopril treatment does not dilate small arterioles in the cremaster muscle of normotensive or renal hypertensive rats.[69] These data suggest that angiotensin II may not contribute significantly to microvascular tone.

The final aspect of the vascular actions of ACE inhibitors concerns their effects on microvascular growth. The angiogenic potential of angiotensin II implies that inhibition of angiotensin II formation is associated with a decreased microvascular growth. Indeed, Wang and Prewitt showed that in renal hypertensive as well as normotensive rats, chronic captopril treatment caused a reduction in microvascular density.[69] Recent studies show comparable results with chronic captopril treatment in a rat model of neovascularization following ischaemia of a hindlimb.[70] The therapeutic consequences of a depressed neovascularization during ACE inhibition are unknown. They may be favourable in certain circumstances, like diabetic retinopathy, but unfavourable in others, like ischaemic heart diseases or peripheral ischaemia.

ACE inhibitors: effects on cardiac structure and function in hypertension

The cardiac muscle is composed of several different cell types. Cardiomyocytes represent approximately 30% of the total cell population. In view of their size, however, they occupy approximately three-quarters of the structural space, and thereby are the cells that contribute most to weight and volume changes. Of the remaining extracellular space, approximately 60% is occupied by vessels, leaving 10% of cardiac structural space for macrophages, fibroblasts and structural proteins such as collagens.[71] This implies that the 'clinical' indicators for cardiac hypertrophy that are normally used (wall thickness and/or mass) relate to cardiomyocyte mass and that the other components contribute relatively little to these measures.

ACE effects on cardiac hypertrophy

Cardiac hypertrophy is a frequent finding in pressure overload. In animal experiments, hypertrophy is usually defined on the basis of cardiac weight or thickening of the ventricular wall. Clinically, indications of hypertrophy may be derived from electrocardiographic changes or from echocardiography, whereby the former detects only approximately 50% of the cases detected with the latter methodology. It is now assumed that left ventricular hypertrophy is present in approximately 20% of patients with mild hypertension, whereas the incidence in severe hypertension is more than 50%.[72] Although hypertrophy has long been regarded a physiological response to the increased cardiac load (which is true to a certain extent), epidemiological surveys have revealed that it is to be regarded as a major independent risk factor for cardiovascular morbidity and mortality.[73,74] In fact, reduction of left ventricular hypertrophy has been suggested as one of the therapeutic goals of antihypertensive therapy.[75,76]

There has been a long-standing debate about whether reduction of blood pressure is a sufficient measure for reduction of left ventricular hypertrophy. It is clear that imposition of increased afterload to the heart triggers increased protein synthesis, evidenced from increased mRNA synthesis as early as 20 minutes after increase of afterload in isolated hearts.[77] Similarly, increased load on isolated ventricular strips results in cardiomyocyte hypertrophy.[78] Also, the opposite has long been recognized, that is that the heart is capable of reducing its weight following unloading. A dramatic demonstration of that phenomenon was described by Korecky and Rakusan, who transplanted hearts heterotopically into rats in such a way that they were perfused but not working.[79] In these animals, a 60% weight reduction of the heterotopic heart was observed.

Confusion arose from studies employing drugs to reduce blood pressure (Table 8.2). Both in animals[30,80,81] and in hypertensive patients[82,83] blood pressure reduction with arteriolar vasodilators like hydralazine and minoxidil were ineffective with respect to reduction of hypertrophy. A similar discrepancy was observed for diuretics.[84,85] In studies comparing the effectiveness of antihypertensive and antihypertrophic actions of the whole range of antihypertensive drugs in animals[86] and humans,[87] the concept that load-reduction per se is not enough for consistent reduction of hypertrophy was further substantiated. Although Devereux and co-workers did observe a significant relationship between reduction of blood pressure and change in cardiac mass, they also noted inconsistent findings with diuretics and arteriolar vasodilators.[88] A recent meta-analysis of 109 studies between 1977 and 1991 has provided further evidence for such discrepant effects of therapy.[89] From that analysis it was evident that blood pressure reduction with β-blockers, calcium antagonists, ACE inhibitors and diuretics all resulted in regression of hypertrophy. However, in contrast to the others which

Drug class	Cardiac hypertrophy	Vascular hypertrophy
Centrally acting sympatholytics	↓	→ ↓
Neurone blockers	→	→
Alpha-blockers	→	↓
Beta-blockers	↓	→
Arteriolar vasodilators	↑	→ ↓
Diuretics	→ ↓	→ ↓
Calcium antagonists	↓	↓
ACE inhibitors	↓ ↓	↓ ↓

Table 8.2
Effect of antihypertensive treatment on cardiac and vascular hypertrophy: studies comprise clinical as well as experimental studies. Adapted from Smits and Struijker Boudier.[253]

all reverse wall hypertrophy, diuretics reduce left ventricular mass through reduction of left ventricular diameter. In the context of this chapter, however, the most important finding was the fact that ACE inhibitors were more effective in reducing hypertrophy than any of the other treatments. This supports the concept that ACE inhibition stands out from other therapeutic options in this respect.

The phenomena of reduction and regression of hypertrophy have been studied more extensively in experimental models of hypertension. ACE inhibition with a variety of ACE inhibitors prevents the development of left ventricular or cardiac hypertrophy in renovascular,[90,91] neurogenic,[92] and spontaneous (SHR) hypertension.[58,93–95] Furthermore, ACE inhibitors reduce cardiac weight in models for experimental hypertension with established hypertrophy (in SHR[30,94–97] and in renovascular hypertension[91,97,98]). ACE inhibition was ineffective in DOCA-salt hypertension.[97] With respect to the coincidence of blood pressure reduction (which in turn was correlated to the degree of activation of the renin-angiotensin system) and reduction of hypertrophy, Michel concluded that hypertrophy regressed during ACE inhibition only if blood pressure was reduced through inhibition of an activated renin-angiotensin system.[97] In two other models, however, the results are quite disparate. In their renovascular hypertensive rats with established hypertension and hypertrophy, Linz and co-workers observed that a very low dose of ramipril, which did not affect plasma angiotensin II levels, nor blood pressure, could completely reverse hypertrophy.[91] This clearly suggests dissociation of the effects. Such dissociation was also observed in hypertrophy caused by supravalvular aortic stenosis.[99,100] In this model, only ventricular intracavitary pressure and not arterial pressure is increased. Quinapril, which did not affect arterial pressure, effectively regressed hypertrophy. These observations by Linz et al.[91] and Kromer et al,[100] indicate interference of ACE inhibitors with a local mechanism. This may be related to the observed increase in cardiac ACE mRNA expression as was observed in pressure-overloaded hypertrophic ventricles of rats.[101] However, quinapril administration at the time of aortic banding did not prevent the normal hypertrophic response.[100] The authors suggest that, although load is the primary trigger for hypertrophy, mechanisms that can be blocked with ACE inhibitors maintain the hypertrophic state of the heart. An alternative explanation might be that, since the aorta was banded at a very young age in these rats, the blockade of the hypertrophic response with ACE inhibitors might have been replaced by a hyperplastic response.[100] Nevertheless, both clinical and experimental data suggest that the reduction of cardiac mass by ACE inhibitors goes beyond, and is, at least partly, independent of their blood pressure reducing properties.

Qualitative effects on myocytes

If one considers the 'overall' parameter of cardiac weight or size as an indicator for cardiac structural changes, several omissions are made. These relate to the cellular composition of the heart as well as to qualitative changes (as opposed to quantitative changes) that occur in the cardiomyocyte in the process of hypertrophy. In rats, hypertrophic cardiomyocytes express a greater proportion of the fetal (slow) myosin isoform V_3 than of the faster adult isoform V_1.[102] ACE inhibitors have been shown to reverse this transformation in spontaneous and in renovascular hypertension[97] and myocardial infarction,[103] but not in DOCA-salt hypertension.[97] As discussed before, there was also no reduction of hypertrophy in the latter model. Although direct proof is lacking, a reversal of the isomyosin shift may contribute to better cardiac contraction.[97]

Cardiac extracellular matrix

As pointed out above, fibroblasts and the associated collagenous extracellular matrix form another important constituent of the cardiac interstitial space. The extracellular matrix, which is synthesized by the fibroblasts, has long been considered an inert framework to suspend cardiomyocytes. Now it is well established that abnormal concentrations of collagen I contribute significantly to diastolic and systolic dysfunction of the hypertrophied left ventricle.

Increased interstitial collagen has been observed in the hearts of adult humans with aortic stenosis[104,105] and in patients with essential hypertension.[106–108] The modulation of the dysbalance between collagen synthesis and degradation has been investigated extensively in experimental hypertension.[71] Increased interstitial collagen deposition has been reported in SHR,[97,109,110] renovascular hypertension,[111,112] and hypertension induced by aldosterone infusion,[113] but not in hypertension induced by infrarenal banding of the aorta.[110] This phenomenon was also observed in another model of ventricular overload, myocardial infarction in rats.[103,114,115] Several lines of evidence from these studies indicate involvement of the renin-angiotensin-aldosterone system. First of all, in hypertension from infrarenal aorta banding, hypertension and left ventricular hypertrophy were observed without any interstitial fibrosis.[110] These animals exhibited normal angiotensin II and aldosterone levels in plasma. Secondly, fibrosis could be prevented in renovascular hypertension[112] and myocardial infarction,[103,114] when animals were treated with ACE inhibitors. Similarly, long-term (12 week) treatment of rats with established genetic hypertension results in regression of fibrosis.[97,109,110] Short-term (4 week) captopril treatment actually increased collagen concentration;[111] this may, however, depend upon the long half-life of cross-linked collagen as opposed to a rapid regression of myocyte hypertrophy.[71] It is, at this stage, not clear whether angiotensin II, aldosterone or both act in concert to increase collagen. Studies by Weber and Brilla show that the aldosterone-antagonist spironolactone is capable of selectively inhibiting fibrosis in the hearts of hypertensive animals.[71] Also the fact that aldosterone infusion (which reduces angiotensin II levels in plasma) induces fibrosis, suggests that aldosterone is the key mediator of fibrosis.[71,113] However, early observations in a murine fibroblast cell line indicate that angiotensin II may induce fibroblast proliferation.[116]

Studies with ACE inhibitors[109,110,112,114] and AT_1-antagonists[115] which all reduce or prevent collagen deposition are not conclusive because these treatments are supposed to either reduce circulating angiotensin II or block its effects and consequently reduce aldosterone. In myocardial infarction where perindopril was shown to prevent fibrosis, however, no reduction of aldosterone was observed,[103] suggesting an important direct role for angiotensin II. The key to this dilemma may be in a recent study where aldosterone was shown to both upregulate angiotensin II receptor numbers and to prevent desensitization in cultured rat vascular smooth muscle cells.[117] Thus, aldosterone might act to potentiate direct effects of angiotensin II. This subject needs further exploration.

The most important consequence of reduction of fibrosis might be the reduction of cardiac stiffness and ensuing ventricular dysfunction. Such an effect was observed in SHR treated with lisinopril[109,110] and renovascular hypertension following captopril.[112] In one of these studies[109] low dose lisinopril was shown to regress fibrosis without effects on blood pressure or hypertrophy. This indicates that careful dosing may selectively inhibit the process that causes myocardial stiffness without affecting the muscular hypertrophy needed to meet with the increased load.

Coronary circulation

During the process of hypertensive myocyte hypertrophy, the coronary circulation undergoes changes. The effect on arterioles is similar to that on the peripheral circulation — an increase in medial thickness of intramural resistance arterioles.[107] These structural changes affect minimal coronary vascular resistance. Obviously neurohumoral mechanisms, including the sympathetic nervous system and the RAS, also have their effect on the coronary vasculature.

Lisinopril at antihypertensive doses effectively increased coronary vascular reserve[109,110] and decreased media thickness of coronary arteries. In contrast, a lower dose that did not affect blood pressure, although it was capable of reducing perivascular fibrosis, did not have an effect on coronary vascular reserve.[109] This suggests that normalization of blood pressure is the key to normalization of coronary vascular reserve in these isolated hearts. In contrast to these in vitro studies, in vivo, neurohumoral mechanisms may contribute to coronary vascular tone. In hypertensive patients, angiotensin II has been suggested to contribute directly to coronary vascular tone since the increase in coronary resistance as observed following furosemide could be blocked with captopril.[118] Similar observations were made in normotensive rats following hydrochlorothiazide and captopril.[118] It is, however, not clear whether captopril has similar effects in conditions of non-stimulated RAS since in this study captopril did not affect coronary vascular resistance in patients receiving placebo instead of furosemide.

As was argued above, ACE inhibition may modulate sympathetic vascular control in a number of tissues. Recently, this phenomenon was investigated at the level of the coronary circulation in patients with coronary heart disease.[119] The authors reported that a single dose of captopril was capable of reducing the coronary vasoconstriction in response to sympathetic stimulation, probably as a consequence of removal of the sympathetic modulatory effects of angiotensin II. Thus, with respect to coronary vascular resistance ACE inhibitors may favourably affect neurohumoral control of vascular tone, besides their effect on the structure of the coronary vessels; the latter effect seems to be related to their overall blood pressure lowering properties.

In general, hypertensive left ventricular hypertrophy is associated with decreased capillary density,[120–122] although in early renovascular hypertension in rats an unchanged density was observed.[123] Although there is no information on the effect of ACE inhibition on cardiac capillary density in hypertension, there is some indirect information from another model of cardiac pressure overload — the model of myocardial infarction in the rat. There it was noted that the hypertrophy of the viable cardiac muscle is associated with a transient increase in interstitial cell proliferation in the first three weeks after infarction.[114] ACE inhibition during this phase completely blocked this proliferative response.[114] These proliferating cells encompass, amongst others, endothelial cells, suggesting an early capillary growth which depends upon an activated RAS. The possible role of the RAS in this type of neovascularization is discussed elsewhere in this chapter. It should be borne in mind, however, that ACE inhibitors would not be expected to block the proliferative response in the myocardium with a quiescent microcirculatory architecture.

A recent study in SHR underlines the importance of regression of hypertrophy in relation to coronary perfusion.[124] These authors reported that treatment with captopril abolished cardiomyocyte vulnerability to hypoxia. Hydralazine, although equally effective in reducing blood pressure, did not affect hypertrophy or myocyte vulnerability. This may form the basis for the observed antiarrhythmic effects of ACE inhibitors. It remains to be established whether other antihypertensives that reduce left ventricular hypertrophy have similar effects on cardiomyocytes.

The role of circulating and local RAS systems in the cardiovascular effects of ACE inhibitors in hypertension

Soon after their introduction it was shown that there is a lack of correlation between the antihypertensive response of ACE inhibitors and the effect on plasma ACE activity.[125] The involvement of local renin-angiotensin systems in the regulation of blood pressure was subsequently demonstrated by a correlation of the antihypertensive effect of ACE inhibitors and the inhibition of local ACE activity in aorta, renal vascular bed and mesenteric vascular bed.[126–128] Furthermore, the local renin-angiotensin system was found to be changed in spontaneously hypertensive rats,[129,130] whereas in renal hypertension (2-kidney–1-clip animal model) elevated local ACE activity instead of plasma renin and ACE activity seems to be responsible for the maintenance of chronic hypertension.[131,132] Additionally, in pressure overload left ventricle hypertrophy,[101] in myocardial infarction[133] and in hypoxia in the lung[134] local ACE is activated and may be involved in the adaptive mechanisms.

Local renin-angiotensin systems

Angiotensinogen and renin

Angiotensinogen is the precursor of the angiotensin peptides and the only known naturally occurring renin substrate. The main product of the reaction between angiotensinogen and the enzyme renin is angiotensin I, which can be converted into angiotensin II by ACE. Synthesis of angiotensinogen occurs predominantly in hepatocytes. Furthermore, synthesis of angiotensinogen has also been proposed in a diversity of tissues, as angiotensinogen mRNA and/or angiotensinogen has been demonstrated in cardiac tissue, vessel wall, adipose tissue, kidney, brain, adrenal gland, lung and testes.[135–140]

Angiotensinogen production can be influenced by angiotensin II. Angiotensin II inhibited the angiotensinogen production in a human hepatoma-derived cell line,[141] whereas an elevated plasma angiotensin II concentration correlated with an increased hepatic angiotensinogen mRNA in rats.[142,143] Angiotensinogen mRNA increases also with a low sodium diet (renin-dependent,[140] and by glucocorticoids and sex steroids, whereas insulin and ACE inhibitors decrease angiotensinogen mRNA.[139,142,144]

Renin derived from the juxtaglomerular cells of the kidney is the main source of circulating renin. In multiple tissues, renin derives from local synthesis and indirectly from kidney by uptake from plasma.[145] This latter has been demonstrated by the lack of angiotensin I and II production in the absence of renin in isolated perfused hearts[138] and the increase of angiotensin I and II production in the presence of renin in perfused hindlimbs.[146,147] Local renin synthesis has been suggested by renin gene expression and renin synthesis in a variety of tissues, such as arterial endothelial and smooth muscle cells,[148] adrenal gland,[149] brain and testes.[150,151] Although in isolated perfused hearts no angiotensin I and II production could be demonstrated in the absence of renin,[138] renin gene expression has been demonstrated in cardiomyocytes of atria and ventricles of the heart.[150,151] Renin release from the kidney is regulated by chloride and sodium ions in the juxtaglomerular cells of the kidney. Additionally, the renin mRNA level can be increased by sodium depletion, ACE inhibition and bradykinin, and decreased by angiotensin II and vasopressin.[143,144,151] Local regulation of the expression of renin mRNA, however, is different among tissues.[139]

The majority of circulating renin is present in the inactive form prorenin.[150,152] Prorenin can be converted into renin by proteases like trypsin, plasmin, kallikrein and cathepsin B.[153] As for renin, the kidney is mainly responsible for the production of blood-borne prorenin. Furthermore, local synthesis of prorenin has been proposed in eyes,[154] reproductive tissues and the adrenal glands.[149,152,155] In these tissues prorenin can exist in high concentrations and has probably intrinsic catalytic activity (without conversion into renin).[152] Plasma levels of prorenin can vary in physiological (menstrual cycle and pregnancy) and pathophysiological conditions (diabetes mellitus and renin-secreting tumours).[153]

Angiotensin converting enzyme

ACE is mainly responsible for the conversion of angiotensin I into angiotensin II as well as for the degradation of the vasodilator bradykinin. It can also cleave other peptide substrates like enkephalins, substance P, neurotensin and the luteinizing hormone-releasing hormone.[156,157] This implies that inhibition of ACE prevents the degeneration of angiotensin I, of bradykinin and also of other peptides. However, the physiological relevance of the cleavage of these latter peptides by ACE is not yet known.

Circulating ACE is produced predominantly in endothelial cells of the lung. ACE has, however, been demonstrated in endothelial and vascular smooth muscle cells of blood vessels.[158–160] ACE is also present in cardiac tissue,[159,161,162] intestine,[163] kidney,[164,165] and brain,[166–168] and is probably localized in the vasculature of these tissues. In testes[169,170] and brain,[168] different isozymes of ACE with similar enzymatic properties are present. Other important angiotensin II generating enzymes are a chymostatin-sensitive angiotensin II generating enzyme (CAGE), kallikrein, tonin, cathepsin G and trypsin.[171] Kallikrein has been demonstrated to play an important role in the generation of angiotensin II in pancreas[172] and heart.[173] In humans, dogs and monkeys, CAGE has been localized in lung, vascular tissues,[158,174,175] and heart.[176] In blood vessels, CAGE is predominantly localized in the adventitia, whereas ACE is mainly in the endothelial cells.[158] This may suggest a distinct functional role for these two enzymes. The alternative angiotensin II generating enzymes may become more important than ACE in cells with an acidic pH due to ischaemia[173] and during inhibition of ACE.

Circulating and tissue ACE activity can be inhibited by ACE inhibitors. Furthermore, the conversion of angiotensin I into angiotensin II in pulmonary artery endothelial cells can be suppressed by atrial natriuretic peptide[177] and bradykinin,[178] whereas platelet-activating factor[173] and aldosterone[179] stimulate ACE activity. In contrast to angiotensinogen and renin, tissue ACE activity is not influenced by salt diets.[180] The different isozyme of ACE in testes is under hormonal control, as it is absent in immature rats and develops with puberty.[156]

Angiotensin II

The generation of angiotensin II has been demonstrated in a variety of tissues, such as heart,[138,173,176] pancreas,[181] brain,[129] kidney,[182] adrenal gland[183] and endothelial and smooth muscle cells of blood vessels.[184–187] Local angiotensin II can exert both autocrine and paracrine influences on local tissue function. Additionally, the locally synthesized angiotensin II can contribute to the amount of circulating angiotensin II, as demonstrated in the vascular bed of rat mesentery and hindlimb,[188] in human umbilical vein[189] and in the vascular bed of human kidney.[190]

Both circulating and local angiotensin II have been proposed to play an important role in long-term blood pressure and flow regulation. Angiotensin II in the vasculature can control blood flow in the individual tissues, whereas angiotensin II in kidney and larger arteries seems to be involved in blood pressure control, as demonstrated by the relation between inhibition of local ACE and blood pressure reduction by ACE inhibitors.[126–128] Besides vasoconstriction through stimulation of angiotensin II receptors, angiotensin II probably also increases arterial blood pressure by activation of the sympathetic nervous system or by potentiation of the circulatory responses to sympathetic stimulation.[191] Furthermore, it regulates blood pressure by cardiac effects and by stimulation of aldosterone secretion in the adrenals and sodium and water retention in the kidney. These aspects are discussed later in this chapter.

Angiotensin II receptors

Angiotensin II can exert its effects by binding to an angiotensin II receptor. With the discovery of potent, highly selective nonpeptide angiotensin II-receptor antagonists, the existence of at least two subtypes of the angiotensin II receptor could be demonstrated, classified as the AT_1 receptor and the AT_2 receptor.[192] Recently, cloning experiments showed the existence of AT_{1A} and AT_{1B} receptors in rat tissues.[193,194] The distribution of the angiotensin II receptor subtypes differs in tissues. AT_{1A} receptor mRNA is expressed primarily in vascular smooth muscle and lung, whereas AT_{1B} receptor mRNA is expressed in adrenal and pituitary gland.[194] The occurrence of the AT_1 and the AT_2 receptor in a variety of tissues in humans is presented in Table 8.3. The distribution of AT_1 and AT_2 receptors is similar

Tissue	AT_1	AT_2
Myocardium (normal)*	+	++
Myocardium (ischaemia)*	+	+++
Renal artery†	+	+
Kidney‡,§	+	–
Adrenal cortex†,‖,#	+/++	+
Uterus†,‖	–	+
Liver#,**	+	?
Lung#	+	?
Vascular smooth muscle cells‖	+	–

– = absence of receptors; +, ++ and +++ = presence of receptors indicating the relative density compared to the other receptor subtype; *Rogg et al[254] and personal communication; †Timmermans et al[196]; ‡Sechi et al;[195] §Chansel et al;[255] ‖Whitebread et al;[256] #Takayanagi et al;[257] **Bergsma et al.[258]

Table 8.3
Occurrence of AT_1 and AT_2 receptors in human tissues

in human, rabbit and rat. The distribution in kidney, adrenal cortex and vascular smooth muscle cells is species-independent,[195,196] whereas in the uterus of rat the same amount or a higher amount of AT_1 receptors than of AT_2 receptors has been demonstrated.[196,197]

The transmembrane signaling pathways of the AT_1 receptor and the AT_2 receptor are not the same. The AT_1 receptor, but not the AT_2 receptor, is linked to phospholipase C, adenylyl cyclase and calcium channels.[198,199] The effector pathway and cellular targets of the AT_2 receptor are still unknown. Recently, the AT_2 receptor has been demonstrated to stimulate protein tyrosine phosphatase (PTP) activity and inhibit particulate guanylate cyclase (pGC).[200] The AT_{1A} and AT_{1B} receptors seem to have a comparable effector pathway.[194] However, the functional difference between these two receptors is not yet known. Evidence for a third AT receptor type, involved in blood vessel growth, was recently published.[201]

Hypertension and local RAS

The involvement of the local RAS in hypertension has been extensively investigated in several experimental hypertension models, whereas little is known about alterations of the local RAS in hypertensive patients. The most common animal models are spontaneously hypertensive rats (SHR), one-kidney–one-clip (1K–1C) rats, two-kidney–one-clip (2K–1C) rats, aorta stenotic rats and DOCA-salt rats. In SHR, plasma renin activity (PRA) is normal or lower compared to the normotensive Wistar-Kyoto rats.[202] In contrast to PRA, renin mRNA in the vessel wall is dependent on age and tissue. In the prehypertensive stage of the SHR, renin mRNA is already increased in kidney, liver, brain, adrenals and heart, whereas mRNA in the aorta is normal.[203] In 1K–1C rats and in the chronic hypertension phase of 2K–1C rats, however, both PRA and vascular renin are normal,[204–206] whereas renin mRNA in the kidney is increased.[207,208] This implies that in these latter hypertension models, circulating and vascular renin are not involved in the maintenance of hypertension.

Plasma ACE activity, which is lower in SHR compared to controls[44,209] and is normal in 1K–1C and 2K–1C rats,[132,205] cannot explain the development of hypertension. In contrast, ACE in the vessel wall may be involved in hypertension, as ACE activity and ACE mRNA in main arteries is increased.[132,205,209,210] In aorta stenotic rats, the pressure overload results in an increased ACE activity, angiotensinogen mRNA and ACE mRNA in the septum and left ventricle of the heart.[101,211] Besides alterations in ACE activity, binding of angiotensin II to the receptors is altered in hypertension.[200] In a low-renin hypertension model, the functional response to angiotensin II receptors was decreased without a change in angiotensin II receptor number.[212] The changes in binding capacity in hypertension are tissue-specific and depend on the experimental model investigated.

ACE inhibitors and local RAS

The ACE inhibitors can be divided in three structurally different groups:[213] the sulfhydryl compounds (for example, captopril, zofenopril); the carboxyalkyl peptides (for example, enalapril, lisinopril, perindopril, ramipril) and the phosphoric acid compounds (for example, fosinopril). One of the factors that differentiates ACE inhibitors is the tissue bioavailability of each drug. Therefore the diffusion characteristics, which depend on the lipophilicity of the drugs, plays an important role in the inhibition of local ACE. Lipid solubility of the drugs is most important for the passage of the barrier to the brain and to the testes.[214,215] Because of a low oral bioavailability, ACE inhibitors (except captopril and lisinopril) are prodrug esters and must be hydrolyzed before inhibitory activity. In general, ester prodrug compounds are more lipid-soluble than the active compound (diacid), which may enable the ester to penetrate tissues more easily. This implies that the penetration of the ACE inhibitor into tissues to contact local ACE depends not only on the diffusion characteristics of the ACE inhibitor, but also on the location of hydrolysis of the prodrug. The major site of de-esterification of the prodrug to the active diacid appears to be in the liver. Other esterase activity can be found in plasma and in a variety of tissues.[213] Several drugs, such as ramipril and enalapril in kidney[216] and zofenopril in a diversity of tissues,[217] are de-esterified rapidly in tissue cells. Those drugs might accumulate in these tissues by entering the tissue as inactive prodrugs, where they are converted to the active drug, which do not leave the tissue easily. Hydrolysis of the ACE inhibitor in specific tissues may imply a selective inhibition of local ACE, whereas hydrolysis in the liver results in an almost non-selective inhibition. Such a lack of regional selective ACE inhibition in heart and kidney has been demonstrated for lisinopril, captopril and zofenopril,[218] whereas a higher inhibitory activity of enalapril and ramipril was found in the kidney.[216]

As the inhibition of local ACE by ACE inhibitors depends primarily on their lipophilicity, there is no major guideline for the choice of drug in the treatment of hypertension. Preferential inhibition of ACE in a specific tissue can only be achieved by administration of a prodrug, which is converted predominantly in the target tissue. At the moment, no such ACE inhibitor is available.

Besides ACE inhibition, scavenging of free radicals is a structure-related characteristic of some ACE inhibitors. Sulfhydryl-containing ACE inhibitors appear to be effective scavengers of non-superoxide free radicals.[219]

This effect is related to the chemical structure of these ACE inhibitors and not to the ACE inhibitory effect, as non-sulfhydryl ACE inhibitors have no radical scavenging properties, and epicaptopril (the non-ACE-inhibitory stereoisomer of captopril) is also a scavenger of free radicals. This characteristic of sulfhydryl-containing ACE inhibitors has been suggested to play a role in myocardial ischaemia and reperfusion injury.[220–222]

Consequences of vascular and cardiac effects of ACE inhibition: mechanism of antihypertensive action

All ACE inhibitors that are clinically employed are effective antihypertensive agents.[223] They lower blood pressure through a reduction of peripheral resistance and have little or no effect on cardiac output or heart rate.[223] This provides an attractive haemodynamic profile for the treatment of essential hypertension.

Several mechanisms have been proposed to explain the ACE inhibitor-induced fall in blood pressure and vascular resistance. These include interaction with the autonomic nervous control of blood pressure, the vasculature and renal blood pressure control. In turn, these may result from blockade of circulating or local ACE resulting in diminished angiotensin II production, or diminished bradykinin breakdown.

The sympathetic nervous system as the primary site of effects

Through their effect on angiotensin synthesis. ACE inhibitors may interact with nervous control of blood pressure. Angiotensin II interferes with sympathetic nervous system responses both in the central nervous system[224] and at the level of the presynaptic postganglionic nerve endings.[225] The question whether, and especially where ACE inhibitors interact with sympathetic neurotransmission is still a matter of debate. Antonaccio and Kerwin obtained evidence in pithed rats that captopril inhibits both pre- and postsynaptic mechanisms in the vasculature during sympathetic stimulation.[219] This observation was not confirmed by some authors in dogs[226] and in hypertensive patients[227] although the majority of studies is at least in part confirmatory. Thus in pithed rats, captopril inhibited both noradrenaline (norepinephrine) overflow and vasoconstriction during nerve stimulation.[228] Also in the human forearm[229] and coronary circulation,[119] evidence has been obtained for presynaptic modulation of sympathetic activity. Although chronic treatment of SHR with ramipril or captopril[230] did not provide biochemical evidence for presynaptic inhibition, an attenuation of pressor responses to sympathetic stimulation but not of the cardiac effects was observed.

The baroreflex as the primary site of effects

One of the earliest and probably most clinically relevant observations with respect to ACE inhibition and the nervous system is the lack of tachycardia compared to other vasodilator therapy.[231] This is associated with an attenuated increase in plasma noradrenaline (norepinephrine) levels, suggesting inhibition of the sympathetic component of the baroreflex. Berecek et al. observed that chronic low-dose intracerebroventricular (ICV) infusion of captopril resulted in a reduced vascular response to posterior hypothalamic stimulation as well as to splanchnic stimulation.[232] This effect was specific for SHR. These observations suggest that captopril might dampen the baroreflex through an effect in the central nervous system and/or the postganglionic fibres. However, cardiac effects were not influenced,[232] which suggests, again, specific interactions with vascular control. Chronic ICV application of enalaprilat, in contrast, enhances baroreflex responses assessed by measuring HR as well as splanchnic nerve activity. This suggests that a central action, if any, might be enhancement rather than attenuation of the baroreflex.[233] In studies in humans with acute captopril administration[234] or chronic cilazapril treatment,[221] no evidence was obtained for changes in baroreflex sensitivity. However, given the lack of heart rate increase, in view of the reduced pressure, there must have been a resetting to a new setpoint.[234] There is now evidence that vagal mechanisms also may contribute to the lack of tachycardia through central mechanisms.[138]

Interaction with kinins and prostanoids

ACE inhibitors may act at several sites to reduce blood pressure. As was argued above, ACE inhibitors not only interact with the conversion of angiotensin I to angiotensin II but also with bradykinin breakdown and prostaglandin synthesis. The potential relevance of these interactions has recently become clear. The antihypertensive effect of ramipril in renovascular hypertensive rats can be significantly attenuated with a bradykinin BK_2 antagonist.[235] In humans, bradykinin exerts an antihypertensive effect upon infusion.[236] The effect of angiotensin II was potentiated by concomitant administration of the antagonist in rats,[237] suggesting functional antagonism between bradykinin and angiotensin II. In addition, captopril has been shown to increase renal prostaglandin release[238] and inhibition of prostaglandin synthesis with indomethacin greatly attenuates renal vasodilatation as induced by captopril.[238,239] Thus, interactions with these mechanisms have to be considered at least as a partial basis for the blood pressure reduction, although more evidence is needed before firm conclusions can be reached.

Vascular effects

ACE-inhibitors reduce peripheral resistance through an effect on the small arteries and arterioles. The vascular effects may also include a venous dilatory component. In rats with heart failure following myocardial infarction, captopril reduced mean circulatory filling pressure by dilatation of capacitance vessels.[240] This effect may depend upon inhibition of angiotensin II formation or upon an effect of bradykinin.[240] The same authors reported decreases of cardiac filling pressures in the same model following the AT_1-receptor antagonist losartan,[241] which suggests an important contribution of angiotensin II to the observed effect.

In heart failure and hypertension, ACE inhibition causes an acute redistribution of blood flow. In hypertension this results in an increase of blood flow through the kidney and liver (Table 8.4). In this respect, ACE inhibitors differ from other antihypertensive agents. Several studies in SHR[18,19] and in hypertensive patients[242] indicate that the effect on the renal circulation is sustained during long-term therapy. The redistribution of blood flow to the kidneys has profound consequences for cardiovascular control. First of all, an acute reduction of resistance to the organ, which receives approximately 20% of total cardiac output, will significantly contribute to total peripheral resistance. Secondly, through an ensuing diuresis[243] (see below) there may be a further reduction of blood pressure, which does, however,

ACE inhibitor	Kidney	Heart	Liver	Brain	Skin	Spleen	Splanchnic organs	Skeletal muscle
Captopril[*,†,‡,§,‖,#,**]	↑	↑	↑	↔	↔	↔	↔	↓ ↔ ↑
Enalapril[‡,§,††,§§]	↑		↑	↔	↔	↔/↑	↔	↔/↑
Lisinopril[‖‖]	↑					↓		
Perindopril[##,***]	↑			↔			↔	↔
Cilazapril[†††,‡‡‡]	↑	↔	↔	↓/↔	↔	↓	↔	↔
Ramipril[§§§]	↑		↑	↔	↔	↑		↔
Quinapril[‖‖‖]	↑	↑		↔	↔		↔	↓

*Paulson and Waldemar,[259] †Smits and Struijker Boudier,[260] ‡Nelissen-Vrancken et al.,[244] §Richer et al.,[261] ‖Magrini et al.,[118] #Ventura et al., [242] **Roberts et al.,[252] ††Strocchi et al.,[263] §§Wang et al.,[264] ‖‖Garavaglia et al.,[265] ##Bray et al.,[266] ***Harrap,[267] †††Clozel,[268] ‡‡‡Clozel et al.,[53] §§§Richer et al.,[18,19] ‖‖‖Richer et al.[269]

Table 8.4
Effects of ACE inhibitors on regional blood flow in hypertension

take time to develop. The relative importance of the renal vascular effects is illustrated by observations with respect to inhibition of ACE; the antihypertensive effect in animals is not correlated with inhibition of circulating ACE, but with inhibition of ACE in the vascular wall and kidney.[126,127]

The mechanism through which regional specificity is attained is still under debate. In a recent study in SHR and rats after myocardial infarction we observed specific reductions of renal vascular resistance at doses of ACE inhibitors that did not lower blood pressure,[244] indicating that the effect is not dependent on haemodynamic changes induced elsewhere in the circulation. Thus the primary site of effect is localized within the renal circulation. Interaction with renal sympathetic nerve endings,[230,245] bradykinin and/or prostaglandins[239,246,247] and the local formation of ACE,[247] as discussed elsewhere in this chapter, have been implied.

Renal effects

The increase in renal blood flow as discussed above is associated with slightly increased or normal GFR.[243] Furthermore, proteinuria is decreased in patients with diabetic nephropathy.[243] These observations are compatible with preferential dilatation of efferent glomerular arterioles and with the observation that angiotensin II exerts its main intrarenal constrictor effect on these efferent arterioles.[248] The ensuing reduction of glomerular pressure (both from efferent arteriolar dilatation and blood pressure reduction) together with the increased flow, maintain glomerular filtration rate around normal. The well-documented exception to this situation is observed in patients with renal artery stenosis, where ACE inhibitors may actually reduce GFR to extremely low values.[249] This probably depends on the fact that GFR in these patients, in the

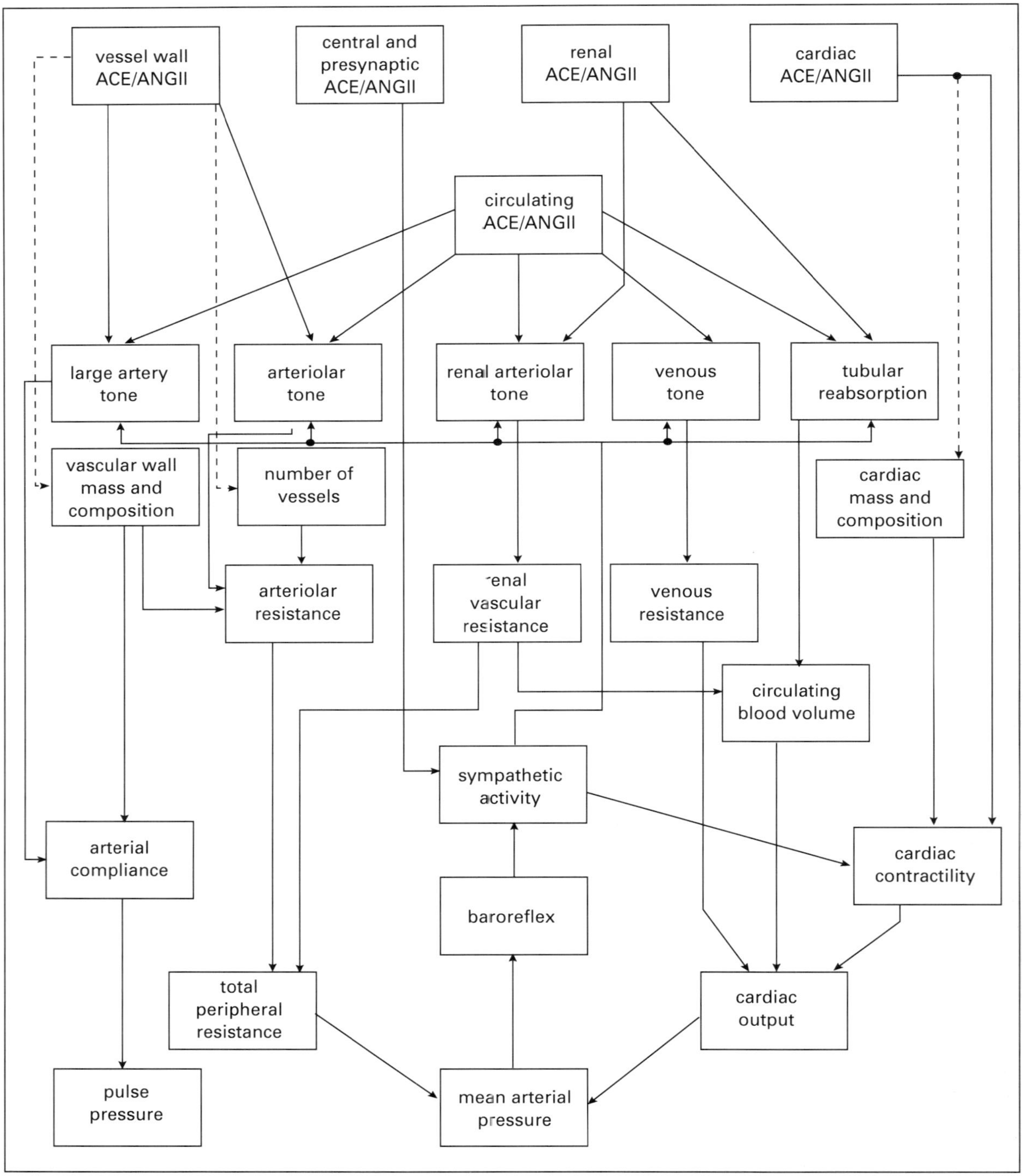

Figure 8.1
Mechanisms of antihypertensive effect of ACE inhibitors. Solid lines represent acute effects: interrupted lines are long-term actions.

face of increased afferent resistance, depends upon strongly increased efferent arteriolar tone by angiotensin II.

In hypertensive patients, acute administration of ACE inhibitors results in diuresis and natriuresis,[243] which, to a large extent, is of tubular origin.[248] This effect may be mediated by inhibition of aldosterone secretion, but the rapidity of onset of natriuresis makes this unlikely.[250] Also, the natriuresis may depend upon increased bradykinin levels.[251] Alternatively, it may depend upon inhibition of formation of angiotensin II which has sodium retaining effects in the proximal tubule.[248] Furthermore, the renal sympathetic nervous system involves innervation of the renal tubules which also promote sodium reabsorption.[252] As was argued above, ACE inhibitors reduce sympathetic effects at different sites in the circulation and, hence, this may contribute to the natriuretic responses as well. Hall put forward that the reduction of efferent arteriolar resistance per se might also explain decreased tubular reabsorption through increasing peritubular capillary flow rate.[248] This remains to be established.

The increase in natriuresis and diuresis in hypertensives following ACE inhibition is not normally maintained. Absence of changes in body fluid volumes suggests a restoration of sodium homeostasis.[243] In view of the concomitant reduction in blood pressure, this indicates that the renal pressure–diuresis relationship is reset by ACE inhibitors.

Conclusions

The evidence reviewed in this chapter allows a preliminary conclusion about the role of different cardiovascular control mechanisms, underlying the antihypertensive effect of ACE inhibitors. We must distinguish between the acute and long-term effects of ACE inhibition. The scheme in Fig. 8.1 is an attempt to integrate the evidence discussed thus far.

With respect to the acute blood pressure lowering effect, the following mechanisms seem to contribute predominantly:

- lowering of peripheral vascular resistance as a consequence of dilatation of resistance arteries and arterioles, especially those in the renal vascular bed;
- a venous vascular effect to reduce venous return of blood to the heart and attenuate a (reflex) rise in cardiac output;
- a central, or more likely, presynaptic peripheral interference with sympathetic nerves to inhibit a (reflex) rise in sympathetic nerve activity and probably even reduce presynaptic noradrenaline release;
- an intrarenal, probably tubular, effect to shift the pressure–diuresis curve to allow for a normal renal fluid and electrolyte excretion, even at a lower level of blood pressure.

These effects are maintained in the long term, although a structural factor then takes over part of the original mechanism of reduction of vascular resistance. In the long term, the structure of the arterial tree as well as the heart adapt to the lower blood pressure and related mechanical factors, such as arterial wall stress and cardiac wall stress. This leads to a gradual remodelling of the arterial tree and the heart. The long-term consequence of this remodelling process is that, in addition to a maintained lowered blood pressure and vascular resistance, the compliance of the large arteries increases. Thus, not only mean arterial pressure remains reduced, but also pulse pressure. Although mechanical factors contribute to the remodelling of the heart and vascular tree, part of this process seems to be mediated by the direct molecular influence of ACE inhibitors on circulating or local renin-angiotensin systems.

References

1 Struijker Boudier HAJ, Smits JFM, Kleinjans JCS, Van Essen H. Hemodynamic actions of diuretic agents. *Clin Exp Hypert Ther Pract* 1983; **A5**:209–223.

2 Van Baak MA, Sruijker Boudier HAJ, Smits JFM. Antihypertensive mechanisms of beta-adrenoceptor blockade: a review. *Clin Exp Theor Pract* 1985; **A7**:1–72.

3 Struijker Boudier HAJ, De Mey JGR, Smits JFM, Nievelstein HMNW. Hemodynamic actions of calcium entry blockers. In: Van Zwieten PA, ed. *Clinical aspects of calcium entry blockers*. Basel: S Karger, 1989:21–66.

4 Struijker Boudier HAJ, Van Bortel LMAB, De Mey JGR. Remodeling of the vascular tree in hypertension: drug effects. *TIPS* 1990;**11**:240–5.

5 Mulvany MJ, Aalkjær C. Structure and function of small arteries. *Physiol Rev* 1990; **70**:921–61.

6 Safar M, Peronneau J, Levenson J, Simon A. Pulsed doppler: diameter, velocity and flow of brachial artery in sustained essential hypertension. *Circulation* 1981; **63**:393–400.

7 Van Merode T, Hick PJJ, Hoeks APG, Rahn KH, Reneman RS. Vessel wall properties of the carotid artery in normotensive and borderline hypertensive male subjects of various ages. *Ultrasound Med Biol* 1988; **14**:563–69.

8 Westerhof N, Huisman RM. Arterial haemodynamics of hypertension. *Clin Sci* 1987; **72**:391–98.

9 Ismard RN, Pannier BM, Laurent S, London GM, Diebold B, Safar ME. Pulsatile diameter and elastic modulus of the aortic arch in essential hypertension: a non-invasive study. *J Am Coll Cardiol* 1989; **13**:399–405.

10 Safar ME, Laurent S, Benetos A, London GM. The common carotid circulation in patients with essential hypertension. *Stroke* 1988; **19**:1198–202.

11 Safar ME, London GM, Bouthier JA, Levenson JA, Laurent S. Brachial artery cross-sectional area and distensibility before and after arteriolar vasodilatation in men with sustained hypertension. *J Cardiovasc Pharmacol* 1987; **9**:734–42.

12 Levy BI, Michel J-B, Salzman J-L, Poitevin P, Camilleri J-P, Safar ME. Arterial effects of angiotensin converting enzyme inhibition in renovascular and spontaneously hypertensive rats. *J Hypert* 1988; **6[Suppl 3]**:S23–25.

13 Levy B, Benessiano J, Poitevin P, Safar ME. Endothelium-dependent mechanical properties of the carotid artery in WKY and SHR. Role of angiotensin converting enzyme inhibition. *Circ Res* 1990; **66**:321–28.

14 Safar ME, Levy BI, Laurent S, London GM. Hypertension and the arterial system: clinical and therapeutic aspects. *J Hypertens* 1990; **8[Suppl 7]**:S113–19.

15 Safar ME, Levy BI, London GM. Arterial structure in hypertension and the effects of angiotensin converting enzyme inhibition. *J Hypertens* 1992; **10[Suppl 5]**:S51–57.

16 Darne B, Girerd X, Safar ME, Cambien F, Guize L. Pulsatile versus steady state component of blood pressure: a cross-sectional analysis and a prospective analysis on cardiovascular mortality. *Hypertension* 1989; **13**:392–400.

17 Safar ME. Therapeutic trials and large arteries in hypertension. *Am Heart J* 1988; **115**:702–10.

18 Richer C, Doussau M-P, Giudicelli J-F. Systemic and regional hemodynamic profile of five angiotensin I converting enzyme inhibitors in the spontaneously hypertensive rat. *Am J Cardiol* 1987; **59**:12D–17D.

19 Richer C, Thuillez C, Giudicelli JF. Perindopril, converting enzyme blockade, and peripheral arterial hemodynamics in the healthy volunteer. *J Cardiovasc Pharmacol* 1987; **9**:94–102.

20 Bouthier JD, Safar ME, Benetos A, Simon AC, Levenson JA, Huguesch M. Hemodynamic effects of vasodilating drugs

on the common carotid and brachial circulations of patients with essential hypertension. *Br J Clin Pharmacol* 1986; **21**:137–42.

21 Simon AC, Levenson JA, Bouthier JD et al. Comparison of oral MK 421 and propranolol in mild to moderate essential hypertension and their effects on arterial and venous vessels of the forearm. *Am J Cardiol* 1984; **53**:781–85.

22 Asmar RG, Pannier B, Santoni JP et al. Reversion of cardiac hypertrophy and reduced arterial compliance after converting enzyme inhibition in essential hypertension. *Circulation* 1988; **78**:941–50.

23 Benetos A, Santoni JP, Safar ME. Vascular effects of intravenous infusion of the angiotensin converting enzyme inhibitor perindoprilat. *J Hypertens* 1990; **8**:819–26.

24 Levy BI, Stefas L, Babalis D, Benetos A. Vascular endothelium, mechanical properties of the arterial wall and local angiotensin converting enzyme inhibition. *J Hypert* 1992; **10[Suppl 5]**:S21–27.

25 Goldschmidt JE, Tallarida RJ. Pharmacological evidence that captopril possesses an endothelium-mediated component of vasodilation: effect of sulfhydryl groups on endothelium-derived relaxing factor. *J Pharmacol Exp Ther* 1991; **257**:1136–45.

26 Rubin B, Laffan RJ, Kotler DG, O'Keefe EH, DeMaio DA, Goldberg ME. SQ 14,255, a novel orally active inhibitor of angiotensin I-converting enzyme. *J Pharmacol Exp Ther* 1978; **204**:271–80.

27 Kerth PA, Vanhoutte PM. Effects of perindoprilat on endothelium-dependent relaxations and contractions in isolated blood vessels. *Am J Hypertens* 1991; **4[Suppl 3]**:226S–34S.

28 Hirooka Y, Imaizumi T, Ando S, Harada S, Momohara M, Takeshita A. Captopril improves impaired endothelium-dependent vasodilation in hypertensive patients. *Hypertension* 1992; **20**:175–80.

29 Clozel M, Kuhn H, Hefti F. Effects of angiotensin converting enzyme inhibitors and of hydralazine on endothelial function in hypertensive rats. *Hypertension* 1990; **16**:532–40.

30 Freslon JL, Giudicelli JF. Compared myocardial and vascular effects of captopril and dihydralazine during hypertension development in spontaneously hypertensive rats. *Br J Pharmacol* 1983; **80**:533–43.

31 Schachter M. Drug-induced modification of vascular structure: effects of antihypertensive drugs. *Am Heart J* 1991; **122**:316–23.

32 Cadilhac M, Giudicelli JF. Myocardial and vascular effects of perindopril, a new converting enzyme inhibitor, during hypertension development in spontaneously hypertensive rats. *Arch Int Pharmacodyn Ther* 1986; **284**:114–26.

33 Lee RMKW, Berecek KH, Tsoporis J, McKenzie R, Triggle CR. Prevention of hypertension and vascular changes by captopril treatment. *Hypertension* 1991; **17**: 141–50.

34 Harrap SB, Nicolaci JA, Doyle AE. Persistent effects on blood pressure and renal hemodynamics following chronic angiotensin converting enzyme inhibition with perindopril. *Clin Exp Pharmacol Physiol* 1986; **13**:753–65.

35 Clozel J-P, Kuhn H, Hefti F. Vascular protection with cilazapril in hypertension. *J Cardiovasc Pharmacol* 1992; **19[Suppl 5]**:S28–33.

36 Levy BI, Michel J-B, Salzmann J-L et al. Effects of chronic inhibition of converting enzyme on mechanical and structural properties of arteries in rat renovascular hypertension. *Circ Res* 1988; **63**:227–39.

37 Schelling P, Fischer H, Ganten D. Angiotensin and cell growth: a link to cardiovascular hypertrophy? *J Hypertens* 1991; **9**:3–15.

38 Naftilan AJ. The role of angiotensin II in vascular smooth muscle cell growth. *J Cardiovasc Pharmacol* 1992; **20[Suppl 1]**:S37–40.

39 Scott-Burden T, Hahn AWA, Bühler FR, Resink TJ. Vasoactive peptides and growth factors in the pathophysiology of hypertension. *J Cardiovasc Pharmacol* 1992; **20[Suppl 1]**:S55–64.

40 Chiu AT, McCall DE, Price WA et al. Nonpeptide angiotensin II receptor antagonists. VII. Cellular and biochemical pharmacology of DuP 753, an orally active antihypertensive agent. *J Pharmacol Exp Ther* 1989; **252**:711–18.

41 Owens GK. Control of hypertrophic versus hyperplastic growth of vascular smooth muscle cells. *Am J Physiol* 1989; **26:**H1755–65.

42 Struijker Boudier HAJ, Messing MWJ, Nelissen HJMG, Huijberts MSP, Le Noble FAC, Smits JFM. Vascular actions of ACE-inhibitors in hypertension: structural and functional consequences. In: MacGregor GA, Sever PS, eds. *Current advances in ACE inhibition*. Edinburgh: Churchill Livingstone, 1991; 26–30.

43 Vanhoutte PM, Auch-Schwelk W, Biondi ML, Lorenz RR, Schini VB, Vidal MJ. Why are converting enzyme inhibitors vasodilators? *Br J Clin Pharmac* 1989; **28:**95S–104S.

44 Grima M, Welsch C, Giesen-Crouse EM, Coquard C, Barthelembs M, Imbs JL. Age-related variations in tissue angiotensin converting enzyme activities: comparison between spontaneously hypertensive and Wistar-Kyoto rats. *J Hypertens* 1990; **8:**697–702.

45 Carmines PK, Fleming JT. Control of the renal vasculature by vasoactive peptides. *FASEB* 1990; **4:**3300–9.

46 Nyborg NC, Nielsen PJ, Prieto D, Benedito S. Angiotensin II does not contract bovine retinal resistance arteries in vitro. *Exp Eye Res* 1990; **50:**469–74.

47 Juul B, Aalkjær Mulvany MJ. Responses of femoral resistance vessels to angiotensin in vitro. *Eur J Pharmacol* 1987; **135:**61–68.

48 Haberl RL, Decker PJ, Einhäupl KM. Angiotensins in degradation products mediate endothelium-dependent dilation of rabbit brain arterioles. *Circ Res* 1991; **68:**1621–27.

49 Urabe M, Lee TJ-F. Pre- and postsynaptic effects of angiotensins in the femoral artery of spontaneously hypertensive and Wistar-Kyoto rats. *Blood Vessels* 1987; **24:**1–10.

50 Yamazaki M, Toda N. Comparison of responses to angiotensin II of dog mesenteric arteries and veins. *Eur J Pharmacol* 1991; **201:**223–29.

51 Christensen KL, Jespersen LT, Mulvany MJ. Development of blood pressure in spontaneously hypertensive rats after withdrawal of long-term treatment related to vascular structure. *J Hypertens* 1989; **7:**83–90.

52 Clozel J-P, Kuhn H, Hefti F. Decreases of vascular hypertrophy in four different types of arteries in spontaneously hypertensive rats. *Am J Med* 1989; **87[Suppl 6B]:**92S–95S.

53 Clozel J-P, Kuhn H, Hefti F. Effects of cilazapril on the cerebral circulation in spontaneously hypertensive rats. *Hypertension* 1989; **14:**645–51.

54 De Mey JGR, Daemen MJAP, Boonen HCM et al. In vivo DNA synthesis is not uniformly increased in arterial smooth muscle of young spontaneously hypertensive rats. *J Hypert* 1991; **9:**695–701.

55 Mulvany MJ. Determinants of vascular structure. *J Cardiovasc Pharmacol* 1992; **19[Suppl 5]:**S1–6.

56 Aalkjær C, Eiskjær H, Mulvany MJ et al. Abnormal structure and function of isolated subcutaneous resistance vessels from essential hypertensive patients despite antihypertensive treatment. *J Hypertens* 1989; **7:**308–10.

57 Mulvany MJ. Changes to the vascular system resulting from hypertension and their effects on response to therapy. *Drugs* 1990; **39[Suppl 1]:**9–20.

58 Harrap SB, Van derMerwe WM, Griffin SA, Macpherson F, Lever AF. Brief angiotensin converting enzyme inhibitor treatment in young spontaneously hypertensive rats reduces blood pressure long term. *Hypertension* 1990; **16:**603–14.

59 Prewitt RL, Hashimoto H, Stacy DL. Structural and functional rarefaction of microvessels in hypertension. In: Lee R, ed. *Blood vessel changes in hypertension: structure and function*. Boca Raton: CRC Press, 1990: 71–90.

60 Struijker Boudier HAJ, Le Noble JLML, Messing MWJ, Huijberts MSP, Le Noble FAC, Van Essen H. The microcirculation and hypertension. *J Hypertens* 1992; **10[Suppl 7]:**S147–45.

61 Adair TH, Gay WJ, Montani J-P. Growth regulation of the vascular system: evidence for a metabolic hypothesis. *Am J Physiol* 1990; **259:**R393–404.

62 Fernandez LA, Twickler J, Mead A. Neovascularization produced by angiotensin II. *J Lab Clin Med* 1985; **195:**141–45.

63 Le Noble FAC, Schreurs N, Van Straaten HWM, Slaaf DW, Smits JFM, Struijker Boudier HAJ. Angiotensin-II induced angiogenesis is not mediated through the AT1 receptor [abstract]. *FASEB* 1992; **6**:A937.
64 Vicaut E, Montalescot G, Hou X, Stucker O, Teisseire B. Arteriolar vasoconstriction and tachyphylaxis with intraarterial angiotensin II. *Microvasc Res* 1989; **37**:28–41.
65 Messing M, Van Essen H, Smith TL, Smits JFM, Struijker Boudier HAJ. Microvascular actions of calcium channel antagonists. *Eur J Pharmacol* 1991; **198**:189–95.
66 Messing MWJ. Antihypertensive drugs and the microcirculation. [Dissertation]. University of Limburg, Maastricht, 1992.
67 Messing MWJ, Van Essen H, Struijker Boudier HAJ. Effects of α-adrenoceptor antagonists on striated muscle microcirculation of conscious rats. *Drugs* 1990; **40[Suppl 4]**:31–33.
68 Struijker Boudier HAJ, Messing MWJ, Van Essen H. Preferential small arteriolar vasodilatation by the potassium channel opener lemakalim in conscious spontaneously hypertensive rats. *Eur J Pharmacol* 1992; **218**:191–93.
69 Wang D-H, Prewitt RL. Captopril reduces aortic and microvascular growth in hypertensive and normotensive rats. *Hypertension* 1990; **15**:68–77.
70 Nelissen-Vrancken HJMG Struijker Boudier HAJ, Daemen MJAP, Smits JFM. Antihypertensive therapy and adaptive mechanisms in peripheral ischaemia. *Hypertension* 1993; **22**:780–8..
71 Weber KT, Brilla CG. Pathological hypertrophy and cardiac interstitium. Fibrosis and renin-angiotensin-aldosterone system. *Circulation* 1991; **83**:1849–65.
72 Giles TD. Angiotensin converting enzyme inhibition, cell growth, and left ventricular hypertrophy in hypertension. *Clin Cardiol* 1991; **14[Suppl 4]**:63–67.
73 Savage DD, Garrison RJ, Kannel WB et al. The spectrum of left ventricular hypertrophy in a general population: The Framingham Study. *Circulation* 1987; **75**:I126–33.
74 Levy D. Left ventricular hypertrophy. Epidemiological insights from the Framingham Heart Study. *Drugs* 1988; **35[Suppl 5]**:1–5.
75 Houston MC. New insights and new approaches for the treatment of essential hypertension: selection of therapy based on coronary heart disease risk factor analysis, hemodynamic profiles, quality of life, and subsets of hypertension. *Am Heart J* 1989; **117**:911–51.
76 Messerli FH. Antihypertensive therapy — going to the heart of the matter. *Circulation* 1990; **81**:1128–35.
77 Schreiber SS, Evans CD, Oratz M, Rothschild MA. Protein synthesis and degradation in cardiac stress. *Circ Res* 1981; **48**:601–11.
78 Cooper G, Kent RL, Mann DL. Load induction of cardiac hypertrophy. *J Mol Cell Cardiol* 1989; **21**:11–30.
79 Korecky B, Rakusan K. Morphological and physiological aspects of cardiac atrophy. In: Alpert NR, ed. *Perspectives in cardiovascular research: Myocardial hypertrophy and failure. vol 7*. New York: Raven Press, 1983: 293–309.
80 Tsoporis J, Leenen FHH. Effects of arterial vasodilators on cardiac hypertrophy and sympathetic activity in rats. *Hypertension* 1988; **11**:376–86.
81 Tsoporis J, Yuan B, Leenen FHH. Arterial vasodilators, cardiac volume load, and cardiac hypertrophy in normotensive rats. *Am J Physiol* 1989; **256**:H876–80.
82 Leenen FHH, Prowse S. Time course of changes in cardiac hypertrophy and pressor mechanisms in two-kidney, one-clip hypertensive rats during treatment with minoxidil, enalapril of after uninephrectomy. *J Hypertens* 1987; **5**:73–83.
83 Julien J, Dufloux M-A, Prasquier R et al. Effects of captopril and minoxidil on left ventricular hypertrophy in resistant hypertensive patients: a 6-month double-blind comparison. *J Am Coll Cardiol* 1990; **16**:137–42.
84 Drayer JIM, Weber MA, Gardin JM, Lipson JL. Effect of long-term antihypertensive therapy on cardiac anatomy in patients with arterial hypertension. *Am J Med* 1983; **75**:116–20.
85 Reichek N. Echocardiographic assessment of left ventricular hypertrophy. *Eur Heart J* 1982; **3**:49–57.

86 Sen S, Tarazi RC, Khairallah PA, Bumpus FM. Cardiac hypertrophy in spontaneously hypertensive rats. *Circ Res* 1974; **35:** 775–81.

87 Strauer BE. Regression of myocardial and coronary vascular hypertrophy in hypertensive heart disease. *J Cardiovasc Pharmacol* 1988; **12[Suppl 4]**:45–54.

88 Devereux RB, Pickering TG, Cody RJ, Laragh JH. Relation of renin-angiotensin system activity to left ventricular hypertrophy and function in experimental and human hypertension. *J Clin Hypertens* 1987; **3:** 87–103.

89 Dahlöf B, Pennert K, Hansson L. Regression of left ventricular hypertrophy — A meta-analysis. *Clin Exp Hypert Theor Pract* 1992; **A14**:173–80.

90 Leenen FHH, Smith DL, Farkas RM, Reeves RA, Marquez-Julio A. Vasodilators and regression of left ventricular hypertrophy. *Am J Med* 1987; **82:**969–78.

91 Linz W, Schölkens BA, Ganten D. Converting enzyme inhibition specifically prevents the development and induces regression of cardiac hypertrophy in rats. *Clin Exp Hypert Theor Pract* 1989; **A11**:1325–50.

92 Julius S, Li Y, Brant D, Krause L, Taylor D. Quinapril, an angiotensin converting enzyme inhibitor, prevents cardiac hypertrophy during episodic hypertension. *Hypertension* 1991; **17:**1161–66.

93 Clozel J-P, Hefti F. Cilazapril prevents the development of cardiac hypertrophy and the decrease of coronary vascular reserve in spontaneously hypertensive rats. *J Cardiovasc Pharmacol* 1988; **11:**568–72.

94 Pfeffer JM, Pfeffer MA, Mirsky I, Braunwald E. Regression of left ventricular hypertrophy and prevention of left ventricular dysfunction by captopril in the spontaneously hypertensive rat. *Proc Nat Acad Sci* USA 1982; **79:**3310–14.

95 Sen S, Tarazi RC, Bumpus FM. Effect of converting enzyme inhibitor (SQ 14,225) on myocardial hypertrophy in spontaneously hypertensive rats. *Hypertension* 1980; **2:**169–76.

96 Pfeffer MA, Pfeffer JM. Pharmacologic regression of cardiac hypertrophy in experimental hypertension. *J Cardiovasc Pharmacol* 1984; **6[Suppl]**:865–69.

97 Michel J-B. Relationship between decrease in afterload and beneficial effects of ACE-inhibitors in experimental cardiac hypertrophy and congestive heart failure. *Eur Heart J* 1990; **11[Suppl D]**:17–26.

98 Gosse P, Grellet J, Bonoron S, Tariosse L, Besse P, Dallocchio M. Effects of perindopril on left ventricular hypertrophy, coronary blood flow, and mechanical properties of cardiac muscle in renovascular hypertensive rats. *Am J Hypert* 1991; **4[Suppl]**:235–39.

99 Kromer EP, Riegger GAJ. Effects of long-term angiotensin converting enzyme inhibition on myocardial hypertrophy in experimental aortic stenosis in the rat. *Am J Cardiol* 1988; **62:**161–63.

100 Kromer EP, Elsner D, Riegger GAJ. Role of neurohumoral systems for pressure induced left ventricular hypertrophy in experimental supravalvular aortic stenosis in rats. *Am J Hypertens* 1991; **4:**521–24.

101 Schunkert H, Dzau VJ, Tang SS, Hirsch AT, Apstein CS, Lorell BH. Increased rat cardiac angiotensin converting enzyme activity and mRNA expression in pressure overload left ventricular hypertrophy. Effects on coronary resistance, contractility, and relaxation. *J Clin Invest* 1990; **86:**1913–20.

102 Bugaisky LB, Gupta M, Gupta MP, Zak R. Cellular and molecular mechanisms of cardiac hypertrophy. In: Fozzard HA, Haber E, Jennings RJ, Katz AM, Morgan HE, eds. *The heart and cardiovascular system, 2nd ed.* New York: Raven Press, 1992: 1621–40.

103 Michel J-B, Lattion A-L, Salzmann J-L et al. Hormonal and cardiac effects of converting enzyme inhibition in rat myocardial infarction. *Circ Res* 1988; **62:**641–50.

104 Hess OM, Schneider J, Koch R, Bamert C, Grimm J, Krayenbuhl HP. Diastolic function and myocardial structure in patients with myocardial hypertrophy. Special reference to normalized viscoelastic data. *Circulation* 1981; **63:**360–71.

105 Krayenbuhl HP, Hess OM, Monard ES, Schneider J, Mall G, Turina M. Left ventric-

ular myocardial structure in aortic valve disease before, intermediate and late after aortic valve replacement. *Circulation* 1989; **79**:744–55.

106 Pearlman ES, Weber KT, Janicki JS, Pietra G, Fischman AP. Muscle fiber orientation and connective tissue content in the hypertrophied human heart. *Lab Invest* 1982; **46**:158–64.

107 Tanaka M, Fujiwara H, Onodera T et al. Quantitative analysis of narrowings of intramyocardial small arteries in normal hearts, hypertensive hearts, and hearts with hypertrophic cardiomyopathy. *Circulation* 1987; **75**:1130–9.

108 Huysman JAN, Vliegen HW, Van der Laarse A, Eulderink F. Changes in non-myocyte tissue composition associated with pressure overload of hypertrophic human hearts. *Pathol Res Pract* 1989; **184**:577–81.

109 Brilla CG, Janicki JS, Weber KT. Impaired diastolic function and coronary reserve in genetic hypertension; role of interstitial fibrosis and medial thickening of intramyocardial coronary arteries. *Circ Res* 1991a; **69**:107–15.

110 Brilla CG, Janicki JS, Weber KT. Cardioreparative effects of lisinopril in rats with genetic hypertension and left ventricular hypertrophy. *Circulation* 1991b; **83**:1771–79.

111 Sen S, Tarazi RC, Bumpus FM. Reversal of cardiac hypertrophy in renal hypertensive rats: medical vs. surgical therapy. *Am J Physiol* 1981; **240**:H408–12.

112 Jalil JE, Janici JS, Pick R, Weber KT. Coronary vascular remodeling and myocardial fibrosis in the rat with renovascular hypertension. Response to captopril. *Am J Hypertens* 1991; **4**:51–55.

113 Brilla CG, Pick R, Tan LP, Janicki JS, Weber KT. Remodeling of the rat right and left ventricles in experimental hypertension. *Circ Res* 1990; **67**:1355–64.

114 Van Krimpen C, Smits JFM, Cleutjens JPM et al. DNA synthesis in the non-infarcted cardiac interstitium after left coronary artery ligation in the rat: effects of captopril. *J Mol Cell Cardiol* 1991; **23**:1245–53.

115 Smits JFM, Van Krimpen C, Schoemaker RG, Cleutjens JPM, Daemen MJAP. Angiotensin II receptor blockade after myocardial infarction in rats: effect on hemodynamics, myocardial DNA synthesis, and interstitial collagen content. *J Cardiovasc Pharmacol* 1992; **20**:772–78.

116 Schelling P, Ganten D, Speck G, Fischer H. Effects of angiotensin II and angiotensin II antagonist saralasin on cell growth and renin in 3T3 and SV3T3 cells. *J Cell Physiol* 1979; **98**:503–14.

117 Ullian ME, Schelling JR, Linas SL. Aldosterone enhances angiotensin II receptor binding and inositol phosphate responses. *Hypertension* 1992; **20**:67–73.

118 Magrini F, Shimizu M, Roberts N, Fouad FM, Tarazi RC, Zanchetti A. Converting enzyme inhibition and coronary blood flow. *Circulation* 1987; **75[Suppl 1]**:168–74.

119 Perondi R, Saino A, Thio RA et al. ACE-inhibition attenuates sympathetic coronary vasoconstriction in patients with coronary artery disease. *Circulation* 1992; **85**:2004–13.

120 Breisch EA, Houser SR, Carey RA, Spann JF, Bove AA. Myocardial blood flow and capillary density in chronic pressure overload of the feline left ventricle. *Cardiovasc Res* 1980; **14**:469–75.

121 Mall G, Mattfeldt T, Hasslacher C, Mann J. Morphological reaction patterns in experimental cardiac hypertrophy — a quantitative stereological study. *Bas Res Cardiol* 1986; **81**:193–201.

122 Tomanek RJ, Palmer PJ, Pfeifer GL, Schreiber KL, Eastham CL, Marcus ML. Morphometry of canine coronary arteries, arterioles and capillaries during hypertension and left ventricular hypertrophy. *Circ Res* 1986; **58**: 38–46.

123 Cimini CM, Weiss HR. Microvascular morphometry and perfusion in renal hypertension-induced cardiac hypertrophy. *Am J Physiol* 1988; **255**:H1384–90.

124 Canby CA, Tomanek RJ. Regression of ventricular hypertrophy abolishes cardiocyte vulnerability to acute hypoxia. *Anat Rec* 1990; **226**:198–206.

125 Waeber B, Brunner HR, Brunner DB, Curtet A, Turini GA, Gavras H. Discrepancy between antihypertensive effect and angiotensin converting enzyme inhibition by captopril. *Hypertension* 1980; **2**:236–42.

126 Cohen ML, Kurz KD. Angiotensin converting enzyme inhibition in tissues from spontaneously hypertensive rats after treatment with captopril or MK-421. *J Pharm Pharmacol* 1982; **220**:63–69.

127 Unger T, Ganten D, Lang RE, Schölkens BA. Is tissue converting enzyme inhibition a determinant of the antihypertensive efficacy of converting enzyme inhibitors? Studies with two different compounds, Hoe498 and MK421, in spontaneously hypertensive rats. *J Cardiovasc Pharmacol* 1984; **6**:872–80.

128 Weishaar RE, Panek RL, Mojor TC, Simmerman J, Rapundalo ST, Taylor DG. Evidence for a functional tissue renin-angiotensin system in the rat mesenteric vasculature and its involvement in regulating blood pressure. *J Pharmacol Exp Ther* 1991; **256**:568–74.

129 Ganten D, Hermann K, Bayer C, Unger T, Lang RE. Angiotensin synthesis in the brain and increased turnover in hypertensive rats. *Science* 1983; **22**:1969–71.

130 Welsch C, Giesen-Crouse EM, Schmidt M, Imbs JL. Tissue converting enzyme activity is low in the kidney of spontaneously hypertensive rats. *J Cardiovasc Pharmacol* 1987; **10[Suppl 7]**:S129–32.

131 Okamura T, Miyazaki M, Inagami T, Toda N. Vascular renin-angiotensin system in two-kidney, one clip hypertensive rats. *Hypertension* 1986; **8**:560–5.

132 Miyazaki M, Okunishi H, Okamura T, Toda N. Elevated vascular angiotensin converting enzyme in chronic two-kidney, one clip hypertension in the dog. *J Hypertens* 1987; **5**:155–60.

133 Hirsch AT, Talsness CE, Schunkert H, Paul M, Dzau VJ. Tissue-specific activation of cardiac angiotensin converting enzyme in experimental heart failure. *Circ Res* 1991; **69**:475–82.

134 King SJ, Booyse FM, Lin P-H, Traylor M, Narkates AJ, Oparil S. Hypoxia stimulates endothelial cell angiotensin-converting enzyme antigen synthesis. *Am J Physiol* 1989; **256**:C1231–38.

135 Campbell DJ, Habener JF. Angiotensinogen gene is expressed and differentially regulated in multiple tissues of the rat. *J Clin Invest* 1986; **78**:31–39.

136 Frederich RC, Kahn BB, Peach MJ, Flier JS. Tissue-specific nutritional regulation of angiotensinogen in adipose tissue. *Hypertension* 1992; **19**:339–44.

137 Kunapuli SP, Benedict CR, Kumar A. Tissue specific hormonal regulation of the rat angiotensinogen gene expression. *Arch Biochem Biophys* 1987; **254**:642–46.

138 Lindpaintner K, Jin M, Niedermaier N, Wilhelm MJ, Ganten D. Cardiac angiotensinogen and its local activation in the isolated perfused beating heart. *Circ Res* 1990; **67**:564–73.

139 Lynch KR, Peach MJ. Molecular biology of angiotensinogen. *Hypertension* 1991; **17**: 263–69.

140 Naftilan AJ, Zuo WM, Inglefinger J, Ryan TJ, Pratt RE, Dzau VJ. Localization and differential regulation of angiotensinogen mRNA expression in the vessel wall. *J Clin Invest* 1991; **87**:1300–11.

141 Coezy E, Darby I, Mizrahi J et al. Inhibition of angiotensinogen production by angiotensin II analogues in human hepatoma cell line. *Am J Physiol* 1989; **257**:C888–95.

142 Iwao H, Nakamura A, Fukui K, Kimura S, Tamaki T, Abe Y. Endogenous angiotensin II regulates hepatic angiotensinogen production. *Life Sci* 1990; **47**:2343–49.

143 Nakamura A, Iwao H, Fukui K et al. Regulation of liver angiotensinogen and kidney renin mRNA levels by angiotensin II. *Am J Physiol* 1990; **258**:E1–6.

144 Dzau VJ, Ingelfinger JR, Pratt RE. Regulation of tissue renin and angiotensin gene expression. *J Cardiovasc Pharmacol* 1986; **8[Suppl 10]**:S11–16.

145 Campbell DJ, Extrarenal renin and blood pressure regulation. An alternative viewpoint. *Am J Hypertens* 1989; **2**:266–75.

146 Hilgers KF, Mann FE, Hilgenfeldt U, Ganten D. Vascular production and regulation of angiotensin. *Blood Vessels* 1991; **28**:201–9.

147 Kuczera M, Hilgers KF, Lisson C et al. Local angiotensin formation in hindlimbs of uremic hypertensive and renovascular hypertensive rats. *J Hypertens* 1991; **9**:41–48.

148 Okamura T, Aimi Y, Kimura H, Murakami K, Toda N. Existence of renin in the endothelium of human artery. *J Hypertens* 1992; **10**:49-53.

149 Inagami T, Mizuno K, Naruse M et al. Active and inactive renin in the adrenal. *J Hypertens* 1989; **2**:311–19.
150 Dzau VJ, Burt DW, Pratt RE. Molecular biology of the renin-angiotensin system. *Am J Physiol* 1988; **255**:F563–73.
151 Hackenthal E, Paul M, Ganten D, Taugner R. Morphology, physiology, and molecular biology of renin secretion. *Physiol Rev* 1990; **70**:1067–116.
152 Sealey JE, Rubattu S. Prorenin and renin as separate mediators of tissue and circulating systems. *Am J Hypertens* 1989; **2**:358–66.
153 Hsueh WA, Baxter JD. Human prorenin. *Hypertension* 1991; **17**:469–79.
154 Danser AHJ, Van den Dorpel MA, Deinum J et al. Renin, prorenin, and immunoreactive renin in vitreous fluid from eyes with and without diabetic retinopathy. *J Clin Endocrinol Metab* 1989; **68**:160–67.
155 Hare GMT, Osmond DH. New renin-angiotensin from plasma prorenin? *Am J Hypertens* 1990; **3**:196–203.
156 Ehlers MRW, Riordan JF. Angiotensin-converting enzyme: new concepts concerning its biological role. *Biochem* 1989; **28**:5311–18.
157 Erdös EG. Angiotensin I converting enzyme and the changes in our concepts through the years. *Hypertension* 1990; **16**:363–70.
158 Okunishi H, Miyazaki M, Okamura T, Toda N. Different distribution of two types of angiotensin II-generating enzymes in the aortic wall. *Biochem Biophys Res Commun* 1987; **149**:1186–92.
159 Schölkens BA, Linz W, Han Y-F. Heart and vascular wall as targets for tissue converting enzyme inhibition. *Clin Exp Theor Pract* 1987; **A9**:427–33.
160 Pipili E, Manopoulos VG, Catravas JD, Maragoudakis ME. Angiotensin converting enzyme activity is present in the endothelium-denuded aorta. *Br J Pharmacol* 1989; **98**:333–35.
161 Fabris B, Jackson B, Cubela R, Mendelsohn FAO, Johnston CI. Angiotensin converting enzyme in the rat heart: studies of its inhibition in vitro and ex vivo. *Cl Exp Pharmacol Physiol* 1989; **16**:309–13.
162 Yamada H, Fabris B, Allen AM, Jackson B, Johnston CI, Mendelsohn FAO. Localization of angiotensin converting enzyme in the rat heart. *Circ Res* 1991; **68**:141–49.
163 Duggan KA, Mendelsohn FAO, Levens NR. Angiotensin receptors and angiotensin I-converting enzyme in rat intestine. *Am J Physiol* 1989; **257**:G504–10.
164 Morin JP, Moulin B, Borghi H, Grise P, Fillastre JP. Comparative regional distribution of angiotensin-I-converting enzyme in the rat, rabbit, dog, monkey and human kidney. *Renal Physiol Biochem* 1989; **12**:96–103.
165 Ikemoto F, Song G-B, Tominaga M, Kanayama Y, Yamamoto K. Angiotensin-converting enzyme in the rat kidney. Activity, distribution, and response to angiotensin-converting enzyme inhibitors. *Nephron* 1990; **55[Suppl 1]**:3–9.
166 Bardelay C, Mach E, Worcel M, Hunt P. Angiotensin-converting enzyme in rat brain and extraneural tissues visualized by quantitative autoradiography using 3H-trandolaprilate. *J Cardiovasc Pharmacol* 1989; **14**:511–18.
167 Correa FMA, Guilhaume SS, Saavedra JM. Autoradiography of angiotensin-converting enzyme in fixed and unfixed rat brain using the specific enzyme inhibitor [I125]351A or a polyclonal antibody and [I125]staphylococcal protein A. *Neurosc Let* 1990; **110**:244–48.
168 Williams TA, Hooper NM, Turner AJ. Characterization of neuronal and endothelial forms of angiotensin converting enzyme in pig brain. *J Neurochem* 1991; **57**:193–99.
169 Ehlers MRW, Fox EA, Strydom DJ, Riordan JF. Molecular cloning of human testicular angiotensin-converting enzyme: the testis isozyme is identical to the C-terminal half of endothelial angiotensin-converting enzyme. *Proc Natl Acad Sci USA* 1989; **86**:7741–45.
170 Ganten D, Unger T, Lang RE. Pharmacological interferences with the renin-angiotensin system. *Arzneim Forsch* 1984; **34(II)**:1391–98.
171 Dzau VJ. Multiple pathways of angiotensin production in the blood vessel wall: evidence, possibilities and hypotheses. *J Hypertens* 1989; **7**:933–36.

172 Maruta H, Arakawa K. Confirmation of direct angiotensin formation by kallikrein. *Biochem J* 1983; **213**:193–200.
173 Gondo M, Maruta H, Arakawa K. Direct formation of angiotensin II without renin or converting enzyme in the ischemic heart. *Jpn Heart J* 1989; **30**:219–29.
174 Okunishi H, Miyazaki M, Toda N. Evidence for a putatively new angiotensin II-generating enzyme in the vascular wall. *J Hypertens* 1984; **2**:277–84.
175 Okamura T, Okunishi H, Ayajiki K, Toda N. Conversion of angiotensin I to angiotensin II in dog isolated renal artery: role of two different angiotensin II-generating enzymes. *J Cardiovasc Pharmacol* 1990; **15**:353–59.
176 Urata H, Kinoshita A, Misono KS, Bumpus FM, Husain A. Identification of a highly specific chymase as the major angiotensin II-forming enzyme in the human heart. *J Biol Chem* 1990; **265**:22348–57.
177 Kawaguchi H, Sawa H, Yasuda H. Effect of atrial natriuretic factor on angiotensin converting enzyme. *J Hypertens* 1990; **8**:749–53.
178 Kawaguchi H, Sawa H, Iizuka K, Yasuda H. Platelet-activating factor stimulates angiotensin converting enzyme activity. *J Hypertens* 1990; **8**:173–77.
179 Dasarathy Y, Stevens J, Lanzillo JJ, Fanburg BL. Elevation of angiotensin converting enzyme in bovine endothelial cells quantitated by an ELISA. *Life Sci* 1990; **47**:883–89.
180 Jackson B, Cubela R, Johnston CI. Tissue angiotensin converting enzyme (ACE) during changes in the renin-angiotensin system. *J Cardiovasc Pharmacol* 1987; **10[Suppl 7]**:S137–40.
181 Chappell MC, Millsted A, Diz DI, Brosnihan KB, Ferrario CM. Evidence for an intrinsic angiotensin system in the canine pancreas. *J Hypertens* 1991; **9**:751–59.
182 Misumi J, Gardes J, Gonzalez M-F, Corvol P, Menard J. Angiotensinogen's role in ANG formation, renin release, and renal hemodynamics in isolated perfused kidney. *Am J Physiol* 1989; **256**:F719–27.
183 Horiba N, Nomura K, Shizume K. Exogenous and locally synthesized angiotensin II and glomerulosa cell functions. *Hypertension* 1990; **15**:190–7.
184 Kifor I, Dzau VJ. Endothelial renin-angiotensin pathway: evidence for intracellular synthesis and secretion of angiotensins. *Circ Res* 1987; **60**:422–28.
185 Englème C, Cressier F, Wood JM. Local formation of angiotensin II in the rat aorta: effect of endothelium. *Br J Pharmacol* 1990; **100**:237–40.
186 Tang S-S, Stevenson L, Dzau VJ. Endothelial renin-angiotensin pathway. *Circ Res* 1990; **66**:103–8.
187 Ziogas J, Story DF. Angiotensin II generation in the rat vena cava: stimulation of local synthesis by beta-adrenoceptor activation. *Naunyn Schmiedeberg's Arch Pharmacol* 1991; **343**:31–36.
188 Mizuno K, Nakamura M, Higashimori K, Inagami T. Local generation and release of angiotensin II in peripheral vascular tissue. *Hypertension* 1988; **11**:223–29.
189 Mizuno K, Tani M, Niimura S et al. Direct evidence for local generation and release of angiotensin II in human vascular tissue. *Biochem Biophys Res Commun* 1989; **165**:457–63.
190 Admiraal PJJ, Danser AHJ, Derkx FHM, Schalekamp MADH. Angiotensin II production in different vascular beds in hypertensive subjects. *J Hypertension* 1991; **9[Suppl 6]**:S208–9.
191 Ball SG. The sympathetic nervous system and converting enzyme inhibition. *J Cardiovasc Pharmacol* 1989; **8[Suppl 10]**:S17–21.
192 Bumpus FM, Catt KJ, Chiu AT et al. Nomenclature for angiotensin receptors. A report of the Nomenclature Committee Council for High Blood Pressure Research. *Hypertension* 1991; **17**:720–1.
193 Iwai N, Inagami T. Identification of two subtypes in the rat type I angiotensin II receptor. *FEBS* 1992; **298**:257–60.
194 Kakar SS, Sellers JC, Devor DC, Musgrove LC, Neill JD. Angiotensin II-type-1 receptor subtype cDNAs: differential tissue expression and hormonal regulation. *Biochem Biophys Res Commun* 1992; **183**:1090–6.
195 Sechi LA, Grady EF, Griffin CA, Kalinyak JE, Schambelan M. Distribution of angiotensin II receptor subtypes in rat and human kidney. *Am J Physiol* 1992; **262**:F236–40.

196 Timmermans PBMWM, Wong PC, Chiu AT, Herblin WF. Nonpeptide angiotensin II receptor antagonists. *TIPS* 1991; **12**:55–62.

197 DeGasparo M, Whitebread S, Mele M et al. Biochemical characterization of two angiotensin II receptor subtypes in the rat. *J Cardiovasc Pharmacol* 1990; **16[Suppl 4]**:S31–35.

198 Crawford KW, Frey EA, Cote TE. Angiotensin II receptor recognized by DuP753 regulates two distinct guanine nucleotide-binding protein signaling pathways. *Mol Pharmacol* 1992; **41**:154–62.

199 Linz W, Martorana PA, Schölkens BA. Local inhibition of bradykinin degradation in ischemic hearts. *J Cardiovasc Pharmacol* 1990; **15[Suppl 6]**:S99–S109.

200 Bottarri SP, King IN, Reichlin S, Dahlström I, Lydon N, DeGasparo M. The angiotensin AT_2 receptor stimulates protein tyrosine phosphatase activity and mediates inhibition of particulate guanylate cyclase. *Biochem Biophys Res Commun* 1992; **183**:206–11.

201 Le Noble FAC, Hekking JWM, Van Straat HWM, Slaaf DW, Struijker Boudier HAJ. Angiotensin II stimulates angiogenesis in the chorio-allantoic membrane of the chick embryo. *Eur J Pharmacol* 1991; **195**:306.

202 Morton JJ, Beattie EC, Griffin SA, MacPherson F, Lyall F, Russo D. Vascular hypertrophy, renin and blood pressure in the young spontaneously hypertensive rat. *Cl Sci* 1990; **79**:523–30.

203 Nakamura Y, Nakamura K, Matsukura T, Nakamura K. Vascular angiotensin converting enzyme activity in spontaneously hypertensive rats and its inhibition with cilazapril. *J Hypertens* 1988; **6**:106–10.

204 Juillière Y, Danchin N, Grentzinger A et al. Role of previous angina pectoris and collateral flow to preserve left ventricular function in the presence or absence of myocardial infarction in isolated total occlusion of the left anterior descending coronary artery. *Am J Cardiol* 1990; **65**:277–81.

205 Salgado MCO, Caldo H. Circulating and tissue renin-angiotensin systems in one-kidney, one clip hypertension. [abstract]. *Hypertension* 1989; **13**:523.

206 Wilson SK, Lynch DR, Ladenson PW. Angiotensin II and atrial natriuretic factor-binding sites in various tissues in hypertension: comparative receptor localization and changes in different hypertension models in the rat. *Endocrinology* 1989; **124**:2799–808.

207 Morishita R, Higaki J, Okunishi H et al. Changes in gene expression of the renin-angiotensin system in two-kidney, one clip hypertensive rats. *J Hypertens* 1991; **9**:187—92.

208 Nishimura M, Milsted A, Block CH, Brosniban KB, Ferrario CM. Tissue renin-angiotensin systems in renal hypertension. *Hypertension* 1992; **20**:158–67.

209 Perich R, Jackson B, Paxton D, Johnston CI. Characterization of angiotensin converting enzyme in isolated cerebral microvessels from spontaneously hypertensive and normotensive rats. *J Hypertens* 1992; **10**:149–53.

210 Shiota N, Miyazaki M, Okunishi H. Increase of angiotensin converting enzyme gene expression in the hypertensive aorta. *Hypertension* 1992; **20**:168–74.

211 Baker KM, Chernin MI, Wixson SK, Aceto JF. Renin-angiotensin system involvement in pressure-overload cardiac hypertrophy in rats. *Am J Physiol* 1990; **259**:H324–32.

212 Socorro L, Vallega G, Nunn A, Moore TJ, Canessa M. Vascular smooth muscle cells from the Milan hypertensive rat exhibit decreased functional angiotensin II receptors. *Hypertension* 1990; **15**:591–99.

213 Ondetti MA. Structural relationships of angiotensin converting-enzyme inhibitors to pharmacologic activity. *Circulation* 1988; **77[Suppl I]**:I74–8.

214 Gohlke P, Urbach H, Schölkens B, Unger T. Inhibition of converting enzyme in the cerebrospinal fluid of rats after oral treatment with converting enzyme inhibitors. *J Pharmacol Exp Ther* 1989; **249**:609–16.

215 Johnston CI, Fabris B, Yamada H et al. Comparative studies of tissue inhibition by angiotensin converting enzyme inhibitors. *J Hypertens* 1989; **7[Suppl 5]**:S11–16.

216 Grima M, Welsch C, Michel B, Barthelmebs M, Imbs JL. In vitro tissue potencies of converting enzyme inhibitors. Prodrug activation by kidney esterase. *Hypertension* 1991; **17**:492–96.

217 Cushman DW, Wang FL, Fung WC, Harvey CM, DeForrest JM. Differentiation of angiotensin-converting enzyme (ACE) inhibitors by their selective inhibition of ACE in physiologically important target organs. *Am J Hypertens* 1989; **2**:294–306.
218 Sun Y, Mendelsohn FAO. Angiotensin converting enzyme inhibition in heart, kidney, and serum studied ex vivo after administration of zofenopril, captopril, and lisinopril. *J Cardiovasc Pharmacol* 1991; **18**:478–86.
219 Chopra M, Besweick H, Clapperton M, Dargie HJ, Smith WE, McMurray J. Antioxidant effects of angiotensin-converting enzyme (ACE) inhibitors: free radical and oxidant scavenging are sulfhydryl dependent, but lipid peroxidation is inhibited by both sulfhydryl- and nonsulfhydryl-containing ACE-inhibitors. *J Cell Physiol* 1992; **19**:330–40.
220 Przyklenk K, Kloner RA. Relationships between structure and effects of ACE-inhibitors: comparative effects in myocardial ischaemic/reperfusion injury. *Br J Clin Pharmacol* 1989; **28**:167S–75S.
221 Grover GJ, Sleph PG, Dzwonczyk S et al. Effects of different angiotensin-converting enzyme (ACE) inhibitors on ischemic isolated rat hearts: relationship between cardiac ACE-inhibition and cardioprotection. *J Pharmacol Exp Ther* 1991; **257**:919–29.
222 Van Gilst WH. Protection of the myocardium against postischemic reperfusion damage. *J Cardiovasc Pharmacol* 1989; **14[Suppl 9]**:S49–54.
223 Mancia G, Perondi R, Saino A et al. Hemodynamic effects of ACE-inhibitors. *Eur Heart J* 1990; **11[Suppl D]**:27–32.
224 Ferrario CM. The renin-angiotensin system: importance in physiology and pathology. *J Cardiovasc Pharmacol* 1990; **15[Suppl 3]**:1–5.
225 Zimmerman BG, Wong PC, Kounenis GK, Kraft EJ. No effect of intrarenal converting enzyme inhibition on canine renal blood flow. *Am J Physiol* 1982; **243**:H277–83.
226 Oliver JA, Sciacca RR, Cannon PJ. Effect of inhibition of angiotensin II with captopril on renal overflow of norepinephrine. *Hypertension* 1982; **4**:257–63.
227 Rosendorff C, Patton J, Radford HM, Kalliatakis B. Alpha-adrenergic and angiotensin II pressor sensitivity in hypertensive patients treated with an angiotensin-converting enzyme inhibitor. *J Cardiovasc Pharmacol* 1992; **19[Suppl 6]**:105–9.
228 Tsuda K, Shima H, Kuchii M, Nishio I, Masuyama Y. Effects of captopril on neurosecretion and vascular responsiveness in hypertension. *Clin Exp Hypert Theor Pract* 1987; **A9**:375–79.
229 Taddei S, Favilla S, Duranti P, Simonini N, Salvetti A. Vascular renin-angiotensin system and neurotransmission in hypertensive persons. *Hypertension* 1991; **18**:266–77.
230 Dominiak P, Elfrath A, Türck D. Effects of chronic treatment with ramipril, a new ACE blocking agent, on presynaptic sympathetic nervous system of SHR. *Clin Exp Theor Pract* 1987; **A9**:369–73.
231 Mohara O, Masuyama Y. Captopril attenuates reflex adrenergic response in essential hypertension. *Life Sci* 1991; **48**:203–7.
232 Berecek KH, Kirk KA, Nagahama S, Oparil S. Sympathetic function in spontaneously hypertensive rats after chronic administration of captopril. *Am J Physiol* 1987; **252**:H796–806.
233 Bunag RD, Eriksson L, Tanabe S. Baroreceptor reflex enhancement by chronic intracerebroventricular infusion of enalapril in normotensive rats. *Hypertension* 1990; **15**:284–90.
234 Giudicelli JF, Berdeaux A, Edouard A, Richer C, Jacolot D. The effect of enalapril on baroreceptor mediated reflex function in normotensive subjects. *Br J Clin Pharmacol* 1985; **20**:211–18.
235 Bao G, Gohlke P, Qadri F, Unger T. Chronic kinin receptor blockade attenuates the antihypertensive effect of ramipril. *Hypertension* 1992; **20**:74–79.
236 Bönner G, Preis S, Schunk U, Toussaint C, Kaufmann W. Hemodynamic effects of bradykinin on systemic and pulmonary circulation in healthy and hypertensive humans. *J Cardiovasc Pharmacol* 1990; **15[Suppl 6]**:46–56.
237 Waeber B, Niederberger M, Gavras H, Nussberger J, Brunner HR. Hemodynamic

effects of a kinin antagonist. *J Cardiovasc Pharmacol* 1990; **15[suppl 6]**:78–82.

238 Oliver JA, Sciacca RR, Cannon PJ. Renal vasodilation by converting enzyme inhibition. Role of renal prostaglandins. *Hypertension* 1983; **5**:166–71.

239 Tobia AJ, Giardino EC. Renal vasodilator responses to captopril in dogs pretreated with indomethacin. *Proc Soc Exp Biol Med* 1981; **167**:242–47.

240 Raya TE, Gay RG, Aguirre M, Goldman S. Importance of venodilatation in prevention of left ventricular dilatation after chronic large myocardial infarction in rats: a comparison of captopril and hydralazine. *Circ Res* 1989; **64**:330–7.

241 Raya TE, Fonken SJ, Lee RW et al. Hemodynamic effects of direct angiotensin II blockade compared to converting enzyme inhibition in rat model for heart failure. *Am J Hypertens* 1991; **4[Suppl]**:334–40.

242 Ventura HO, Frohlich ED, Messerli FH, Kobrin I, Kardon MB. Cardiovascular effects and regional blood flow distribution associated with angiotensin converting enzyme inhibition (captopril) in essential hypertension. *Am J Cardiol* 1985; **55**:1023–26.

243 Bauer JH. Angiotensin converting enzyme inhibitors. *Am J Hypertens* 1990; **3**:331–37.

244 Nelissen-Vrancken HJMG, Leenders PJA, Struijker Boudier HAJ, Smits JFM. Renal hemodynamic effects of non-hypotensive doses of ACE-inhibitors in hypertension and heart failure in rats. *J Cardiovasc Pharmacol* 1992; **19**:163–68.

245 Antonaccio MJ, Kerwin L. Pre- and postjunctional inhibition of vascular sympathetic function by captopril in SHR. *Hypertension* 1981; **3[Suppl 1]**:54–62.

246 Clappison BH, Anderson WP, Johnston CI. Renal hemodynamics and renal kinins after angiotensin-converting enzyme inhibition. *Kidney Int* 1981; **20**:615–20.

247 Zimmerman BG, Raich PC, Vavrek RJ, Stewart JM. Bradykinin contribution to renal blood flow effect of angiotensin converting enzyme inhibitor in the conscious sodium-restricted dog. *Circ Res* 1990; **66**:234–40.

248 Hall JE. Intrarenal actions of converting enzyme inhibitors. *Am J Hypertens* 1989; **2**:875–84.

249 Mimran A, Ribstein J, DuCailar G. Converting enzyme inhibitors and renal function in essential and renovascular hypertension. *Am J Hypert* 1991; **4[Suppl]**:7–14.

250 Hollenberg NK, Swartz SL, Passan DR, Williams GH. Increased glomerular filtration rate after converting enzyme inhibition in essential hypertension. *New Engl J Med* 1979; **301**:9–12.

251 Simamoto K, Matsuki T, Iimura O. Angiotensin converting enzyme inhibitors and the kallikrein-kinin system. *J Cardiovasc Pharmacol* 1990; **15[Suppl 6]**:83–90.

252 DiBona GF. Neural control of renal tubular sodium reabsorption in the dog. *Fed Proc* 1978; **37**:1214–17.

253 Smits JFM, Struijker Boudier HAJ. Effects of drug treatment on cardiovascular structure. *Neth J Cardiol* 1991; **3**:87–92.

254 Rogg H, De Gasparo M, Grädel E, Stulz P, Erne P. Identification of angiotensin-II-receptor subtypes in human atrial tissue [abstract]. *Eur Heart J* 1991; **12[Suppl]**:224.

255 Chansel D, Czekalski S, Pham P, Ardaillou R. Characterization of angiotensin II receptor subtypes in human glomeruli and mesangial cells. *Am J Physiol* 1992; **262**: F432–41.

256 Whitebread S, Mele M, Kamber B, DeGasparo M. Preliminary biochemical characterization of two angiotensin II receptor subtypes. *Biochem Biophys Res Commun* 1989; **163**:284–91.

257 Takayanagi R, Ohnaka K, Sakai Y et al. Molecular cloning, sequence analysis and expression of a cDNA encoding human type-1 angiotensin II receptor. *Biochem Biophys Res Commun* 1992; **183**:910–6.

258 Bergsma DJ, Ellis C, Kumar C et al. Cloning and characterization of a human angiotensin II type 1 receptor. *Biochem Biophys Res Commun* 1992; **183**:989–95.

259 Paulson OB, Waldemar G. ACE-inhibitors and cerebral blood flow. *J Hum Hyperten* 1990; **4[Suppl. 4]**:69–73.

260 Smits JFM, Struijker Boudier HAJ. Systemic and regional hemodynamics following acute inhibition of angiotensin I-converting enzyme in the conscious spontaneously hypertensive rat. *Prog Pharmacol* 1984; **5**:39–49.

261 Richer C, Doussau M-P, Giudicelli J-F. Effects of captopril and enalapril on regional vascular resistance and reactivity in spontaneously hypertensive rats. *Hypertension* 1983; **5**:312–20.

262 Roberts DH, Tsao Y, Grimmer SFM, Winstanley PA, Orme MLE, Breckenridge A. Haemodynamic effects of atenolol, labetolol, pindolol and captopril: a comparison in hypertensive patients with special reference to changes in blood flow, heart rate and left ventricular function. *Br J Clin Pharmacol* 1987; **24**:163–72.

263 Strocchi E, Luigi Malini P, Ciavarella A et al. The effect of ACE-inhibition on peripheral hemodynamics in normotensive and hypertensive patients with type II diabetes. *J Clin Pharmacol* 1991; **31**:140–3.

264 Wang Y-X, Gavras I, Wierzba T, Gavras H. Comparison of systemic and regional hemodynamic effects of a diuretic, an angiotensin II receptor antagonist, and an angiotensin-converting enzyme inhibitor in conscious renovascular hypertensive rats. *J Lab Clin Med* 1992; **119**:267–72.

265 Garavaglia GE, Messerli FH, Nunez BD, Schmieder RE, Frohlich ED. Immediate and short-term cardiovascular effects of a new converting enzyme inhibitor (lisinopril) in essential hypertension. *Am J Cardiol* 1988; **62**:912–16.

266 Bray L, Lartaud I, Muller F, Atkinson J, Capdeville C. Effects of the angiotensin I converting enzyme inhibitor perindopril on cerebral blood flow in awake hypertensive rats. *Am J Hypertens* 1991; **4**:246S–52S.

267 Harrap SB. Angiotensin converting enzyme inhibitors, regional vascular hemodynamics, and the development and prevention of experimental genetic hypertension. *Am J Hypertens* 1991; **4**:212S–16S.

268 Clozel J-P. Effects of nitrendipine and cilazapril alone or in combination on hemodynamics and regional blood flows in conscious spontaneously hypertensive rats. *J Cardiovasc Pharmacol* 1988; **12**:600–7.

269 Richer C, Mulder P, Fornes P, Richard U, Camilleri JP, Giudicelli JF. Hemodynamic and morphological effects of quinapril during genetic hypertension development. *J Cardiovasc Pharmacol* 1991; **18**:631–42.

9

ACE inhibitors and atherosclerosis

Gunnar Aberg, Mark C Kowala and Christopher J Molloy

Hypertension as a risk factor for atherosclerosis

There is overwhelming evidence that hypertension accelerates coronary heart disease and atherosclerosis.[1–3] Thus normalizing the blood pressure with antihypertensive drugs should delay the atherosclerotic process. Two recent clinical trials have demonstrated that the formation of new coronary lesions was diminished as a result of blood pressure reductions with calcium channel blockers.[4,5] All modern antihypertensive drugs significantly lower blood pressure, however as described below, drugs that interfere with the renin-angiotensin system may have additional, direct and beneficial effects on both the progression and the regression of atherosclerosis. The purpose of this chapter is not to discuss the benefits of blood pressure reduction with antihypertensive agents, but rather to review the potential 'direct' effects of ACE inhibitors on atherogenesis.

Effects of ACE inhibitors on plasma lipids

It has been suggested that captopril may reduce atherosclerosis by lowering total serum cholesterol and triglyceride concentrations in hyperlipidemic patients.[6] Sasaki et al did not report any significant changes in total, high density lipoprotein (HDL) — or low density lipoprotein (LDL) — cholesterols or total triglyceride levels in patients given captopril (37.5–50 mg/day) or enalapril (5–10 mg/day).[7] However during three months of therapy with either ACE inhibitor there were increases in the antiatherogenic HDL_2-cholesterol fraction, in apo A-I, and in apo A-II as well as decreases in the atherogenic apoB and apoB/A-I ratio.

Since the results of Sasaki et al. have not been repeated, it was of considerable interest that in the hyperlipidemic hamster, fosinopril had beneficial effects on the plasma lipid profile.[8] Male Golden Syrian hamsters (F_1B hybrid strain) fed chow supplemented with 0.05% cholesterol and 10% coconut oil become mildly hyperlipidemic, where the low-density lipoprotein (LDL) fraction carried most of the plasma cholesterol.[9] In control hamsters, three weeks of feeding this atherogenic diet produced total, LDL plus very low density lipoproteins (VLDL) and high-density lipoprotein [HDL]) cholesterols and total triglycerides averaging at 236, 161, 75 and 428 mg/dl, respectively. Compared to controls, three weeks of fosinopril treatment at 50 mg/kg/day reduced total cholesterols by 17%, LDL plus VLDL cholesterols fell by 27% and total triglycerides were reduced by 45%. Captopril (50 mg/kg/day) only reduced HDL cholesterol

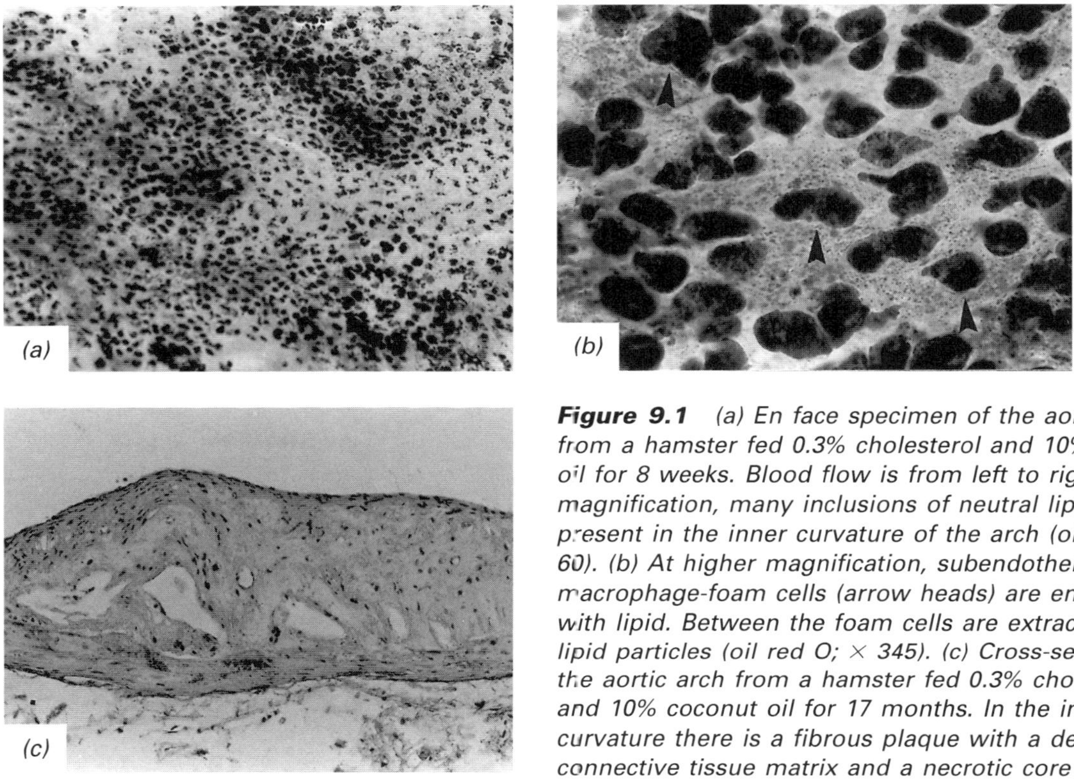

Figure 9.1 *(a) En face specimen of the aortic arch from a hamster fed 0.3% cholesterol and 10% coconut oil for 8 weeks. Blood flow is from left to right. At low magnification, many inclusions of neutral lipid are present in the inner curvature of the arch (oil red O; × 60). (b) At higher magnification, subendothelial macrophage-foam cells (arrow heads) are engorged with lipid. Between the foam cells are extracellular lipid particles (oil red O; × 345). (c) Cross-section of the aortic arch from a hamster fed 0.3% cholesterol and 10% coconut oil for 17 months. In the inner curvature there is a fibrous plaque with a dense connective tissue matrix and a necrotic core of cholesterol crystals (hematoxylin stain; × 75).*

by 20% at three weeks.[8] The minimum effective dose of either fosinopril or captopril in the hamster was 50 mg/kg, and subsequent plasma or blood concentrations of each drug were within the range found in humans.[8]

To determine the mechanism behind fosinopril's hypolipidemic effect, HepG2 cells (a cell model of hepatocytes) were treated with 10–50 μg/ml of fosinopril. To measure LDL receptor activity, the uptake of fluorescent LDL (dioctadecyltetramethylindocarbocyanine (DiI) perchlorate labeled LDL) into HepG2 cells was quantified. Fosinopril at 40 and 50 μg/ml increased the uptake of DiI-LDL by 83% and 138%, respectively (M Rosser and S O'Conner, unpublished observations). Thus it was feasible that fosinopril reduced the LDL plus VLDL cholesterols by upregulating hepatic LDL receptor activity. This dramatic effect of fosinopril (the pro-drug) may be unrelated to ACE inhibition since fosinoprilat (the active drug that inhibits ACE) had no effect on plasma lipids in the hamster or on DiI-LDL uptake in HepG2 cells. Although intriguing, the effect of fosinopril on plasma lipids in hamsters may be specific for this species, since fosinopril had not altered plasma lipids in cholesterol-fed rats or pigs or in humans. However, in the pig study, a potent atherogenic diet produced very high cholesterol levels, where hepatic LDL receptor activity was possibly overwhelmed by the high concentrations of the ligand LDL. Thus severe hypercholesterolemia may have produced unfavorable conditions for demonstrating an elevation of LDL receptor activity in previous studies in pigs and monkeys. As for humans, a double-blind study of the effect of fosinopril on plasma lipids has yet to be undertaken.

Effects of ACE inhibitors on early atherosclerosis

In early atherosclerosis, the monocyte derived macrophage is the major participant in the development of fatty streaks. Hypercholesterolemia promotes the adhesion and migration of monocytes into the artery wall, where the tissue macrophages scavenge modified LDL, esterify the intracellular cholesterol forming droplets of cholesteryl esters and transform into macrophage-foam cells. Foam cells accumulate in the intima to form fatty streaks and they are also present in mature atherosclerotic plaques.[10–12]

The hamster model

The F_1B hybrid strain of Golden Syrian hamster has been used to study the development of the fatty streak in vivo. After only several weeks of feeding chow plus 0.05% cholesterol and 10% coconut oil, macrophage-foam cells rapidly accumulated in the subendothelial space along the inner curvature of the aortic arch.[8,13] If the dietary cholesterol concentration was increased to 0.2% or 0.3%, plasma total cholesterol levels become severely elevated ranging from 700 to 2600 mg/dl, with most of the cholesterol in the VLDL fraction. With the more potent diet, macrophage-foam cell accumulation was accelerated by four-fold compared to the 0.05% cholesterol diet. After 12 months, fibro-fatty plaques developed in the inner curvature of the arch, and their structure was similar to that of human lesions. The hamster plaques had a fibrous cap of smooth muscle cells, a dense connective tissue matrix, a few macrophage-foam cells, and a necrotic core of cholesterol crystals (Fig. 9.1).

The hamster model has been used to investigate the effect of two ACE inhibitors, fosinopril and captopril, on early atherosclerosis. Initially the short-term effect of these ACE inhibitors on blood pressure was determined. Normotensive hamsters were treated with 50 mg/kg of fosinopril or 50 mg/kg of captopril. Fosinopril produced a slight decrease in mean arterial pressure only at six hours after the first dose, whereas captopril was ineffective during the first 24 hours. After eight days of chronic dosing neither ACE inhibitor lowered blood pressure; thus there were no sustained alterations of blood pressure or heart rate with either drug after oral dosing.[8] (Higher doses of fosinopril or captopril decreased mean arterial pressure.) The pressor response following an intravenous injection of angiotensin I was inhibited by 50 mg/kg of either drug, indicating that ACE activity was impaired in vivo.[8]

In an atherosclerosis study, hamsters were fed an atherogenic diet of normal hamster chow plus 0.05% cholesterol and 10% coconut oil. The animals were divided into three groups. One group were controls, the second group received 50 mg/kg/day of fosinopril and the third group was treated with 50 mg/kg/day of captopril. Animals were treated for three

	Mean arterial pressure (mmHg)	Heart rate (beats/min)	Total cholesterol (mg/dl)	HDL cholesterol (mg/dl)	Total triglycerides (mg/dl)	n
1K–1C-control	138 ± 14	317 ± 14	244 ± 43	27 ± 7	12 ± 12	10
1K–1C-fosinopril 50 mg/kg/day	122 ± 18	291 ± 19	260 ± 73	35 ± 10	16 ± 16	9

Table 9.1
Mean arterial pressure, heart rate and plasma lipids (mean ± SD) of 1K–1C-control and treated rats at 4 weeks.

weeks, on day 18 fasted hamsters were bled for plasma lipid measurements, and on day 21 mean arterial pressure was monitored. The aortic arch was fixed by perfusing formalin, stained with oil red-O and examined *en face* by light microscopy. The number of macrophage-foam cells per square millimeter was counted, their average size was determined and the total area of the foam cells (fatty streak area) was measured, as described elsewhere.[13]

In the three-week atherosclerosis study, the fosinopril and captopril treated groups had similar blood pressures and heart rates as the controls (136 mmHg; 354 beats/minute). In the aortic arch, controls had 72 macrophage-foam cells/mm^2, an average foam cell size of 109 μm^2 and a fatty streak area of 113 000 μm^2. Fosinopril reduced the number of foam cells/mm^2, foam cell size and fatty streak area by 85%, 35% and 90%, respectively. Captopril treatment reduced these parameters by 44%, 16% and 53%.[8] In conclusion, captopril impeded the progression of early atherosclerosis without altering plasma LDL cholesterol and blood pressure. The data suggests that inhibiting ACE (or kininase II) directly suppressed the formation of the fatty streak. Fosinopril probably further suppressed atherogenesis by lowering plasma lipids.

Hypertensive-hypercholesterolemic rat models

Normotensive rats fed cholesterol have been used to study the pathogenesis of early atherosclerosis[14] and hypertensive rats were investigated to determine the effect of elevated blood pressure on the structure of arteries.[15] Previously, the influence of ACE inhibitors on atherosclerosis was studied in animal models with only one risk factor, either hypercholesterolemia or hypertension. In this section we will discuss the effect of ACE inhibitors on accelerated atherosclerosis in rat models with two risk factors, that is hypertension and hypercholesterolemia. Hypertensive rats fed with cholesterol were used for this study, and blood pressure was elevated by either renin-dependent or renin-independent mechanisms. Fosinopril was first tested in the 1-kidney–1-clip (1K–1C) rat, which is a model of acute renin-dependent hypertension. In this model, atherosclerosis was accelerated by four to six times compared to normotensive rats.[16] Two

	Subendothelial macrophages/mm²	Subendothelial macrophage-foam cells/mm²	n
1K–1C-control	97 ± 76	376 ± 184	10
1K–1C-fosinopril 50 mg/kg/day	37 ± 35*	137 ± 72*	9

*$P < 0.05$ compared to the 1K–1C-control group (Students' t test).

Table 9.2
Subendothelial macrophages and macrophage-foam cells/mm² (mean ± SD) in 1K–1C-control and treated rats at 4 weeks.

	Mean arterial pressure (mmHg)	Heart rate (beats/min)	Total cholesterol (mg/dl)	HDL cholesterol (mg/dl)	Total triglycerides (mg/dl)	n
DOCA-salt control	115 ± 18*	318 ± 39*	365 ± 231	15 ± 4	63 ± 18	8
DOCA-salt-fosinopril 42 mg/kg/day	119 ± 15	286 ± 51	347 ± 112	23 ± 10	72 ± 14	8

*n = 7

Table 9.3
Mean arterial pressure, heart rate and plasma lipids (mean ± SD) in DOCA-salt-control and treated rats.

groups were established, 1K–1C-controls and 1K–1C rats treated with fosinopril. The dose of fosinopril used in this study was 50 mg/kg/day in the drinking water and the diet was normal chow plus 4% cholesterol and 1% cholic acid. Unilateral nephrectomy, clipping the contralateral renal artery, cholesterol feeding and drug treatment started simultaneously. Blood pressure, plasma lipids and atherosclerosis were measured four weeks later.

In normotensive rats, mean arterial pressure averaged at 100 mmHg.[15,16] Rats in the 1K–1C group were hypertensive with an average mean arterial pressure of 138 mmHg. Surprisingly, fosinopril only slightly reduced this parameter by 12% while heart rate was unchanged (Table 9.1). Plasma cholesterol in rats fed chow averaged at 60 mg/dl, whereas the 4% cholesterol and 1% cholic acid diet increased this lipid fraction up to 400 mg/dl.[14] In this study

	Subendothelial macrophages/mm²	Subendothelial macrophage-foam cells/mm²	n
DOCA-salt-control	19 ± 10	357 ± 246	8
DOCA-salt-fosinopril 42 mg/kg/day	32 ± 16	393 ± 145	8

Table 9.4
Subendothelial macrophages, macrophage-foam cells/mm² (mean ± SD) in DOCA-salt-control and treated rats.

the plasma total cholesterols were similarly elevated in the 1K–1C-control and treated groups, indicating that fosinopril was unable to reduce cholesterol levels in rats.

Early atherosclerosis parameters were measured in the thoracic aorta adjacent to the first and second intercostal branches. The aortae were stained with hematoxylin and oil red-O[14] and examined *en face* by light microscopy. Subendothelial macrophages and macrophage-foam cells were counted and divided by the area of the specimen. Compared to 1K–1C controls, fosinopril decreased the accelerated accumulation of macrophages by 62% and macrophage-foam cells by 64% in rats with renin-dependent hypertension (Table 9.2). In the same hypertensive 1K–1C model, captopril (28 mg/kg/day) also impeded the progression of accelerated atherosclerosis at one and two weeks, however in that study, elevated mean arterial pressure was reduced to normal levels.[16]

In a second experiment, fosinopril was tested in a low-renin model of mineralocorticoid hypertension. In this model, atherosclerosis was accelerated three to nine times faster compared to normotensive animals (Mark C Kowala, unpublished observations). Two groups were established. The first group were two-kidney rats treated with 30 mg/kg/week of deoxycorticosterone acetate (DOCA) for four weeks; they were fed the 4% cholesterol, 1% cholic acid diet and drank 1% sodium chloride (DOCA-salt). The second group of DOCA-salt rats fed with cholesterol received fosinopril at 42 mg/kg/day in the 1% sodium chloride drinking solution. Cholesterol feeding, DOCA-salt treatment and drug dosing started on the same day.

At four weeks, mean arterial pressure in DOCA-salt rats averaged at 115 mmHg (Table 9.3) which was higher than the usual 100 mmHg of normotensive rats.[15,16] Thus DOCA-salt treatment induced mild hypertension, however blood pressure and heart rate were not changed by fosinopril treatment. In both groups, total cholesterols were similarly elevated and thus fosinopril failed to alter the plasma lipid profile.

Early atherosclerosis in the thoracic aorta was quantified as described earlier. Fosinopril failed to decrease the accelerated accumulation of subendothelial macrophages and macrophage-foam cells when compared to DOCA-salt controls (Table 9.4). Thus in a low-renin model of hypertension there was no significant effect of ACE inhibition on accelerated atherogenesis. From these rat studies, it appeared that

Aortic region	Control (% intimal surface) N = 8	Captopril (% intimal surface) N = 14
Total aorta	47.9 ± 3.6	29.7 ± 3.9
Ascending and arch	76.3 ± 4.7	76.3 ± 3.7
Descending thoracic	48.9 ± 5.2	15.3 ± 3.9
Abdominal	28.6 ± 2.5	26.2 ± 4.4

Table 9.5
Effects of captopril on aortic surface atherosclerosis in Watanabe Heritable Hyperlipidemic Rabbits

ACE inhibitors prevented aggravated atherosclerosis in renin-dependent hypertension, but not in renin-independent models. The results suggested that ACE inhibitors reduced accelerated atherosclerosis by antagonizing the renin-angiotensin system. (Whether fosinopril also inhibited kininase II in the 1K–1C rat remains to be determined.) It may prove useful to keep this fact in mind when designing protocols for further preclinical and clinical studies since there is no reason to believe that ACE inhibitors, renin inhibitors or angiotensin-receptor antagonists should prevent atherosclerosis under conditions where the renin-angiotensin system is not operating.

Effects of ACE inhibitors on the progression of atherosclerosis

The formation of complex atherosclerotic lesions occur when monocyte/macrophages infiltrate regions where smooth muscle cells have already accumulated, or when macrophages and endothelial cells recruit medial smooth muscle cells into the intima and induce smooth muscle cell proliferation. As the processes of leukocyte infiltration, smooth muscle migration and proliferation continue, a complex plaque ultimately forms, which has a fibrous cap of smooth muscle cells, macrophage-foam cells, a dense connective tissue matrix and a necrotic core of cholesterol crystals.[17,18] Several animal models have been used to investigate drug interactions with the formation of complex arterial lesions. In particular we will focus on studies that determined the role of the renin-angiotensin system on the progression of atherosclerosis.

Rabbit models

Both adrenergic β-blockers[19] and calcium antagonists[20] have been shown to inhibit atherogenesis in cholesterol-fed animals, but the same drugs have failed to influence the development of atherosclerosis in the Watanabe heritable hyperlipidemic (WHHL) rabbit.[21,22] It was therefore of considerable interest when Chobanian et al. showed that the ACE inhibitor captopril reduced aortic atherosclerosis in WHHL rabbits.[23] In this study, captopril was added to the diet in doses (25–50 mg/kg) which reduced serum ACE activity and treatment was continued for 3–12 months. A significant reduction in atherosclerosis was observed in the

Group	Carotid bifurcation	Common carotid	Thoracic aorta	Abdominal aorta
Control	2.6 ± 1.7	0.2 ± 0.1	0.0 ± 0.0	0.1 ± 0.1
High cholesterol	55.7 ± 10.8†	13.6 ± 3.6†	5.9 ± 2.5	9.0 ± 2.3†
Low captopril	32.7 ± 9.4†	2.7 ± 1.1*	5.2 ± 2.0	4.7 ± 1.9
High captopril	18.8 ± 7.7*	0.6 ± 0.4*	3.7 ± 2.1	2.9 ± 1.1*

Mean ± SEM; N = 6
*Significantly different from the high cholesterol group ($P < 0.05$)
†Significantly different from the control group ($P < 0.05$)

Table 9.6
Effects of captopril on per cent intimal lesions in carotid arteries and aortic segments in monkeys given a high-cholesterol diet and treated with captopril for 6 months.

descending aortae of the captopril-treated animals compared to control WHHL rabbits (Table 9.5). The intimal cholesterol content was also reduced in the descending aorta. There was a marked reduction in arterial blood pressure, but it was unlikely that the reduction in blood pressure was the only reason for the reduced atherosclerosis since other blood pressure lowering agents did not have beneficial effects on atherosclerosis in the WHHL rabbit.[21–22]

Non-human primate models

The effects of captopril on atherosclerosis in cholesterol-fed cynomolgus monkeys has been investigated.[24] Twenty-four monkeys were divided into four groups:

- a control group given a normal monkey diet and placebo medication;
- a high cholesterol group given a cholesterol diet and placebo medication;
- a low-dose captopril group given the cholesterol diet and 25 mg/kg of captopril twice daily;
- and a high-dose captopril group given the cholesterol diet and 50 mg/kg of captopril twice daily.

The doses seem high, but there was a low bioavailability of captopril in these monkeys and the plasma levels of captopril were approximately equivalent to those seen in patients after high clinical doses of the drug. There was no effect of captopril on the levels of total serum cholesterol, HDL cholesterol, or triglycerides. The total cholesterol/HDL ratio was also unaffected by captopril. Blood pressures and heart rates were assessed before the start of drug administration and at three and six months into the study. After six months of treatment, the progression of atherosclerosis was assessed by gross pathology, histopathology, and biochemical methods.

Group	Right coronary artery	Left circumflex artery	Left anterior descending artery
Control	0.0 ± 0.0	0.0 ± 0.0	0.0 ± 0.0
High cholesterol	5.9 ± 2.3†	13.4 ± 4.6†	6.3 ± 1.9†
Low captopril	4.1 ± 2.6	5.5 ± 3.7	1.6 ± 1.3*
High captopril	0.0 ± 0.0*	0.9 ± 0.6*	0.1 ± 0.1*

Mean ± SEM; N = 6
*Significantly different from the high cholesterol group ($P < 0.05$).
†Significantly different from the control group ($P < 0.05$).

Table 9.7
Effects of captopril on per cent intimal lesions in coronary arteries of monkeys given a high-cholesterol diet and treated with captopril for 6 months.

There was a statistically significant reduction of blood pressure by the high dose of captopril; the low dose only reduced the diastolic pressure and only after six months treatment. As shown in Table 9.6, the intimal lesion areas were significantly reduced in the carotid arteries and in the abdominal aorta. Few animals demonstrated atherosclerosis in the thoracic aorta. The most obvious antiatherosclerotic effects of captopril in the cynomolgus monkeys were seen in the coronary arteries (Table 9.7), which were essentially free of atherosclerosis in the animals treated with the high dose of captopril.

It is not believed that the reduction in blood pressure was of major importance for the antiatherosclerotic effect of captopril in this study since nifedipine reduced blood pressure to a similar degree in a previous study without having any effect on atherosclerosis in the monkeys.[25]

Mini-pig model

In this study[26] atherosclerosis was produced in mini-pigs of the Gottingen strain by adding 11% egg yolk and 1% cholesterol to the diet for 52 weeks (total cholesterols averaging 615 mg/dl). The animals were divided into three groups. The first group was fed the atherogenic diet alone and served as the control. The animals in the second group were treated with 80 mg/kg/day of captopril which was added to the atherogenic diet on an individual basis. The animals in the third group were treated with 8 mg/kg/day of fosinopril. Both drugs produced a significant reduction in serum ACE activity associated with a reactive rise in plasma renin activity and a slight fall in serum aldosterone concentration. The drug treatment had no significant effects on plasma lipids and lipoproteins.

After 52 weeks of the treatment, the aorta, the carotid and the coronary arteries were removed and examined for atherosclerosis. Atherosclerotic plaques had developed in the abdominal aorta whereas fatty streaks were present in the thoracic aorta and coronary arteries. The extent of atherosclerosis was determined by gross pathology, histopathology, and biochemical methods. The per cent visible atherosclerosis was calculated from color photographs of the intimal aspect of the abdominal aorta and is shown in Table 9.8.

Group	N	% surface with lesions
Control	11	19.1 ± 4.4
Captopril	11	4.5 ± 1.4*)
Fosinopril	10	5.1 ± 1.0*)

Table 9.8
*Gross pathology of abdominal aortas from minipigs fed a high cholesterol diet. Two groups of animals were treated with captopril and fosinopril, respectively. The areas of the abdominal lumen surface covered with fatty streaks and plaques were quantified and the surface area with lesions is expressed as per cent of the total abdominal aortic area. *$P < 0.01$.*

Effects of ACE inhibitors on regression of atherosclerosis

Hypertensive patients treated with ACE inhibitors probably have preexisting atherosclerosis in the coronary arteries. In this section we will describe a study which determined whether the ACE inhibitors fosinopril and captopril could induce the regression of early atherosclerosis in hamsters fed with cholesterol. First, a separate progression experiment established that feeding F_1B hamsters with chow plus 0.05% cholesterol and 10% coconut oil for 12 weeks produced a continuous increase in the number of macrophage-foam cells accumulating in the inner curvature of the aortic arch.[27]

In the regression study, three groups of hamsters were fed the same atherogenic diet used in the progression experiment. The first group of control hamsters had their plasma lipids, blood pressure and atherosclerosis parameters quantified by four weeks. The second and third groups were also fed the atherogenic diet for four weeks, then these two groups were gavaged with either 100 mg/kg/day of fosinopril or 100 mg/kg/day of captopril for another six weeks, while still receiving the atherogenic diet. By week 10 plasma lipids, blood pressure and atherosclerosis were determined in the drug-treated groups, and these parameters were compared to those of the four-week controls. Control hamsters had total, LDL plus VLDL, and HDL cholesterols of 339, 269 and 70 mg/dl, respectively. Total triglycerides were 444 mg/dl. By week 10, fosinopril had significantly reduced total, LDL plus VLDL, and HDL cholesterols by 58%, 70%, and 16%, while total triglycerides fell by 56%. Captopril at week 10 failed to change the LDL plus VLDL cholesterols and total triglyceride levels, however HDL and total cholesterols decreased by 23% and 28%, respectively.[27] These results confirm the hypolipidemic effect of fosinopril in the hamster model.[8]

Captopril reduced mean arterial pressure by 9% and both ACE inhibitors lowered heart rate by 8% to 11%, when compared to the controls (137 mmHg; 384 beats/min). Early atherosclerosis in the aortic arch was quantified as described previously. Control hamsters at four weeks had 87 macrophage-foam cells/mm^2, an average foam cell size of 113 μm^2, a fatty streak area of 125 000 μm^2, and 144 000 μm^2 of extracellular lipid in the subendothelial space (surrounding the foam cells). At week 10, fosinopril decreased these parameters by 53%, 32%, 94%, and 63%, respectively, compared to controls. Captopril reduced foam cell number by 40%, decreased foam cell size by 21% and fatty streak area declined by 56%, however the area of the extracellular lipid increased by 9%.[27]

In hamsters, foam cell accumulation was a continuous process during the first 12 weeks of cholesterol feeding. The fact that, compared to four-week controls, fosinopril and captopril had reduced fatty streak formation after treatment was initiated at week four and continued to week 10, suggested that these ACE inhibitors induced the regression of early atherosclerosis. Fosinopril reversed all the parameters of atherosclerosis while captopril reversed the accumulation and formation of macrophage-foam cells. A reduction of plasma LDL plus VLDL cholesterol, blood pressure and heart rate as well as inhibition of the renin-angiotensin system (or kininase II) were the probable mechanisms behind the effects of these ACE inhibitors. The results of this study indicated that the regression of early arterial lesions was feasible, however it remains to be determined whether ACE inhibitors are capable of promoting the regression of advanced atherosclerotic plaques.

Effects of ACE inhibition on monocyte-macrophage function

Captopril decreased atherosclerosis in hamsters without affecting plasma LDL cholesterol or blood pressure, which suggested that ACE inhibitors directly impeded the formation and accumulation of macrophage-foam cells. Endothelial cells, monocyte-macrophages and smooth muscle cells expressed renin and ACE and produce angiotensin II.[28–34] This peptide was chemotactic for monocytes,[35] it promoted monocyte adhesion to endothelial cells[36] and inhibited macrophage migration.[37] It is possible that during atherogenesis, arterial angiotensin II recruits monocytes to the artery wall and immobilizes subendothelial macrophages, leading to an accumulation of foam cells. ACE inhibitors may reduce the production of arterial angiotensin II, thereby decreasing monocyte adhesion and chemotaxis and enhancing the efflux of arterial macrophages. The result is fewer arterial foam cells. A simultaneous reduction of monocyte chemotaxis and an elevation of macrophage efflux, suggests that the residence time for arterial macrophages to scavenge modified LDL was reduced. This would account for the small size of the macrophage-foam cells with captopril treatment.

Since a selective angiotensin type I receptor antagonist failed to alter the development of the fatty streak in cholesterol-fed rabbits,[38] which suggested that this receptor and its ligand, angiotensin II, were not crucial for the formation of early macrophage-rich lesions. Thus, ACE inhibitors may impede atherogenesis independently of the renin-angiotensin system. ACE is identical to kininase II, the enzyme that inactivates bradykinin.[39] When the ACE inhibitor ramiprilat was added to endothelial cells in vitro, there was an increase in the levels of bradykinin, prostacyclin and nitric oxide.[40] Prostacyclin inhibits monocyte chemotaxis[41] and macrophage activation,[42] and stable prostacyclin agonists retard the progression of atherosclerosis in hamsters.[13] Nitric oxide decreased monocyte chemotaxis and monocyte adhesion to endothelial cells[41] and nitric oxide reduces atherosclerosis in rabbits.[43] Therefore ACE inhibitors could have prevented atherosclerosis by elevating arterial prostacyclin and nitric oxide levels. A summary of this hypothesis is shown in Fig. 9.2.

Effects of angiotensin II on vascular smooth muscle cell proliferation

Proliferation of vascular smooth muscle cells has been implicated as a critical event in the

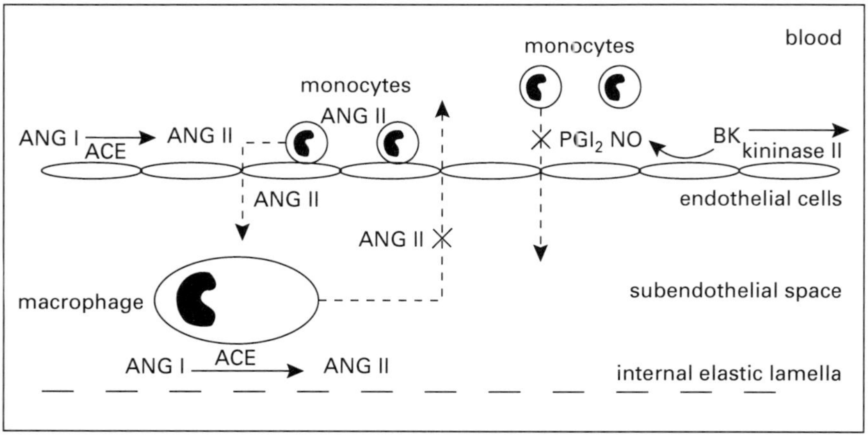

Figure 9.2
Potential pathways for ACE inhibitors to alter the progression of early atherosclerosis. Angiotensin I (ANG I) is cleaved by endothelial ACE to angiotensin II (ANG II), where ANG II promotes monocyte adhesion to the arterial endothelium and monocyte chemotaxis into the subendothelial space. Arterial macrophages synthesize ANG II to further enhance monocyte recruitment, and ANG II impedes macrophage efflux from the artery wall. ACE inhibitors decrease the production of ANG II and reduce these cellular events. In addition, inhibiting kininase II increases bradykinin (BK) levels and elevates the production of prostacyclin (PGI_2) and nitric oxide (NO), both of which reduce monocyte adhesion and chemotaxis to the artery wall.

development of fibrous atherosclerotic plaques and is also associated with the vascular restenosis that often follows balloon angioplasty. However, the cellular mechanisms responsible for smooth muscle cell proliferation are not yet defined. Several lines of evidence suggest a role for the renin-angiotensin system in the marked smooth muscle cell growth observed as a consequence of hypertension, vascular injury, and atherosclerosis.[44–47] For example, in the spontaneously hypertensive rat, inhibition of ANG II production by ACE inhibitors causes a decrease in overall smooth muscle cell mass.[48] Furthermore, recent studies demonstrate that treatment of rats with ACE inhibitors or an angiotensin-receptor antagonist significantly decreases myointimal proliferation in rats following acute arterial injury with a balloon catheter,[49,50] and infusion of ANG II in rats prior to balloon injury greatly exacerbates the myoproliferative lesions.[51]

ANG II is known to bind to specific high-affinity receptors on the surface of responsive cells and distinct ANG II receptor genes have been identified.[52–54] In vascular smooth muscle cells, activation of the AT_1 receptor stimulates phosphatidylinositol-specific phospholipase C activity through the activation of at least one closely coupled G-protein.[55] This results in the generation of the second messengers inositol trisphosphate and diacylglycerol, leading to intracellular calcium mobilization and protein kinase C (PKC) activation, respectively. ANG II also stimulates rapid protein tyrosine phosphorylation in cultured rat aortic smooth muscle cells, and the activation of other serine, threonine protein kinases, including Raf-1 kinase and the 'mitogen-activated' protein kinases (MAP kinases) that are implicated in mitogenic signal transduction.[56] ANG II also stimulates rapid increases in the expression of the growth-associated nuclear proto-oncogenes c-*myc*, c-*fos*, and c-*jun*.[57–59] These studies imply that ANG II can induce rapid changes in gene expression that may ultimately lead to increased cell growth.

In cell culture growth studies, ANG II has been reported to stimulate vascular smooth muscle cell growth directly or enhance the growth response of cells to serum.[60–64] However, these findings have conflicted with other studies that conclude that ANG II stimulates cellular hypertrophy but not proliferation.[65,66] A possible resolution of these discrepancies is provided in a recent study.[67] To evaluate the effects of ANG II on vascular cell growth, mitogenic assays were performed using quiescent cultures of rat aortic smooth muscle cells (Fig. 9.3). Cells were incubated without (serum-free) or with fetal bovine serum (FBS, 10% v/v) or ANG II (1 μM). As shown in Fig. 9.3, although FBS treatment led to a 10–15 fold increase in DNA synthesis after 24 hours, ANG II addition caused only minor (two-fold) stimulation. However, when mitogenic assays were extended to 48 hours ANG II was found to produce a consistent five- to eight-fold stimulation of DNA synthesis in comparison to quiescent controls. Thus ANG II is capable of stimulating a delayed mitogenic effect in certain cultured vascular smooth muscle cell lines. Furthermore, ANG II-stimulated DNA synthesis was inhibited in a concentration dependent manner by losartan, a specific AT_1-receptor antagonist.

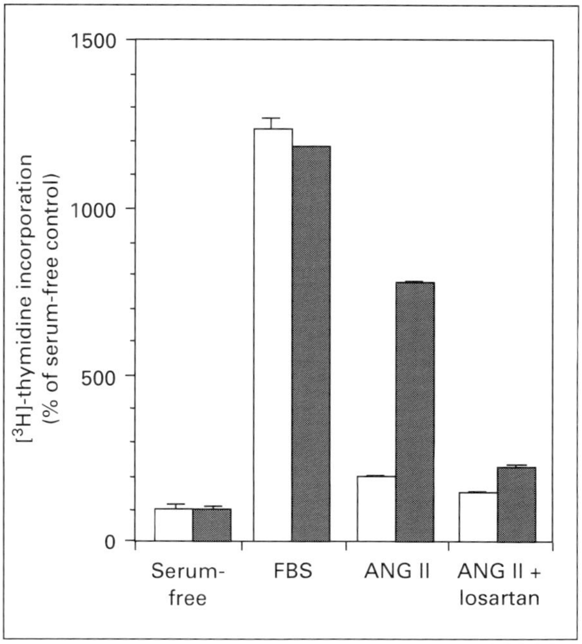

Figure 9.3
ANG II stimulates delayed mitogenesis in cultured RASM cells. Cells were incubated for 24 h (open bars) or 48 h (closed bars) with [^{3}H]-thymidine in SF-DMEM, FBS (10% v/v), ANG II (1 μM) or ANG II plus losartan (5 μM). Results are expressed as % of the unstimulated serum-free control. Bars are the mean ± SD of three treatment wells from a representative experiment. Reproduced from J Clin Invest 1994; ***93:****788–98,*[67] *by copyright permission of the American Society for Clinical Investigation.*

Delayed mitogenesis stimulated by ANG II in cultured rat aortic smooth muscle cells correlates with the expression of endogenous growth factors

One possible explanation for the delayed mitogenesis observed following ANG II stimulation of the RASM cells is that increased expression of new gene products is required prior to the initiation of DNA synthesis. In support of this concept, several studies have demonstrated that ANG II stimulates the expression of several growth factors, including PDGF and TGF-β, such that one or more of these may act as 'proximal' mitogens that are ultimately responsible for the ANG II stimulated cell growth.[57,63,64,67] Furthermore, increased local tissue expression of these factors is described following arterial balloon injury in animal models of restenosis.[68] Thus the effects of ANG II on vascular smooth muscle cell growth in vitro appear to correlate with certain animal models of smooth muscle cell proliferation.

To provide direct evidence for autocrine growth factor production, Weber et al.[67] performed experiments to detect growth factors secreted by cultured rat aortic smooth muscle cells following ANG II stimulation. For these experiments, conditioned media from untreated (control) and ANG II-stimulated cells was added to quiescent cultures of NIH/3T3 fibroblasts. As a control, NIH/3T3 cultures were treated with ANG II and recombinant forms of PDGF-AA, -BB, or basic FGF for 10 minutes. Protein extracts of NIH/3T3 cells were then analysed by immunoblotting with anti-phosphotyrosine antibodies. By comparing the characteristic 'biochemical signatures' of endogenous protein tyrosine phosphorylation induced by specific growth factors in the fibroblasts to those observed following stimulation with rat aortic smooth muscle cell conditioned media samples, secreted growth factors induced by ANG II were identified.

As shown in Fig. 9.4, pure growth factors induced characteristic patterns of protein tyrosine phosphorylation. For example, PDGF-AA and -BB each stimulated rapid tyrosine phosphorylation of proteins with bands at 180–185 kD, 145, 120, and several bands between 30 and 46 kD. The 180–185 kD bands correspond to activated PDGF receptors, which rapidly undergo tyrosine autophosphorylation upon binding to a ligand. Basic FGF stimulated tyrosine phosphorylation of a prominent 90 kD protein, thought to be a major substrate of the activated receptor kinase. In contrast, NIH/3T3 cells treated with ANG II did not show any significant change in protein tyrosine phosphorylation.

By comparison, conditioned media samples from untreated and ANG II-treated RASM cells stimulated complex patterns of protein tyrosine phosphorylation in the NIH/3T3 fibroblasts (Fig. 9.4B). For example, PDGF activity was detected in conditioned media samples from rat aortic smooth muscle cells treated with ANG II for between 6 to 48 hours. As shown in Fig. 9.4B, these samples stimulated tyrosine phosphorylation of a 180–185 kD protein, which migrated to the same position as the autophosphorylated PDGF receptors. To confirm that this activity represented PDGF, conditioned media from cells treated with ANG II was incubated with a PDGF neutralizing antiserum for 30 minutes prior to stimulation of the NIH/3T3 cells. The resulting pattern of protein tyrosine phosphorylation lacked the 180/185 kD PDGF receptors, although tyrosine phosphorylation of the 150 kD band was observed. Additional support for the induction of PDGF by ANG II was derived from Northern blot

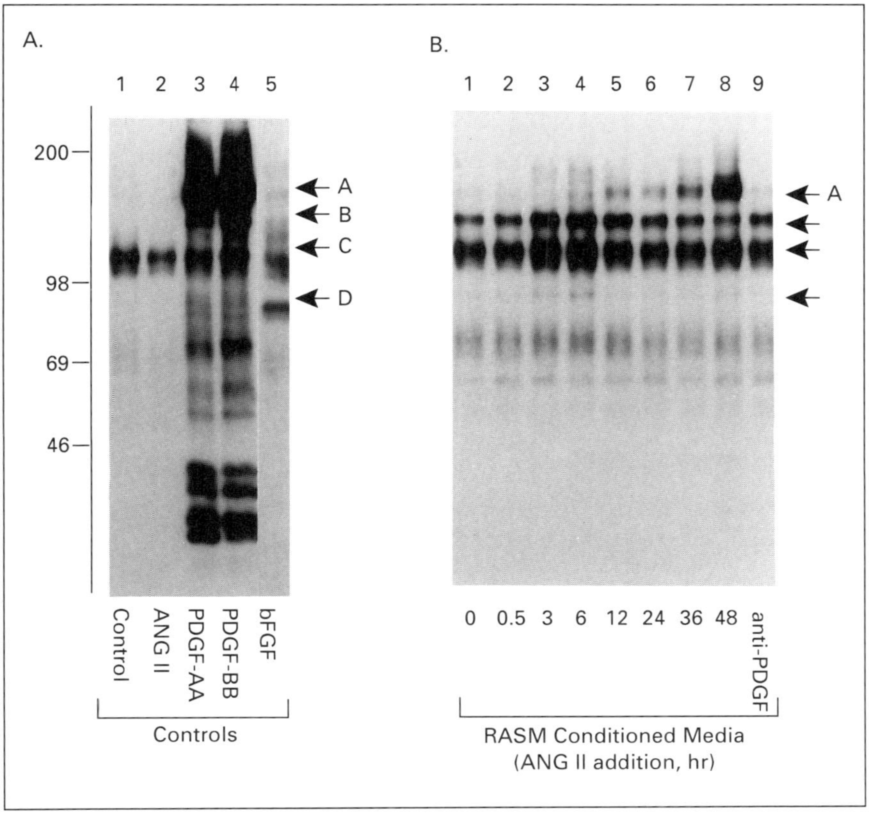

Figure 9.4

*Detection of ANG II-stimulated growth factors from RASM cell conditioned medium by anti-phosphotyrosine immunoblot analysis of challenged NIH/3T3 fibroblasts. **A**: Anti-phosphotyrosine immunoblot analysis of NIH/3T3 cell lysates from quiescent cells (lane 1) and cells stimulated for 10 min with ANG II (1 μM, lane 2), PDGF-AA (10 ng/ml, lane 3), PDGF-BB (10 ng/ml, lane 4) or basic FGF (10 ng/ml, lane 5). Arrows indicate the activated PDGF receptors at 180–185 kD, (A); major tyrosine-phosphorylated proteins of 145 kD (B) and 120 kD (C); and the basic FGF receptor substrate (90 kD) (D). **B**: Anti-phosphotyrosine immunoblot of anti-phosphotyrosine-recovered cell proteins from serum-starved NIH/3T3 cells (lane 1) and cells stimulated for 10 min with conditioned media samples from ANG II (1 μM)-treated RASM cells (lanes 2–9). Individual cell lysate samples (0.5 mg) were first immunoprecipitated with anti-phosphotyrosine antibodies, and immunoprecipitated proteins were resolved by SDS-PAGE. Lanes 1–8 list at the bottom the duration of ANG II treatment used to produce the conditioned media. Note that ANG II-treated RASM conditioned media stimulated the tyrosine phosphorylation of PDGF receptors (←A) (lanes 4–8, compare with panel A, lanes 3,4). The ability of 24-h ANG II-stimulated RASM conditioned media to activate PDGF receptors was blocked by PDGF neutralizing antiserum (lane 9). Arrows below the PDGF receptor band indicate the 150 kD 120 kD and 90 kD tyrosine-phosphorylated proteins induced by the RASM conditioned media samples. Reproduced from J Clin Invest 1994; **93:**788–98,[67] by copyright permission of the American Society for Clinical Investigation.*

analyses of ANG II-stimulated RASM cells. Hybridization of these blots using labeled probes for either PDGF-A or PDGF-B chains showed that the RASM cells exclusively expressed the A chain.

It should be noted, however, that it is still unclear as to the exact roles of the specific ANG II-stimulated autocrine growth factors in vascular smooth muscle cell proliferation. For example, specific neutralizing antibodies, directed against PDGF or other known polypeptide growth factors (including TGF-β_1, EGF, basic FGF, etc.), have failed to significantly decrease ANG II-stimulated DNA synthesis.[67] Furthermore, a PDGF-neutralizing antiserum did not inhibit the mitogenic activity present in the conditioned media derived from ANG II-stimulated RASM cells.[67] Thus, productive autocrine ligand-receptor interactions, if required for ANG II-stimulated cell growth, may occur 'intracellularly', that is, in a compartment inaccessible to neutralizing antibodies. Alternatively, the results of Weber et al.[67] potentially implicate novel autocrine growth factors in the control of vascular smooth muscle cell proliferation. Additional studies, aimed at characterizing the specific autocrine vascular smooth muscle cell mitogens, are required to resolve the mechanisms underlying smooth muscle cell growth control modulated by the renin-angiotensin system.

ANG II and endothelin in hyperproliferative vascular diseases

The endothelins are a family of potent pressor peptides derived from three different genes.[69] These 21-amino-acid peptides, designated ET-1, ET-2 and ET-3, are highly homologous and bind to two distinct receptors, termed ET_A[70] and ET_B.[71] Smooth muscle cells primarily express ET_A receptors which bind ET-1 and ET-2 with equal high affinity but have a lower affinity for ET-3. Endothelins are synthesized in endothelial cells and act on neighboring smooth muscle cells in a paracrine fashion to induce contractility and regulate vascular tone.[69] Consequently, like the renin-angiotensin system, endothelins have been postulated to play important roles in the control of blood pressure, possibly in vascular cell growth control.

An interrelationship between the renin-angiotensin system and endothelins has been suggested by recent studies in which ANG II has been shown to stimulate the release of ET-1 from endothelial cells.[72] Furthermore, since the selective ET receptor antagonist BQ-123 causes shifts in the concentration–response curve of isolated rabbit aortic strips to both ET-1 and ANG II, but not to noradrenaline (norepinephrine) or KCl (S Moreland, personal communication) it is possible that ET is involved in the hypertensive effect of ANG II.

A recent report has also demonstrated that circulating and tissue endothelin (ET) reactivity is increased in patients with advanced atherosclerosis.[73] Like many pressor peptides found to be weak mitogens on smooth muscle cells,[74] ET-1 has been investigated for its ability to stimulate DNA synthesis and cell division. In this regard, ET-1 has been reported to stimulate DNA synthesis in Swiss 3T3 fibroblasts,[75] a model system for studying growth control and signaling pathways. However, in these cells, ET-1 is a relatively weak mitogen, producing only a two-fold increase in DNA synthesis. ET-1 is capable also of producing a four-fold increase in DNA synthesis in the Swiss 3T3 cells when co-administered with insulin-like growth factor.[76] These data correlate with other studies using vascular smooth muscle cells which report that ET-1 is not

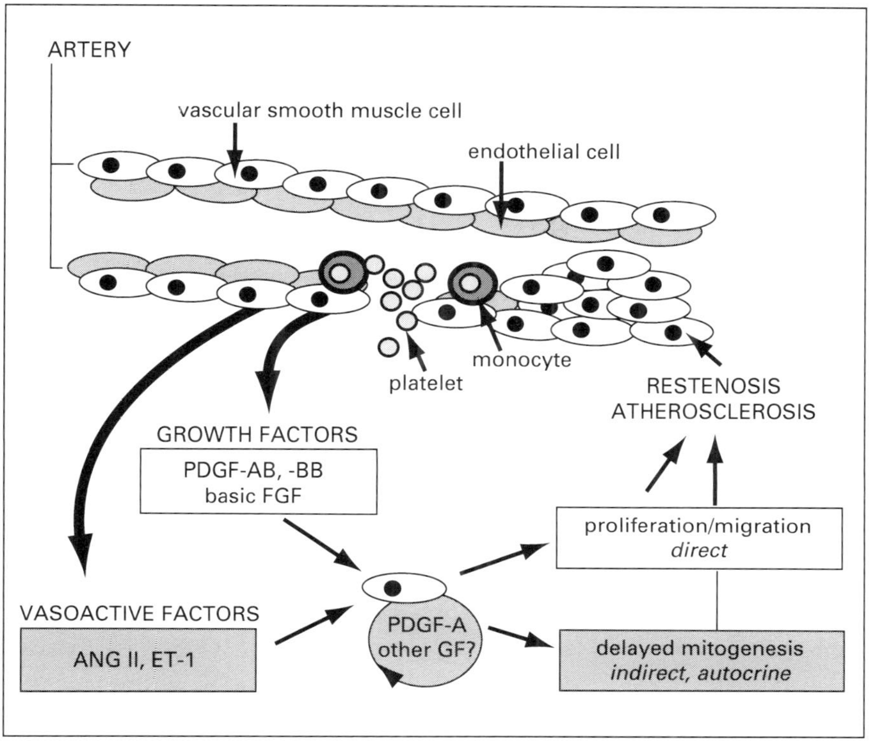

Figure 9.5
Possible autocrine/paracrine mechanisms involved in atherosclerosis/vascular injury. At the top of the figure, individual cell types present in an 'injured' blood vessel are noted. These include endothelial cells, adherent monocytes, platelets, and vascular smooth muscle cells. Soluble paracrine factors produced by these cells in response to injury include the polypeptide growth factors listed: PDGFs, basic fibroblast growth factor (basic FGF). These directly stimulate vascular smooth muscle cells (bottom) to proliferate and/or migrate to lesion site. Local tissue vasoconstrictors, including angiotensin II or endothelin-1, may modulate these processes by indirectly stimulating vascular smooth muscle cell growth through the autocrine production of growth factors such as PDGF-AA.

mitogenic when added alone, but is co-mitogenic when added with insulin,[77,78] or other growth factors.[79]

In studies using cultured rat aortic smooth muscle cells, Weber et al.[80] have demonstrated that ET-1, like ANG II, induces delayed mitogenic stimulation of cultured rat aortic smooth muscle cells. Furthermore, ET-1 exerted an additive mitogenic effect when added to cells along with maximally stimulating concentrations of ANG II. Thus, the co-mitogenic effects of ANG II and ET-1 support a model in which vasoactive peptides exert 'weak' indirect mitogenic stimuli on vascular smooth muscle cells that may then result in a concerted effect on localized cellular proliferation and under conditions where these peptides are oversecreted (Fig. 9.5). For example, Owen and Villereal suggest that contractile agents are generally elevated in cardiovascular diseases resulting in vessel wall thickening.[81] Therefore, an additive effect of contractile agents serving as weak mitogens could produce a proliferation of vascular smooth muscle cells with resulting intimal wall thickening. This concept is also supported by their studies using cultured human fibroblasts in which the mitogenic effect of Lys-bradykinin was shown to be additive with epidermal growth factor, vasopressin, and insulin. In combination, these factors stimulated DNA synthesis to a similar extent as 10% FBS.[81]

Thus potent vasoactive peptides such as ANG II and ET-1 can stimulate DNA synthesis and promote cellular proliferation in a cultured vascular smooth muscle cell systems. These responses appear to correlate with the expression of endogenous growth factor genes, including PDGF and TGF-β_1. Since the effects of vasoconstrictors on the expression of endogenous growth factors in vascular smooth muscle cell cultures mimics similar changes observed in neonintimal formation following arterial injury in vivo[68] these data imply that the autocrine induction of specific growth factors likely accounts for the proliferative effects of ANG II and ET-1 in cardiovascular diseases. These results predict that inhibition of ANG II or ET-1 in vivo, through the use of ACE inhibitors, ANG II-receptor antagonists, or ET-receptor antagonists may inhibit the localized expression of potent polypeptide growth factors and consequently limit vascular smooth muscle cell proliferation. In the future, identification of the precise smooth muscle cell autocrine factor(s) responsible for the proliferative effects of vasoconstrictors should lead to insights into the pathophysiology and treatment of hyperproliferative vascular diseases, including atherosclerosis.

Summary

Inhibitors of ACE decrease the progression of early and advanced atherosclerotic lesions in five animal models, as well as induce the regression of fatty streaks in hamsters. In models with two risk factors, such as hypertension and hypercholesterolemia, ACE inhibitors prevent the development of accelerated atherosclerosis when blood pressure elevation is dependent on the renin-angiotensin system. Pharmacological inhibition of ACE and kininase II decreases angiotensin II production and elevates bradykinin, prostacyclin and nitric oxide levels. Angiotensin II, prostacyclin and nitric oxide modulate the behaviour of monocyte/macrophages, where angiotensin II promotes inflammation, while prostacyclin and nitric oxide act in an opposite manner by suppressing leukocyte activity. Thus ACE inhibitors probably modify monocyte/macrophage activity through several vasoactive factors and prostanoids.

ACE inhibitors, by limiting the localized formation of ANG II in vascular tissues, may also act to reduce vascular smooth muscle cell proliferation, an important component in the progression of fibroproliferative plaques. For example, in animal cell culture models, ANG II stimulation of vascular smooth muscle cells leads to increased cell proliferation closely correlated with the expression of autocrine growth factors. One or more of these endogenous polypeptide growth factors may act as the 'proximal' stimulus for cell growth/tissue remodeling initiated by vasoactive factors such as ANG II. Thus pharmacologic regulation of the renin-angiotensin system may directly affect atherogenesis by altering endothelial cell, monocyte/macrophage and smooth muscle cell function. Future studies, conducted both in vivo and in cell culture studies, should help define the role of the renin-angiotensin system in the complex pathophysiology of atherosclerosis.

References

1 Kannel WB, McGee D, Gordon T. A general cardiovascular risk profile. The Framingham Study. *Am J Cardiol* 1976; **38**:46–51.

2 Robertson WB, Strong JP. Atherosclerosis in persons with hypertension and diabetes mellitus. *Lab Invest* 1968; **18**:538–51.

3 Solberg LA, Strong JP, Holme I, Helgeland A, Hjermann I, Leren P, Mogensen SV. Stenoses in the coronary arteries. Relation to atherosclerotic lesions, coronary heart disease, and risk factors. The Oslo study. *Lab Invest* 1985; **53**:648–55.

4 Waters D, Lesperance J, Francetich M, Causet D, Theroux P, Chiang YK, Hudon G, Lembarbe L, Reitman M, Joyal M et al. A controlled clinical trial to assess the effect of a calcium channel blocker on the progression of coronary atherosclerosis. *Circulation* 1990; **82**:1940–53.

5 Lichtlen PR, Hugenholtz PG, Rafflenbeul W, Hecker H, Jost S, Deckers JW. Retardation of angiographic progression of coronary artery disease by nifedipine. Results of the international nifedipine trial on antiatherosclerotic therapy (INTACT). *Lancet* 1990; **335**:1109–13.

6 Costa FW, Borghi C, Mussi A, Ambrosioni E. Use of captopril to reduce serum lipids in hypertensive patients with hyperlipidemia. *Am J Hypertens* 1988; **1**:221S–3S.

7 Sasaki J. Arakawa K. Beneficial effects of ACE inhibitors (capto-pril and enalapril) on lipoprotein metabolism in mild essential hypertensives. Presented at the 8th Symposium of Atherosclerosis, Rome, 1988. Abstract: 150.

8 Kowala MC, Grove RI, Aberg G. Inhibitors of angiotensin converting enzyme decrease early atherosclerosis in hyperlipidemic hamsters. Fosinopril reduces plasma cholesterol and captopril inhibits macrophage-foam cell accumulation independently of blood pressure and plasma lipids. *Atherosclerosis* 1994. [In press]

9 Kowala MC, Nunnari JJ, Durham SK, Nicolosi RJ. Doxazosin and cholestyramine similarly decrease fatty streak formation in the aortic arch of hyperlipidemic hamsters. *Atherosclerosis* 1991; **91**:35–49.

10 Ross R. The pathogenesis of atherosclerosis – an update. *N Eng J Med* 1986; **314**:488–500.

11 Munro JM, Cotran RS. Biology of disease. The pathogenesis of atherosclerosis: atherogenesis and inflammation. *Lab Invest* 1988; **58**: 249–61.

12 Steinberg D, Parthasarathy S, Carew TE, Khoo JC, Witzium JL. Beyond cholesterol. Modification of low density lipoprotein that increase atherogenicity. *N Eng J M ed* 1989; **320**:915–24.

13 Kowala MC, Mazzucco CE, Hartl KS, Seiler SM, Warr GA, Abid S, Grove RI. Prostacyclin agonists reduce early atherosclerosis in hyperlipidemic hamsters. Octimibate and BMY 42393 suppress monocyte chemotaxis, macrophage cholesteryl ester accumulation, scavenger receptor activity and tumor necrosis factor production. *Arteriosc l Thromb* 1993; **13**:435–44.

14 Joris I, Zand T, Nunnari JJ, Krolikowski FJ, Majno G. Studies on the pathogenesis of atherosclerosis. I. Adhesion and emigration of mononuclear cells in the aorta of hypercholesterolemic rats. *Am J Pathol* 1983; **113**:341–58.

15 Kowala MC, Cuénoud HF, Joris I, Majno G. Cellular changes during hypertension: a quantitative study of the rat aorta. *Exp Mol Pathol* 1986; **45**:323–335.

16 Kowala MC. Captopril decreases accelerated atherosclerosis in hypertensive one kidney one clip rats fed cholesterol. *Drug Devel Res* 1993; **29**:100–7.

17 Masuda J, Ross R. Atherogenesis during low level hypercholesterolemia in the nonhuman primate. II. Fatty Streak conversion. *Arteriosclerosis* 1990; **10**:178–87.

18 Davies MJ, Woolf N, Rowles PM, Pepper J. Morphology of the endothelium over atherosclerotic plaques in human coronary arteries. *Br Heart J* 1988; **60**:459–64.

19 Chobanian AV, Brecher P, Chan C. Effects of propranolol on atherogenesis in the cholesterol-fed rabbit. *Circ Res* 1985; **56**:755–62.

20 Henry PD, Bentley KI. Suppression of atherogenesis in cholesterol-fed rabbit treated with nifedipine. *J Clin Invest* 1981; **68**:1366–69.

21 Lichtenstein AH, Drago R, Nickerson C, Prescot MF, Lee SQ, Chobanian AV. Effects of propranolol on atherogenesis in the Watanabe heritable hyperlipidemic rabbit. *J Vasc Med Biol* 1989; **1**:248–54.

22 Van Niekerk JLM, Hendriks Th, DeBoer HHM, Van'tLaar A. Does nifedipine suppress atherogenesis in WHHL rabbits? *Atherosclerosis*. 1984; **53**:91–98.

23 Chobanian AV, Haudenschild CC, Nickerson C, Drago R. Antatherogenic effect of captopril in the Watanabe heritable hyperlipidemic rabbit. *Hypertension* 1990; **15**:327–31.

24 Aberg G, Ferrer P. Effects of captopril on atherosclerosis in cynomolgus monkeys. *J Cardiovasc Pharm* 1990; **15[Suppl 5]**:S65–S72.

25 Ferrer P, Aberg G. Progression of atherosclerosis in cholesterol-fed cynomolgus monkeys during treatment with nifedipine. *FASEB* 1990; **4**(4):A1151.

26 Jacobsson L, Persson K, Andersson R, Karlberg B, Olsson A, Ferrer P, Valentine M, Aberg G. Antiatherosclerotic effects of the angiotensin converting enzyme inhibitors captopril and fosinoprile in hypercholesterolemic min-pigs. *J Cardiovasc Pharmacol* (in press).

27 Kowala MC, Recce R, Beyer S, Aberg G. Regression of early atherosclerosis in hyperlipidemic hamsters by fosinopril and captopril. *J Cardiovasc Pharmacol* 1994 [submitted].

28 Lilly LS, Pratt RE, Alexander RW et al. Renin expression by vascular endothelial cells in culture. *Circ Res* 1985; **57**:312–18.

29 Re R, Fallon JT, Dzau V, Quay SC, Haber E. Renin synthesis by canine aortic smooth muscle cells in culture. *Life Sci* 1982; **30**:99–106.

30 Caldwell PRB, Seegal BC, Hsu KC, Das M, Soffer RL. Angiotensin-converting enzyme: Vascular endothelial localization. *Science* 1976; **191**:1050–51.

31 Ryan US, Ryan JW, Whitaker C, Chiu A. Localization of angiotensin converting enzyme (kininase II). II. Immunocytochemistry and immunofluorescence. *Tiss Cell* 1976; **8**: 125–45.

32 Johnson AR, Erdos EG. Metabolism of vasoactive peptides by human endothelial cells in culture. Angiotensin I converting enzyme (kininase II) and angiotensinase. *J Clin Invest* 1977; **59**:684–95.

33 Kifor I, Dzau VJ. Endothelial renin-angiotensin pathway: evidence for intracellular synthesis and secretion of angiotensins. *Circ Res* 1987; **60**:422–28.

34 Padgett RC, Vasquez R, Brooks RM, Heistad DD. Human leukocytes release angiotensin II. *Hypertension* 1992; **20**:408.

35 Goetzl EJ, Klickstein LB, Watt KWK, Wintroub BU. The preferential human mononuclear leukocyte chemotactic activity of the substituent peptides of angiotensin II. *Biochem Biophys Res Commun* 1980; **97**:1097–1102.

36 Kim JA, Berliner JA, Nadler JL. Angiotensin-II increases monocyte binding to aortic endothelial-cells. *Clin Res* 1994; **42**:A15.

37 Weinstock JV, Blum AM. Isolated liver granulomas of murine *Schistosoma mansoni* contain components of the angiotensin system. *J Immunol* 1983; **131**:2529–32.

38 Schuh JR, Blehm DJ, Frierdich GE, McMahon EG, Blaine EH. Differential effects of renin-angiotensin blockade on atherogenesis in cholesterol-fed rabbits. *J Clin Invest* 1993; **91**:1453–58.

39 Erdos EG. Some old and some new ideas on kinin metabolism. *J Cardiovasc Pharmacol* 1990; **15[Suppl 6]**:S20–S24.

40 Wiemer G, Scholkens BA, Becker RHA, Busse R. Ramiprilat enhances endothelial autocoid formation by inhibiting breakdown of endothelium-derived bradykinin. *Hypertension* 1991; **18**:558–63.

41 Bath PMW, Hassall DG, Gladwin A-M, Palmer RMJ, Matin JF. Nitric oxide and prostacyclin: divergence of inhibitory effects on monocyte chemotaxis and adhesion to endothelium in vitro. *Arterioscl Thromb* 1991; **11**:254–260.

42 Bonta IL, Adolfs MJP, Fieren MWJA. Cyclic AMP levels and their regulation by prostaglandins in peritoneal macrophages of rats and humans. *Int J Immunopharmacol* 1984; **6**:547–555.

43 Cooke JP, Singer AH, Tsao P, Zera P, Rowan RA, Billingham ME. Antiatherogenic effects of

L-arginine in the hypercholesterolemic rabbit. *J Clin Invest* 1992; **90**:1168–72.
44 Peach MJ, Dostal DE. The angiotensin II receptor and the actions of angiotensin II. *J Cardiovasc Pharmacol* 1990; **16[Suppl 4]**:S25–S30.
45 Gibbons GH, Dzau VJ. Angiotensin converting enzyme inhibition and vascular hypertrophy in hypertension. *Cardiovasc Drugs Ther* 1990; **4**:237–42.
46 Osterrieder W, Muller RKM, Powell JS, Clozel J-P, Hefti F, Baumgartner HR. Role of angiotensin II in injury-induced neointima formation in rats. *Hypertension* 1991; **18[Suppl II]**:II60–II64.
47 Paul M, Bachmann J, Ganten D. The tissue renin-angiotensin systems in cardiovascular disease. *Trends Cardiovasc Med* 1992; **2**:94–99.
48 Owens GK. Influence of blood pressure on development of aortic medial smooth muscle hypertrophy in spontaneously hypertensive rats. *Hypertension* 1987; **9**:178–87.
49 Powell JS, Clozel JP, Muller RKM, Kuhn H, Hefti F, Hosang M, Baumgartner HR. Inhibitors of angiotensin-converting enzyme prevent myointimal proliferation after vascular injury. *Science* 1989; **245**:186–88.
50 Kauffman RF, Bean JS, Zimmerman KM, Brown RF, Steinberg MI. Losartan, a nonpeptide angiotensin II (Ang II) receptor antagonist, inhibits neointima formation following balloon injury to rat carotid arteries. *Life Sci* 1991; **49**:PL223–28.
51 Daemen MJAP, Lombardi DM, Bosman FT, Schwartz SM. Angiotensin II induces smooth muscle cell proliferation in the normal and injured rat arterial wall. *Circ Res* 1991; **68**:450–56.
52 Sasaki K, Yamano Y, Bardhan S, Iwai N, Murray JJ, Hasegawa M, Matsuda Y, Inagami T. Cloning and expression of a complementary DNA encoding a bovine adrenal angiotensin II type-1 receptor. *Nature* 1991; **351**:230–33.
53 Murphy TJ, Alexander RW, Griendling KK, Runge MS, Bernstein KE. Isolation of a cDNA encoding the vascular type-1 angiotensin II receptor. *Nature* 1991; **351**:233–36.
54 Sandberg K, Hong J, Clark AJL, Shapira H, Catt KJ. Cloning and expression of a novel angiotensin II receptor subtype. *J Biol Chem* 1992; **267**:9455–58.
55 Griendling KK, Rittenhouse SE, Brock TA, Ekstein LS, Gimbrone MA, Alexander RW. Sustained diacylglycerol formation from inositol phospholipids in angiotensin II-stimulated vascular smooth muscle cells. *J Biol Chem* 1986; **261**:5901–6.
56 Molloy CJ, Taylor DS, Weber H. Angiotensin II stimulation of rapid protein tyrosine phosphorylation and protein kinase activation in rat aortic smooth muscle cells. *J Biol Chem* 1993; **268**:7338–45.
57 Naftilan AJ, Pratt RE, Dzau VJ. Induction of platelet-derived growth factor A-chain and c-myc gene expressions by angiotensin II in cultured rat vascular smooth muscle cells. *J Clin Invest* 1989; **83**:1419–24.
58 Taubman MB, Berk BC, Izumo S, Tsuda T, Alexander RW, Nadal-Ginard B. Angiotensin II induces c-fos mRNA in aortic smooth muscle. *J Biol Chem* 1989; **264**:526–30.
59 Naftilan AJ, Gilliland GK, Eldridge CS, Kraft AS. Induction of the proto-oncogene c-jun by angiotensin II. *Mol Cell Biol* 1990; **10**:5536–40.
60 Campbell-Boswell M, Robertson AL. Effects of angiotensin II and vasopressin on human smooth muscle cells in vitro. *Exper Mol Pathol* 1981; **35**:265–76.
61 Lyall F, Morton JJ, Lever AF, Cragoe EJ. Angiotensin II activates Na^+–H^+ exchange and stimulates growth in cultured vascular smooth muscle cells. *J Hypertens* 1988; **6[Suppl 4]**:S438–441.
62 Scott-Burden T, Hahn AWA, Resink TJ, Buhler FR. Modulation of extracellular matrix by angiotensin II: stimulated glycoconjugate synthesis and growth in vascular smooth muscle cells. *J Cardiovasc Pharmacol* 1990; **16[Suppl 4]**:S36–41.
63 Stouffer GA, Owens GS. Angiotensin II-induced mitogenesis of spontaneously hypertensive rat-derived cultured smooth muscle cells is dependent on autocrine production of transforming growth factor-β. *Circ Res* 1992; **70**:820–28.
64 Gibbons GH, Pratt RE, Dzau VJ. Vascular smooth muscle cell hypertrophy vs. hyperplasia. *J Clin Invest* 1992; **90**:456–61.

65 Geisterfer AAT, Peach MJ, Owens GK. Angiotensin II induces hypertrophy, not hyperplasia, of cultured rat aortic smooth muscle cells. *Circ Res* 1988; **62**:749–56.
66 Owens GK. Control of hypertrophic versus hyperplastic growth of vascular smooth muscle cells. *Am J Physiol* 1989; **257**:H1755–65.
67 Weber H, Taylor DS, Molloy CJ. Angiotensin II induces delayed mitogenesis and cellular proliferation in rat aortic smooth muscle cells: correlation with the expression of specific endogenous growth factors and reversal by suramin. *J Clin Invest* 1994; **93**:788–98.
68 Schwartz SM, Heimark RL, Majesky MW. Developmental mechanisms underlying pathology of arteries. *Pharmacol Rev* 1990; **70**:1177–1209.
69 Masaki T, Kimura S, Yanagisawa M, Goto K. Molecular and cellular mechanism of endothelin regulation. *Circulation* 1991; **84**:1457–68.
70 Arai H, Hori S, Aramori I, Ohkubo H, Nakanishi S. Cloning and expression of a cDNA encoding an endothelin receptor. *Nature* 1990; **348**:730–32.
71 Sakurai T, Yanagisawa M, Takuwa Y, Miyazaki H, Kimura S, Goto K, Masaki T. Cloning of a cDNA encoding a non-isopeptide-selective subtype of the endothelin receptor. *Nature* 1990; **348**:732–35.
72 Emori T, Hirata Y, Ohta K, Kanno K, Eguchi S, Imai T, Shichiri M, Marumo F. Cellular mechanism of endotheliun-1 release by angiotensin and vasopressin. *Hypertension* 1991; **18**:165–70.
73 Lerman A, Edwards BS, Hallett JW, Henblein DM, Sandberg SM, Burnett JC. Circulating and tissue endothelin immunoreactivity in advanced atherosclerosis. *NEJM*; **325**:997–1001.
74 Owens GK. Role of contractile agonists in growth regulation of vascular smooth muscle cells. In: Cox RH, ed. *Cellular and molecular mechanisms in hypertension*. Plenum: New York, 1991: 71–79.
75 Takuwa N, Takuwa Y, Yanagisawa M, Yamashita K, Masaki T. A novel vasoactive peptide endothelin stimulates mitogenesis through inositol lipid turnover in Swiss 3T3 fibroblasts. *J Biol Chem* 1989; **264**:7856–61.
76 Brown KD, Littlewood CJ. Endothelin stimulates DNA synthesis in Swiss 3T3 cells: Synergy with polypeptide growth factors. *Biochem* 1989; **263**:977–80.
77 Nakaki T, Nakayama M, Yamamoto S, Kato R. Endothelin-mediated stimulation of DNA synthesis in vascular smooth muscle cells. *Biochem Biophys Res Commun* 1989; **158**:880–83.
78 Serradeil-Le Gal C, Herbert JM, Garcia C, Boutin M, Maffrand JP. Importance of the phenotypic state of vascular smooth muscle cells on the binding and mitogenic activity of endothelin. *Peptides* 1991; **12**:575–79.
79 Weissberg PL, Witchell C, Davenport AP, Hesketh TR, Metcalfe JC. The endothelin peptides ET-1, ET-2, ET-3 and sarafotoxin S6b are co-mitogenic with platelet-derived growth factor for vascular smooth muscle cells. *Atherosclerosis* 1990; **85**:257–62.
80 Weber H, Webb M, Serafino R, Taylor DS, Moreland S, Norman J, Molloy CJ. Endothelin-1 and angiotensin II stimulate delayed mitogenesis in cultured rat aortic vascular smooth muscle cells: evidence for common signaling mechanisms. *Mol Endocrinology* 1994; **8**:148–58.
81 Owen NE, Villereal ML. Lys-Bradykinin stimulates Na^+ influx and DNA synthesis in cultured human fibroblasts. *Cell* 1983; **32**:979–85.

10

ACE inhibitors in diabetes

Stephen C Bain and Anthony H Barnett

Introduction

The current indications for use of ACE inhibitors in diabetic subjects are the same as those for the general population, namely treatment of essential hypertension, renovascular hypertension and cardiac failure. Individual ACE inhibitors may, however, have specific benefits for diabetics compared with other agents. This chapter will consider the suitability of ACE inhibitors for use in diabetics, review the evidence to support specific advantages over other agents and examine potential therapeutic uses for the future.

To be acceptable first-line therapy for diabetic patients, drugs must have neutral or beneficial effects on the metabolic abnormalities associated with diabetes mellitus and must not aggravate other complications. ACE inhibitors are well tolerated[1,2] and, in contrast to β-blockers and thiazide diuretics, they do not worsen glucose tolerance or lipid profiles when given to type 1 or type 2 diabetics.[3,4] Indeed, there are reports of a reduction in blood glucose levels in patients receiving ACE inhibitors: McMurray and Fraser reported two cases where the initiation of enalapril led to a reduction in hypoglycaemic treatment.[5] Ferriere and colleagues described a similar scenario following the use of captopril in four diabetic subjects and Rett et al. described two cases with prolonged hypoglycaemic symptoms after initiation of captopril.[6,7] In the latter report, both patients were chronically hyperglycaemic type 2 diabetics receiving sulphonylurea and biguanide medication. As long as physicians are aware of this possibility, the hypoglycaemic action of ACE inhibitors should be regarded as a therapeutic advantage rather than an 'astonishing' side-effect.[8] In addition, they do not aggravate peripheral vascular insufficiency or impotence and are not known to adversely affect retinopathy or neuropathy.

Hypertension

Hypertension in diabetic subjects can be broadly classified into renovascular and essential hypertension. These cosegregate to a large extent with the metabolically distinct forms of disease — type 1 or insulin-dependent, and type 2 or non-insulin-dependent diabetes. About one-third of type 1 diabetics develop the clinical syndrome of diabetic nephropathy, of which hypertension is a part. In contrast, 50% of type 2 diabetics already have essential hypertension at the time of diagnosis and more than 30% have coronary artery disease. The major aim of antihypertensive treatment in the two groups therefore differs; that is, to prevent the onset of end-stage renal failure in type 1 diabetics and to reduce the elevated cardiovascular risk in type 2 subjects.

Diabetic nephropathy

Approximately 35% of patients with type 1 diabetes develop persistent proteinuria (defined as an albumin excretion rate (AER) greater than 200 μg/min), increased arterial blood pressure and a decline in glomerular filtration rate; these signs collectively constitute the clinical syndrome of diabetic nephropathy[9,10] (Figs 10.1, 10.2). Comparable studies have not been performed in type 2 diabetes but a population-based study in North America reported a prevalence of 24.6% after twenty years' disease duration.[11] The WHO multinational study of over 4000 diabetic patients found that 23% of type 2 diabetics had slight or moderate proteinuria at baseline and 6.5% went on to develop renal failure over the mean follow-up period of 8.4 years.[12] In contrast, 27% of type 1 diabetics had increased urinary protein and 8.8% become nephropathic.

Diabetic nephropathy is the main cause of the increased morbidity and mortality in type 1 diabetic subjects.[13] This is due to an excess of cardiovascular disease and renal failure: in the United Kingdom alone there are an estimated 600 cases per year of end-stage renal failure as a consequence of diabetes mellitus.[14] The aetiology of diabetic nephropathy is unknown, but associated risk factors have been identified[15] (Table 10.1). Hypertension has also been implicated as a cause as well as a consequence of diabetic nephropathy.[16]

Longitudinal studies have shown that the subgroup of type 1 diabetics at risk of diabetic nephropathy can be identified fairly accurately by the detection of microalbuminuria.[17,18] During this phase, known as incipient nephropathy, urinary albumin excretion is raised but below the level of clinical proteinuria (AER $\geq$ 20 μg/min but $\leq$ 200 μg/min). A small increase in arterial blood pressure (although still within the 'normal range') is also a common finding. It is thought that prevention or retardation of the progression of microalbuminuria may delay or retard the risk of end-stage renal failure. Detection of evolving nephropathy before the stage of

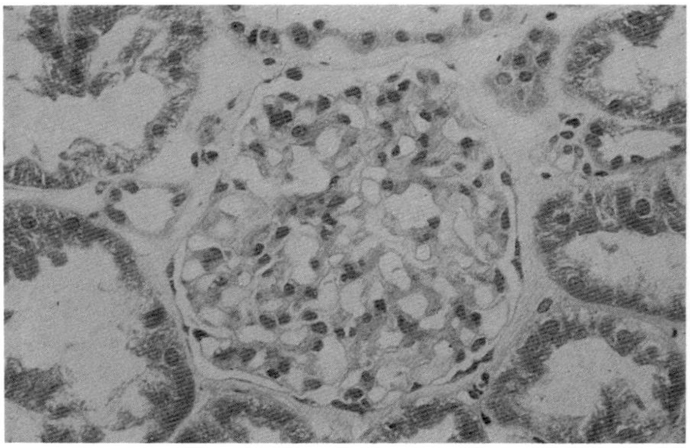

Figure 10.1
Haemotoxylin and eosin stain of normal kidney biopsy with normal glomerular structure.

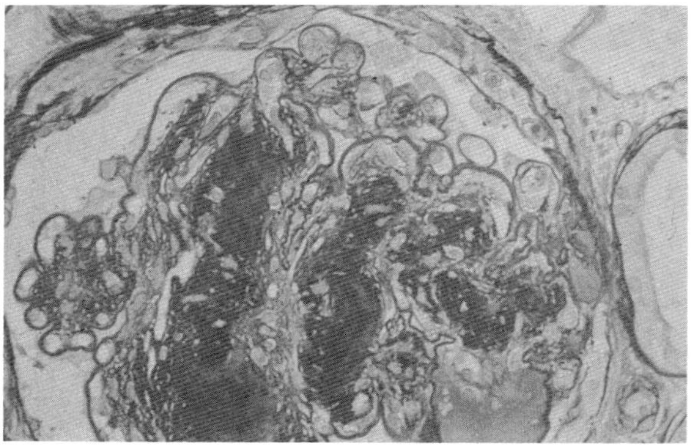

Figure 10.2
Haemotoxylin and eosin stain of kidney biopsy from longstanding diabetic showing the changes of diabetic glomerulosclerosis. The glomerulus is histologically abnormal with replacement of the normal architecture and basement membrane thickening.

Male gender
Early age at onset
Insulin-dependence
Cigarette smoking
Poor long-term glycaemic control
Racial origin
Parental history of hypertension

Table 10.1
Factors associated with development of diabetic nephropathy.

microalbuminuria is the aim of a number of prospective studies. Elevation of glomerular filtration rate (GFR) and renal plasma flow (RPF) is present in a high proportion of patients with type 1 diabetes, even at diagnosis. The prognostic significance of these findings is unclear, although one small retrospective study did correlate the presence of glomerular hyperfiltration early in diabetes with raised AER and reduced GFR a decade later.[19]

Once established, prolonged tight glycaemic control[20] and dietary protein restriction[21,22] may have some influence on progression of overt nephropathy. Treatment of hypertension is, however, the only manoeuvre which has been shown to consistently delay renal insufficiency.[13,23–26] The rate of decline of renal function in patients with established diabetic nephropathy may be slowed by use of a variety of antihypertensive agents including diuretics, vasodilators, β-blockers, calcium channel blocking drugs and ACE inhibitors.[26,27] ACE inhibitors may represent an advance in the hypertensive care of diabetic patients. Specifically, it has been claimed that this class of drugs preserve glomerular function by lowering intraglomerular pressure independently of their systemic antihypertensive action. The background to this hypothesis is as follows: in animal models, a rise in systemic blood pressure is transmitted to the glomerular capillary network resulting in glomerular hypertension. ACE inhibitors act primarily on the efferent renal arteriole and reduce intraglomerular pressure; this in turn reduces filtration pressure across the glomerular capillary basement membrane and so reduces protein filtration.

In diabetic Munich-Wistar rats, ACE inhibition with enalapril reduces glomerular hypertension, restricts the increase in urinary albumin excretion and slows the development of glomerular structural lesions.[28] This protection is afforded despite pronounced hyperglycaemia and elevated levels of glycosylated haemoglobin in these animals. A further study by Anderson et al. compared the effects of enalapril with triple hypotensive drug therapy (reserpine, hydralazine and hydrochlorothiazide) in groups of Munich-Wistar rats after removal of one kidney and segmental infarction of two-thirds of the other.[29] Treatment with enalapril prevented both systemic hypertension and a rise in glomerular transcapillary pressure. Despite an equivalent effect on systemic blood pressure, triple therapy failed to control glomerular hypertension. This was associated with progressive proteinuria and glomerular lesions comparable to those seen in untreated animals.

In man, the acute effects of captopril on systemic and renal haemodynamics and albuminuria were studied in nine normotensive type 1 diabetics with overt nephropathy.[30] Blood pressure and GFR were not significantly altered but AER fell from 403 (interquartile range 812 μg/min) to 333 (707) μg/min ($P < 0.01$). This is consistent with the fall in protein excretion being mediated by a reduction in intraglomerular capillary pressure. Taguma and colleagues investigated the short-term effect of captopril on heavy proteinuria in 10 patients with advanced diabetic nephropathy.[31] Following an observation period of two to four weeks, subjects received 37.5 mg captopril per day (in divided doses) for eight weeks. Urinary protein fell from 10.66 ± 2.2 g/day (mean ± SEM) to 6.1 ± 1.4 g/day with no significant change in systemic blood pressure.

The major questions that need to be answered concerning the use of ACE inhibitors in diabetic renal disease are:

- Do ACE inhibitors represent an advantage over conventional antihypertensive agents in slowing the progression of established nephropathy?
- Can ACE inhibitors postpone or prevent progression from incipient nephropathy to overt nephropathy and, if so, are they more effective in this action than other antihypertensive agents?

To address these questions requires randomized, prospective studies comparing the effects of ACE inhibitors and other antihypertensive agents in a situation where comparable degrees of blood pressure control are achieved in both treatment groups. It must be emphasized, however, that the process that begins with glomerular hyperfiltration takes, on average, 30 years to progress through microalbuminuria and proteinuria to end-stage renal failure. Trials of intervention therefore require end-points other than end-stage renal failure. Reduction of the fall in GFR is likely to postpone end-stage renal failure and this is accepted as a satisfactory end-point, provided it is measured for more than two to three years. Stabilization or reduction of microalbuminuria and albuminuria is also accepted as a surrogate end-point as this seems to be associated with preservation of GFR.[32]

Studies of patients with overt diabetic nephropathy (macroalbuminuria)

To date, there have been no prospective intervention studies which have assessed the effect of ACE inhibitors (or any antihypertensive treatments) on mortality due to diabetic nephropathy. Retrospective analysis of the effect of ACE inhibitors on mortality is not yet available since this group of agents are relatively new.[13,33]

In 1986, Bjorck reported the effects of captopril in 14 insulin-dependent patients with diabetic nephropathy and hypertension (mean age 34 years, mean duration of diabetes 22 years).[34] During two years' treatment with captopril (in addition to existing antihypertensive therapy) the mean arterial blood pressure fell by 5 mmHg ($P < 0.005$) whilst the deterioration in GFR decreased from 10.3 to 2.4 ml/min/year ($P < 0.005$). There was no correlation between the fall in blood pressure and the reduction in GFR. The effect of captopril on mean urinary protein excretion was not significant but proteinuria was reduced in eight out of the fourteen patients. The authors considered it unlikely that the beneficial effect of captopril could be explained by the reduction in blood pressure alone. Interestingly, Ahmed subsequently divided the patients studied by Bjorck et al. into two groups on the basis of the amount of blood pressure (BP) reduction (>4 mmHg or < 4 mmHg).[35] The group with the larger fall in BP had a significant reduction in proteinuria and this correlated with the blood pressure change ($R = 0.58$).

Parving et al. treated 18 hypertensive insulin-dependent diabetic patients with nephropathy (mean age 33 years) with 37.5–100 mg captopril and 98 mg (mean) frusemide (10 patients) or 4 mg (mean) bendrofluazide (seven patients).[36] An untreated control group of 13 hypertensive nephropaths (mean age 32 years) from a prospective study were also observed for the study period of two and a half years. Arterial blood pressure, albuminuria and GFR were the end-points of the study; baseline values were identical in the treated and untreated groups. Captopril produced a significant improvement in mean arterial BP (–8.7 mmHg versus +6.6 mmHg, $P < 0.001$), AER (–592 µg/min versus +431 µg/min, $P < 0.001$) and in the rate of decline in GFR compared with the control subjects. The rate of decrease in GFR and mean arterial BP were significantly correlated ($R = 0.37$, $P < 0.05$). The mean rate of decline of GFR at the end of the study in the captopril group was 5.8 (0.7) ml/min/year compared with 10.0 (1.3) ml/min/year in the controls. This improvement in GFR is, however, no better than has been reported in trials using older antihypertensive agents; 14.7–5.9 ml/min/year,[24] 10.92–4.7 ml/min/year[25] and 10.68–2.64 ml/min/year.[26]

Parving et al. performed an open, follow-up study of normotensive insulin-dependent diabetics with nephropathy, randomized to receive treatment with captopril or no therapy.[37] Thirty-two patients were recruited, of which 15 received captopril 25–100 mg/day for 12 months. Albuminuria, arterial BP and GFR were the main outcome measures. In captopril-treated patients, mean arterial BP fell by 3 (SE 2) mmHg and albuminuria declined by 11%. In contrast, the mean arterial BP in control subjects rose by 6(1) mmHg and albumin excretion increased by 55%. The GFR fell by 3.1(2.8) ml/min/1.73 m^2 with captopril and by 6.4(3.1) ml/min/1.73 m^2 in controls over the 12-month period.

In 1990, Bjorck reported a prospective, open, randomized study of 40 type 1 diabetic patients with diabetic nephropathy and reduced renal function.[38] They had a mean age of 42 years (range 21–58 years), mean age at onset 17 years (range 3–39) and mean duration of diabetes 25 years (range 13–14 years). All patients had diabetic retinopathy and only two were not receiving antihypertensive therapy prior to study. Twenty-two patients were randomized to receive enalapril (E) for eight weeks and 18 to metoprolol (M), usually in combination with frusemide. The main outcome measures were arterial BP and urinary excretion of albumin and protein. After eight weeks' therapy, there was no difference in mean BP between the two groups (E:135/82 mmHg, M:136/86 mmHg). Protein and albuminuria were similar before randomization but fell to 0.7 g/24 hours (95% confidence interval 0.5–1.2) in the enalapril group compared with 1.6 g/24 hours (95% confidence interval 1.11–2.5) in patients given metoprolol ($P < 0.02$). Proteinuria was 1.1 (0.7–1.7) g/24 hours with enalapril and 2.4(1.6–3.6) g/24 hours with metoprolol ($P < 0.02$). The authors concluded that ACE inhibitors reduce proteinuria by a mechanism independent of their effect on systemic blood pressure. As to whether such an effect leads to better preservation of kidney function (as they had indicated four years earlier), the authors conceded that further studies were necessary.

Holdaas and colleagues performed a short, open, cross-over study to assess the effects of lisinopril and nifedipine on albumin excretion, renal haemodynamics and segmental tubular reabsorption in overt diabetic nephropathy.[39] Twelve patients with type 1 diabetes and albuminuria, mild-to-moderate hypertension and serum creatinine < 200 µg/min were recruited. The study consisted of two 3-week periods of active treatment separated by a 4-week wash-out period. Lisinopril reduced albumin excretion from 1343(±337) µg/min to 879(±299) µg/min ($P < 0.01$) whilst nifedipine had little effect (1436[±336] µg/mm versus 1319[±342] µg/min). No differences were observed between the two treatments with regard to their effect on renal haemodynamic parameters, including GFR.

Bjorck et al. performed a prospective study to determine whether inhibition of ACE reduced the rate of decline in kidney function more than lowering of blood pressure with other antihypertensive agents.[40] This was an open, randomized trial lasting a mean of 2.2 years, performed in three outpatient nephrology clinics. Forty patients with insulin-dependent diabetes and diabetic nephropathy with reduced renal function received enalapril or metoprolol, usually combined with frusemide. The rate of decline in GFR was measured by chromium-51 edetic acid clearance. GFR declined a mean of 2.0 (SD3.2) ml/min/year in the group given enalapril and 5.6 (5.9) ml/min/year in the control group (the physiological fall in GFR in middle age is approximately 1 ml/min/year compared with 10 ml/min/year in diabetic patients with albuminuria who have not been treated with antihypertensive drugs). Mean arterial BP during the study was 102(5) mmHg in the enalapril group and 103(5) mmHg in the metoprolol patients. Urinary albumin excretion during treatment with enalapril was 60% lower than during treatment with metoprolol. The authors concluded that enalapril has an antiproteinuric action independent of its effect on blood pressure and that treatment with enalapril reduces the rate of decline in kidney function more than equally effective antihypertensive treatment with metoprolol. This points to a specific renal protective effect of ACE inhibitors in diabetic nephropathy.

These conclusions should, however, be interpreted with caution. Parving et al. reported on ten years' experience with conventional antihypertensive therapy in 11 type 1 patients with diabetic nephropathy.[41] Using metoprolol, hydralazine and frusemide over a 9.7-year period (range 2.8–10.4 years), the rate of decline in GFR fell from 10.7 (5.3–17.5) ml/min/year to 2.5 (0.5–4.8) ml/min/year. This is similar to the results achieved by Bjorck and colleagues using enalapril and much lower than those that they achieved using metoprolol. Furthermore, Gronhagen-Riska et al. reported similar rates of fall in GFR comparing the same treatments as Bjorck.[42]

A prospective, randomized, double-blind study has recently been conducted to compare the effects of lisinopril and nifedipine on renal function and blood pressure in 28 hypertensive, macroproteinuric patients with type 1 and type 2 diabetes.[43] Both drugs reduced systemic blood pressure to the same extent. Protein excretion was reduced in the lisinopril group but the nifedipine-treated group showed a slight increase at the end of the study. GFR fell in both groups but the data in the lisinopril-treated group were heavily skewed by one patient's results. Fractional albumin clearance fell in the lisinopril-treated group despite the reduction in GFR. Although the fall in GFR in the nifedipine-treated group was of comparable magnitude there was an increase in fractional albumin clearance. This suggests that lisinopril may exert a specific intra-renal effect on protein handling independent of alterations in renal haemodynamics.

In summary, although ACE inhibitors seem to be effective in slowing the progression of decline in renal function in patients with diabetic nephropathy, there is little evidence to support the contention that they provide an advantage over conventional antihypertensive agents. It seems likely that they do exert a specific intra-renal effect, but whether this is of therapeutic benefit remains to be proven. Large, long-term prospective, randomized, double-blind comparisons of treatments are required to address this issue. Ideally, type 1 and type 2 diabetics with nephropathy should be examined separately until it is proven that these two conditions are identical.

Studies of patients with incipient nephropathy (microalbuminuria)

Marre and colleagues assessed the effects of enalapril on renal function in 20 normotensive diabetic patients who had persistent microalbuminuria (30–300 mg albumin/24 hours).[44] Two groups of 10 patients were randomly assigned to receive enalapril 20 mg/day or placebo for six months. Before treatment, the enalapril group had a median AER of 124 mg/24 hours and mean arterial BP of 100(SD 8) mmHg), similar to the placebo group. Following treatment with enalapril, the mean arterial BP decreased to 90(10) mmHg, median AER fell to 37 mg/24 hours and the GFR rose from 130(23) ml/min/1.73 m^2 to 141(24) ml/min/1.73 m^2. Albumin excretion increased in the placebo group whilst mean arterial BP was unchanged.

Winnocour et al. reported a six weeks' uncontrolled study of captopril in eight insulin-dependent diabetic patients.[45] Subjects all had newly diagnosed hypertension (systolic or diastolic pressures greater than the 95th centile norm for age and sex) and urinary protein excretion less than 200 mg/day. Although mean BP improved over the study period, there was no improvement in AER or in renal haemodynamics.

In 1988 Insua et al. reported the results of a comparison of captopril, nifedipine and placebo on urinary albumin excretion in normotensive patients with incipient diabetic nephropathy.[46] Twenty-two patients with insulin-dependent diabetes of 8–26 years' duration were studied. All had diastolic BP less than 95 mmHg and average baseline UAER of 15–200 µg/min on three 24-hour collections. Patients were randomized to receive slow release nifedipine 20 mg bd (n = 7), captopril 25 mg bd (n = 8), or placebo (n = 7). Placebo had no effect on any of the parameters that were assessed. Captopril and nifedipine resulted in a similar fall in arterial BP — (–9/–4 mmHg, and –7/–3 mmHg respectively). The UAER, however, decreased in all patients treated with captopril (mean 86±37 µg/min to 51±19 µg/min, $P < 0.05$) whilst it increased in all the nifedipine patients (mean 86±70 µg/min to 122±96 µg/min, $P < 0.05$). There was no correlation between the changes in UAER and arterial BP in either group.

Marre and colleagues reported a further 6 months' follow-up on those subjects described in 1987.[47] Patients remained in the same treatment groups and received the same therapy (enalapril 20 mg/day or placebo), now in a single-blind protocol. Albumin excretion decreased linearly in the enalapril group but not in the placebo group. In the last three months of the trial, three placebo-treated patients had an albumin excretion rate of more than 300 mg/24 hours compared with none in the enalapril group. Five enalapril-treated patients had normal albumin excretion (less than 30 mg/24 h, $P = 0.005$). Mean arterial pressure was reduced by enalapril throughout the study ($P < 0.005$) but increased linearly with placebo ($P < 0.05$); this rise was positively related to the increase in albumin excretion in the control group. GFR decreased significantly in the placebo group but not with enalapril. The conclusion that inhibition of ACE prevented development of nephropathy in normotensive diabetics with persistent microalbuminuria had to be tempered by the fact that seven of the patients under study had initial systolic blood pressure above 140 mmHg and therefore could be considered hypertensive.

Baba and colleagues reported a randomized cross-over trial comparing the renal effects of enalapril and nicardipine in eight hypertensive type 2 diabetic patients with microalbuminuria.[48] Subjects were aged 45–65 years, had normal serum creatinine and increased urinary albumin excretion rate (41–105 mg/24 h).

Patients received enalapril (10–20 mg/day) for four weeks followed by nicardipine (60–120 mg/day) or vice versa. There was no wash-out period between the treatment phases to avoid any rebound phenomenon. Seven patients who completed the study were analysed. Both enalapril and nicardipine significantly reduced systolic and diastolic BP without altering the GFR. UAER was significantly lowered by each treatment compared with the control phase (control: 32.6(20.4–47.4) μg/min; enalapril: 22.1(16.3–31.6) μg/min, $P < 0.05$; nicardipine: 18.5(6.3–31) μg/min, $P < 0.01$).

The Melbourne Diabetic Nephropathy Study Group compared the effects of perindopril with the calcium antagonist nifedipine on urinary albumin excretion in hypertensive and normotensive diabetic patients over a period of 12 months.[49] Fifty patients were enrolled of whom 43 completed the study: 30 normotensive and 13 hypertensive patients. Nineteen patients had type 1 diabetes and 24 had type 2 diabetes; 20 patients received perindopril and 23 were given nifedipine. The main outcome measures were UAER, BP and GFR. Overall, both drugs significantly reduced blood pressure to a similar extent and there was no significant difference with respect to albumin excretion. There was a significant correlation between mean BP and albuminuria and also between the reduction in mean BP and the decrease in albuminuria during treatment. In hypertensive patients, both drugs caused significant decreases in mean BP and albuminuria. In normotensive patients, mean BP fell by 4 mmHg over the twelve-month period but there was no significant reduction in albuminuria with either regime. (Alternatively, both drugs can be said to have prevented an increase in albuminuria in normotensive patients.) Within the total cohort, there was a wide variation in GFR and this was not significantly altered by treatment. In the 21 patients with a baseline GFR of more than 135 ml/min/1.73m^2 (defined as hyperfiltration), the GFR fell significantly with both treatments (from 186(167–204) ml/min/1.72 m^2 to 161(135–187) ml/min/1.73 m^2). In the remaining patient group with a GFR of 60–135 ml/min/1.73 m^2, there was no significant change with either therapy.

In a four-year prospective, randomized, controlled trial Mathieson et al. evaluated the effect of captopril on the progression of microalbuminuria in normotensive insulin-dependent diabetic patients.[50] Forty-four subjects were studied, 21 of whom were initially given captopril 25 mg/24 h. The dose was increased to 100 mg/24 h during the first 16 months of treatment and thiazide was added after 30 months. The remaining 23 patients were untreated. Urinary albumin excretion fell from 82(66–106) mg/24 h to 57(39–85) mg/24 h (geometric mean [95% confidence interval]) in the captopril group compared with an increase from 107(77–153) mg/24 h to 166(83–323 mg/24 h in the control group ($P < 0.05$). Seven of the untreated patients progressed to diabetic nephropathy, whereas none of the captopril-treated patients did so. Systemic BP and GFR did not differ between the two groups.

The authors have recently performed a double-blind placebo-controlled single-centre study of lisinopril in normotensive microalbuminuric diabetic subjects.[51] Twenty-seven patients with type 1 and type 2 diabetes were studied, of whom twelve received low-dose lisinopril (10 mg/day) for 48 weeks. The main outcome measures were changes in UAER and GFR. Seven lisinopril-treated patients became normoalbuminuric and five remain microalbuminuric; none developed nephropathy. Three placebo-treated patients became normoalbuminuric, nine remained microalbuminuric and three developed nephropathy. The reduction in proteinuria in the lisinopril-treated group was independent of alterations in systemic BP. GFR fell in both groups, the fall being greater in the lisinopril-treated patients. We could demonstrate no benefit of ACE inhibition over placebo on any other aspect of renal haemodynamics.

In summary, there is now a body of evidence to support the contention that ACE inhibitors prevent the progression from microalbuminuria to overt proteinuria in patients with diabetes. It is likely that progression to end-stage renal failure will also be slowed or prevented by this therapy. Whether this effect is due to the lowering of systemic blood pressure (even in those patients not formally defined as 'hypertensive') or is related to specific intra-renal effects is unknown. Studies of conventional antihypertensive agents have shown similar benefits in terms of albuminuria,[52] and the few comparative studies suggest that ACE inhibitors provide little additional benefit. Once again, large, long-term, prospective, randomized, double-blind comparisons of treatments are required, ideally with type 1 and type 2 diabetes examined separately.

Essential hypertension

Hypertension is approximately twice as common in diabetic patients as in age-matched controls.[53] Part of this increase is related to high levels of obesity and renal disease. However, even when these variables have been taken into account, the association remains. Both hypertension and diabetes are associated with increased cardiovascular morbidity and mortality and epidemiological studies have demonstrated that these risks are additive.

The effect of treatment on the consequences of hypertension has been extensively investigated in non-diabetics by large multi-centre trials.[54] Unfortunately, the diagnosis of diabetes has been a specific exclusion criterion in almost all major trials and recommendations have been derived by extrapolation from investigation of non-diabetics. Although high blood pressure is a well recognized risk factor for coronary artery disease (CAD), it has been difficult to demonstrate that treatment of essential hypertension leads to improved cardiovascular morbidity and mortality. Similarly, there is little evidence that good glycaemic control in patients with type 2 diabetes reduces the development of CAD, a major cause of premature death in this group.[55] Reaven proposed that these findings may be explained if the conditions are seen in the setting of a syndrome. He suggested the term 'syndrome X' should be applied to a series of related variables important in the genesis of CAD (Table 10.2) and proposed that insulin resistance was causal in this syndrome.[56] The term 'insulin resistance' refers to the level of insulin-stimulated glucose uptake measured in vivo. Resistance to insulin-stimulated glucose uptake is a common phenomenon and a characteristic finding in patients with type 2 diabetes and impaired glucose tolerance (IGT).[57–59] The degree of insulin resistance in

Insulin resistance
Hyperinsulinaemia
Glucose intolerance
Increased very low density lipoprotein triglyceride
Decreased high density lipoprotein cholesterol
Hypertension

Table 10.2
Components of syndrome X (the chronic cardiovascular risk factor

these conditions, however, bears little relation to fasting glucose levels and 25% of obese subjects with normal glucose tolerance show a similar degree of insulin resistance.[60] Several reports indicate that patients with raised blood pressure also have abnormalities in glucose and insulin metabolism.

There is indirect evidence that insulin resistance and hyperinsulinaemia may be relevant in the aetiology of hypertension. For example, there is a correlation between plasma insulin concentration and the level of blood pressure.[61] Further, experimental data provide mechanisms whereby hyperinsulinaemia may raise blood pressure:

- insulin is known to stimulate the sympathetic nervous system, resulting in a significant elevation in plasma catecholamine concentrations;[62,63]
- the sodium retaining properties of insulin[64,65] could increase total exchangeable body sodium, a consistent finding in type 2 diabetics[66] and in some studies of patients with essential hypertension.[67]

Mechanisms by which elements of syndrome X could promote the formation of coronary atheroma are not clear. Insulin resistance, genetically acquired and/or environmentally induced, could be the true aetiological factor and other components of the syndrome simply epiphenomena. Alternatively, hyperinsulinaemia may be of crucial importance; insulin is a potent stimulus to endothelial cell growth[68] and also regulates LDL-receptor activity.[69,70] Similarly, glucose intolerance is associated with abnormalities of platelet function, glycation of lipoproteins and the formation of lipid peroxides, all of which might promote atherogenesis.[71–73] The concept of a cardiovascular risk factor complex has important therapeutic implications. Agents used to correct one variable, for example hypertension, should not adversely affect another since the beneficial effects of treatment may then be negated. Indeed, if Reaven's hypothesis is completely accepted then the most important aspect of therapy is that it does not worsen the insulin resistant state. The disappointing results, in terms of cardiovascular mortality, from trials of antihypertensive therapy may be a consequence of the adverse effects of the agents used on other components of syndrome X. In this respect, thiazide diuretics appear to be the least attractive of the antihypertensive drugs for use in type 2 diabetics. These agents promote the development of glucose intolerance[74–76] and have been shown to increase insulin resistance.[4] Beta-blockers are also known to worsen glucose tolerance, especially in combination with diuretics,[74] and they cause a rise in triglyceride levels in association with a fall in HDL-C concentration.[77] Atenolol and metoprolol have now been shown to worsen insulin resistance in clamp studies.[78] ACE inhibitors have potential advantages over diuretics and β-blockers; they do not raise serum insulin levels and have neutral effects on lipoproteins. Captopril has been shown to reduce insulin resistance in euglycaemic hyperinsulinaemic clamp studies: Pollare and colleagues performed a randomized, double-blind, cross-over study to compare the effecs of captopril and hydrochlorothiazide on carbohydrate and lipid metabolism.[4] Fifty patients with essential hypertension were studied over two four-month treatment periods. Captopril increased the insulin-mediated disposal of glucose, compared with placebo, from 5.7±2.4 mg/kg to 6.3±2.5 mg/kg body weight per minute ($P < 0.05$). Hydrochlorothiazide caused a decrease from 6.4±2.0 mg/kg to 5.7±1.9 mg/kg body weight per minute ($P<0.01$). Captopril had no effect upon the basal insulin concentration but it decreased the late insulin response to glucose and increased the early insulin peak. Hydrochlorothiazide increased the basal insulin concentration and the late insulin response to glucose. There was no significant change in serum lipid or lipoprotein levels during treatment with captopril. The authors concluded that captopril caused a decrease in insulin resistance whilst this was increased with hydrochlorothiazide.

Whilst the association of glucose intolerance, hyperinsulinaemia, lipid abnormalities and hypertension is now well described, there are aspects of the syndrome X hypothesis which are yet to be proven. Although currently available trials of antihypertensive medication (mainly diuretics and β-blockers) have been disappointing with regard to CAD, they have demonstrated a consistent reduction in stroke morbidity and mortality of approximately 30%.[54] Indeed, a recent meta-analysis suggested that coronary deaths are reduced by 12–15%[79] and there is no evidence that these drugs are associated with increased coronary mortality, as might be expected from their effects on insulin resistance. Whilst newer antihypertensive agents, such as ACE inhibitors are theoretically attractive, there is no evidence that these agents offer any substantial advantage in terms of cardiovascular or stroke mortality. It would therefore be premature to abandon the use of diuretics and β-blockers for treatment of essential hypertension in both diabetic and non-diabetic individuals.[80]

Free-radical scavenging

Captopril exhibits anti-inflammatory properties which are independent of its action on ACE and may be due to its possession of a sulphydryl group.[81] The postulated mechanism is related to the free-radical scavenging ability of this moiety.[82] Free radicals (FR) are unstable chemical species, produced during the course of normal metabolism, which have an unpaired electron in their structure.[83] They are highly reactive (oxidizing adjacent molecules within microseconds) and hence are difficult to measure directly. A role for FR in disease is inferred from the characteristic changes they induce in polyunsaturated lipids and serum proteins. Elevated levels of oxidative products of lipids (total diene conjugates, thiobarbituric acid reactive substances) and proteins (fluorescent IgG and albumin) have been demonstrated in diabetic patients with retinopathy.[84] Diabetics with complications also have significantly reduced levels of ascorbic acid (AA) and reduced glutathione (GSH), compounds which act as natural antioxidants.[85] It has therefore been hypothesized that FR mechanisms may be involved in the generation of diabetic complications.[86] Further, FR mechanisms have also been implicated in the pathogenesis of hypertensive vascular disease[87] and atherogenesis.[88]

Captopril is an effective FR scavenger in vitro[82] and in vivo.[89,90] A similar, but reduced effect, is seen in vitro with non-sulphydryl-containing ACE inhibitors.[89,90] It has been demonstrated that captopril inhibits the in vitro formation of fluorescent IgG and albumin.[91] This effect is not seen with enalaprilat, the active metabolite of enalapril, which does not contain a sulphydryl group (although in the same study enalapril inhibited fluorescence development in albumin to a lesser degree). The authors also investigated the effects on markers of FR activity of three months' therapy with captopril versus enalapril in type 2 diabetic subjects with hypertension.[92] Thirty-two patients were randomized to receive captopril or enalapril in an open study, the doses adjusted according to BP response. During three months' treatment, both captopril and enalapril were associated with a fall in levels of two markers of FR activity, namely

TBA reactivity and fluorescent IgG. Other markers were not significantly affected and there was no difference between the two drug treatments. These findings posed two questions: why did captopril not demonstrate an advantage over enalapril, as seen in in vitro studies, and why did both ACE inhibitors lower the levels of two markers of FR activity? Most in vitro studies have assessed the effect of captopril at concentrations in the order of 10–100 μmol.[89–91] In vivo, captopril exists in the blood in the free form, as a disulphide dimer and as mixed disulphides with endogenous thiol compounds such as cysteine.[93] Only free sulphydryl groups act as FR scavengers and so it is the free drug concentration which is relevant.

Previous studies suggest that the levels achieved in the patients in the authors' study were lower than 10 μm. In a study of 12 healthy males, aged 65–76 years, a single 100 mg dose resulted in a mean maximum concentration of 4.15 μmol/l (803 ng/ml) free captopril;[94] in hypertensive patients given single doses of 1 mg/kg captopril peak free levels have been reported at 5.56–6.03 μmol/l.[95,96] Further, free captopril rapidly disappears from the blood and does not accumulate on repeated dosing; in hypertensive patients treated chronically with captopril 25 mg tds, levels of total captopril were 0.69 μmol/l eight hours after dosing, whereas free captopril was only 0.028 μmol/l.[97] Only three patients in the authors' study required more than 12.5 mg captopril bd for control of blood pressure, implying that the levels of free drug in this patient group were even lower than those previously reported. At such concentrations, the FR scavenging activity of the sulphydryl group of captopril may be clinically insignificant. Consideration of plasma levels is especially important when comparing results of in vivo studies. For example, Scheider et al.[98] reported an increase in GSH levels in hypertensive patients after captopril treatment, in contrast to the lack of change that the authors demonstrated. In that study, however, subjects received one 25 mg dose of captopril sublingually and measurements were performed over 120 minutes only. Whilst plasma levels of GSH rose significantly at 30 and 60 minutes after drug administration (when free drug levels would have been high), by two hours the mean GSH had fallen back to the baseline level.

There are various explanations as to why both ACE inhibitors were associated with a reduction in levels of fluorescent IgG and TBA reactivity. For example, hypertension per se may represent a state of increased FR activity. Simply reducing blood pressure, by any means, may lead to reduced levels of FR and hence a reduction in TBS and fluorescent IgG. The literature provides some support for this contention. Uysal et al. described raised TBA reactivity in non-diabetic subjects with essential hypertension[99] and Prabha et al. have reported that control of blood pressure using β-blockers/calcium antagonists decreased plasma TBA reactivity.[100] It is also possible that the reduction in FR markers represents a specific ACE inhibitor effect. In vivo studies have shown that ACE inhibitors without a sulphydryl group scavenge FR by a mechanism that is blocked by indomethacin.[89,101] It has been proposed that stimulation of prostacyclin production is important in this respect. A comparison of patients in whom equivalent blood pressure control was achieved using an ACE inhibitor and an agent with an alternative antihypertensive action is needed to resolve these questions. The most likely reason, however, for the fall in levels of fluorescent IgG and TBA reactivity are that they are non-specific changes which do not represent FR scavenging. Taken together, the seven markers monitored in this study probably give a good indication of FR activity. Individually, they are indirect and non-specific indicators, for example, fluorescent IgG and fluorescent albumin development occur as a result of glycation in diabetics and it has yet to be convincingly proved that the in-vivo mechanism involves free radicals.

In summary, the authors have found that captopril and enalapril, used in therapeutic doses for the treatment of hypertension in diabetic subjects, are associated with similar changes in levels of putative markers of FR activity. This finding is at variance with in vitro studies and may reflect the low concentrations of free captopril achieved by standard dosage regimes. To date, the notion that captopril provides a therapeutic advantage over other classes of antihypertensive drugs (or other ACE inhibitors) by means of FR scavenging has yet to be proven.

Conclusions

This chapter examined the indications for ACE inhibitor therapy in diabetics with renovascular and essential hypertension, and have also discussed their possible advantages based on the syndrome X hypothesis. The authors conclude that ACE inhibitors are suitable as first-line antihypertensive agents in diabetic patients based on efficacy and side-effect profiles, which in many respects appear to be superior to the older more established therapies. Evidence is still lacking that these drugs reduce long-term cardiovascular or nephrological morbidity or mortality in diabetics. Such confirmation (or otherwise) is awaited with interest. In the meantime, ACE inhibitors have established themselves as a further first-line treatment for the management of hypertension in diabetes.

References

1 Helgeland A, Strommen R, Hagelund CH, Tretli S. Enalapril, atenolol and hydrochlorothiazide in mild to moderate hypertension. *Lancet* 1986; **i**:872–75.

2 Croog SH, Levine S, Terta MA et al. The effects of antihypertensive therapy on the quality of life. *N Engl J Med* 1986; **34**:1657–64.

3 Ames RP. The effects of antihypertensive drugs on serum lipids and lipoproteins. *Drugs* 1986; **32**:335–57.

4 Pollare T, Lithell H, Berne C. A comparison of the effects of hydrochlorothiazide and captopril on glucose and lipid metabolism in patients with hypertension. *N Engl J Med* 1989; **321**:868–73.

5 McMurray J, Fraser DM, Captopril, enalapril, and blood glucose. *Lancet* 1986; **i**:1035.

6 Ferriere M, Lachkar H, Richard J, Bringer J, Orsetti A, Mirouze J. Captopril and insulin sensitivity. *Ann Int Med* 1985; **102**:134–35.

7 Rett K, Wicjlmayr M, Dietze GJ. Hypoglycaemia in hypertensive diabetic patients treated with sulphonylureas, biguanides, and captopril. *N Engl J Med* 1988; **319**:1609.

8 Sawicki PT, Muhlhauser I, Baba T, Berger M. Do angiotensin converting enzyme inhibitors represent a progress in hypertension care in diabetes mellitus? *Diabetologia* 1990; **33**:121–24.

9 Andersen AR, Sandahl Christiansen J, Andersen JK, Kreiner S, Deckert T. Diabetic nephropathy in type 1 (insulin dependent) diabetes: An epidemiological study. *Diabetologia* 1983; **25**:496–501.

10 Parving HH, Hommel E, Mathiesen E, Skott P, Edsberg B, Bahnsen M et al. Prevalence of microalbuminuria, arterial hypertension, retinopathy and neuropathy in patients with insulin dependent diabetes. *BMJ* 1988; **296**:156–160.

11 Ballard DJ, Humphrey LL, Melton J, Frohnert PP, Chu C-P, O'Fallon WM et al. Epidemiology of persistent proteinuria in type 2 diabetes mellitus. *Diabetes* 1988; **37**:405–12.

12 Stephenson J, Kenny S, Lee E, Fuller J. Diabetic renal failure in the WHO multinational study. *Diabetes* 1992; **41(Suppl 1)**:128A.

13 Parving HH, Hommel E. Prognosis in diabetic nephropathy. *BMJ* 1989; **299**:230–3.

14 Joint Working Party on Diabetic Renal Failure of the British Diabetic Association, Renal Association and Research Unit, Royal College of Physicians. Renal failure in diabetics in the United Kingdom: Deficient provision of care in 1985. *Diabetic Med* 1988; **5**:79–84.

15 McMillan MA. Diabetic nephropathy – progress but not cure. *Hospital Update* 1990; **16**:947–8.

16 Viberti GC, Keen H, Wiseman MJ. Raised arterial pressure in parents of proteinuric insulin dependent diabetics. *BMJ* 1987; **295**:515–7.

17 Viberti GC, Hill RD, Jarret RJ, Argyropoulos A, Mahmud U, Keen H. Microalbuminuria as a predictor of clinical nephropathy in insulin-dependent diabetes mellitus. *Lancet* 1982; **i**:1430–2.

18 Mogensen CE, Christensen CK. Predicting diabetic nephropathy in insulin-dependent patients. *N Engl J Med* 1984; **311**:89–93.

19 Drummond K, Levy-Marchal C, Laborde K et al. Enalapril does not alter renal function in normotensive, normoalbuminuric, hyperfiltering type 1 (insulin-dependent) diabetic children. *Diabetologia* 1989; **32**:255–60.

20 Feldt-Rasmussen B, Mathiesen ER, Jensen T, Lauritzen T, Deckert T. Effect of improved metabolic control on loss of kidney function in Type 1 (insulin-dependent) diabetic patients: an update of the Steno studies. *Diabetologia* 1991; **34**:164–70.

21 Cohen D, Dodds R, Viberti GC. Effect of protein restriction in insulin dependent diabetics at risk of nephropathy. *BMJ* 1987; **294**:795–98.

22 Evanoff GV, Thompson CS, Brown J, Weinman EJ. The effect of dietary protein restriction on the progression of diabetic nephropathy. *Arch Intern Med* 1987; **147**:492–95.

23 Mogensen CE. Progression of nephropathy in long term diabetics with proteinuria and effect of initial antihypertensive treatment. *Scand J Lab Invest* 1976; **36**:383–88.

24 Mogensen CE. Long term antihypertensive treatment inhibiting progression of diabetic nephropathy. *BMJ* 1982; **285**:685–88.

25 Parving HH, Andersen AR, Smidt UM, Svendsen PA. Early aggressive antihypertensive treatment reduces rate of decline in kidney function in diabetic nephropathy. *Lancet* 1983; **i**:1175–79.

26 Parving HH, Andersen AR, Smidt U, Hommel E, Mathiesen ER, Svendsen PA. Effect of antihypertensive treatment on kidney function in diabetic nephropathy. *BMJ* 1987; **294**:1443–52.

27 Mogensen CE. Long term antihypertensive treatment inhibiting progression of diabetic nephropathy. *BMJ* 1982; **285**:685–88.

28 Zatz R, Dunn R, Meyer T, Anderson, Rennke H, Brenner B. Prevention of diabetic glomerulopathy by pharmacological amelioration of glomerular capillary hypertension. *J Clin Invest* 1986; **77**:1925–30.

29 Anderson S, Rennke H, Brenner B. Therapeutic advantage of converting enzyme inhibition in arresting progressive renal disease associated with systemic hypertension. *J Clin Invest* 1986; **77**:1993–2000.

30 Elving LD, Wetzels JMF, de Nobel E, Hoitsma AJ, Berden JHM. Captopril reduces albuminuria probably through lowering intraglomerular pressure: acute effects in patients with diabetic nephropathy. *Postgrad Med J* 1988; **64[Suppl 3]**:34–35.

31 Taguma Y, Kitamoto Y, Futaki G et al. Effect of captopril on heavy proteinuria in azotemic diabetics. *N Engl J Med* 1985; **313**:1617–20.

32 Mogensen CE. Angiotensin converting enzyme inhibition and diabetic nephropathy. *BMJ* 1992; **304**:327–28.

33 Mathiesen ER, Borch-Johnsen K, Jensen DV, Deckert T. Improved survival in patients with diabetic nephropathy. *Diabetologia* 1989; **32**:884–86.

34 Bjorck S, Nyberg G, Mulec H, Granerus G, Herlitz H, Aurell M. Beneficial effects of angiotensin converting enzyme inhibition on renal function in patients with diabetic nephropathy. *BMJ* 1986; **293**:471–74.

35 Ahmad S. Beneficial effects of angiotensin converting enzyme inhibition on renal function in patients with diabetic nephropathy. *BMJ* 1986; **293**:1028.

36 Parving HH, Hommel E, Smidt UM. Protection of kidney function and decrease in albuminuria by captopril in insulin dependent diabetics with nephropathy. *BMJ* 1988; **297**:1086–91.

37 Parving HH, Hommel E, Nielsen MD, Giese J. Effect of captopril on blood pressure and kidney function in normotensive insulin dependent diabetics with nephropathy. *BMJ* 1989; **299**:533–36.

38 Bjorck S, Mulec H, Johnsen SA, Nyberg G, Aurell M. Contrasting effects of enalapril and metoprolol on proteinuria in diabetic nephropathy. *BMJ* 1990; **300**:904–7.

39 Holdaas H, Hartmann A, Lien MG, Nilsen L, Jervell J, Fauchald P, Endresen L, Djoseland O, Berg KJ. Contrasting effects of lisinopril and nifedipine on albuminuria and tubular transport functions in insulin dependent diabetics with nephropathy. *J Int Med* 1991; **229**:163–70.

40 Bjorck S, Mulec H, Johnsen SA, Nordn G, Aurell M. Renal protective effect of enalapril in diabetic nephropathy. *BMJ* 1992; **304**:339–43.

41 Parving HH, Smidt UM, Mathiesen ER, Hommel E. Ten years' experiences with antihypertensive treatment in diabetic nephropathy. *Diabetologia* 1991; **34(Suppl 2)**:A38.

42 Gronhagen-Riska S. ACE inhibitors' effect on kidney function. *Nordisk Medicin* 1990; **105**:219–22.

43 O'Donnell MJ, Rowe BR, Lawson N, Horton A, Gyde OHB, Barnett AH. Comparison of the effects of an angiotensin converting enzyme inhibitor and a calcium antagonist in hypertensive, macroproteinuric diabetic patients: a randomised double-blind study. *J Hum Hypertension* 1993; **7**:333–9.

44 Marre M, Leblanc H, Suarez L, Guyenne T, Menard J, Passa P. Converting enzyme inhibition and kidney function in normotensive diabetic patients with persistent microalbuminuria. *BMJ* 1987; **294**:1448–52.

45 Winnocour PH, Waldek S, Anderson DC. Converting enzyme inhibition and kidney function in normotensive diabetic patients with persistent microalbuminuria. *BMJ* 1987; **295**:391.

46 Insua A, Ribstein J, Mimran A. Comparative effect of captopril and nifedipine in normotensive patients with incipient diabetic nephropathy. *Postgrad Med J* 1988; **64[Suppl 3]**:59–62.
47 Marre M, Chatellier G, Leblanc H, Guyene TT, Menard J, Passa P. Prevention of diabetic nephropathy with enalapril in normotensive diabetics with microalbuminuria. *BMJ* 1988; **297**:1092–95.
48 Baba T, Murabayashi S, Takebe K. Comparison of the renal effects of angiotensin converting enzyme inhibitor and calcium antagonist in type 2 (non-insulin dependent) diabetic patients with microalbuminuria; A randomised controlled trial. *Diabetologia* 1989; **32**:40–4.
49 Melbourne Diabetic Nephropathy Study Group. Comparison between perindopril and nifedipine in hypertensive and normotensive diabetic patients with microalbuminuria. *BMJ* 1991; **302**:210–16.
50 Mathiesen ER, Hommel E, Giese J, Parving HH. Efficacy of captopril in postponing nephropathy in normotensive insulin dependent diabetic patients with microalbuminuria. *BMJ* 1991; **303**:81–87.
51 O'Donnell MJ, Rowe BR, Lawson, N, Horton A, Gyde OHB, Barnett AH. Placebo controlled trial of lisinopril in normotensive diabetic patients with incipient nephropathy. *J Hum Hypertension* 1993; **7**:327–32.
52 Friedman PJ, Dunn PJ, Jury DR. Metoprolol and albumin excretion in diabetes. *Lancet* 1986; **88**:1042–43.
53 Working Group on Hypertension in Diabetes. Statement on Hypertension in diabetes: final report. *Arch Intern Med* 1987; **147**:830–42.
54 Swales JD, Ramsey LE, Cooper JR et al. Report of the British Hypertension Society working party. Treating mild hypertension. *BMJ* 1989; **298**:694–98.
55 Multi-centre study. UK prospective study of therapies of maturity-onset diabetes. 1. Effect of diet, sulphonylurea, insulin or biguanide therapy on fasting plasma glucose and body weight over one year. *Diabetologia* 1983; **24**:404–11.
56 Reaven GM. Role of insulin resistance in human disease. Banting Lecture. *Diabetes* 1988; **37**:1595–1607.
57 Shen S-W, Reaven GM, Farquhar J. Comparison of impedance to insulin-mediated glucose uptake in normal subjects and in subjects with latent diabetes. *J Clin Invest* 1970; **49**:2151–60.
58 Ginsberg H, Kimmerling G, Olefsky JM, Reaven GM. Demonstration of insulin resistance in untreated adult onset diabetic subjects with fasting hyperglycaemia. *J Clin Invest* 1975; **55**:454–61.
59 Ginsberg J, Olefsky JM, Reaven GM. Further evidence that insulin resistance exists in patients with chemical diabetes. *Diabetes* 1974; **23**:674–78.
60 Hollenbeck C, Reaven GM. Variations in insulin-stimulated glucose uptake in healthy individuals with normal glucose tolerance. *J Clin Endocrinol Metab* 1987; **64**:1169–73.
61 Manicardi V, Camellini L, Belloidi G, Coscelli C, Ferrannini E. Evidence for an association of high blood pressure, and hyperinsulinaemia in obese man. *J Clin Endocrinol Metab* 1986; **62**:1302–4.
62 Rowe JW, Young JB, Minaker KL, Stevens AL, Pallotta J, Landsberg L. Effect of insulin and glucose infusions on sympathetic nervous system activity in normal mean. *Diabetes* 1981; **30**:219.
63 Christensen NJ, Gundersen HJG, Hegedus L et al. Acute effect of insulin on plasma noradrenaline and the cardiovascular system. *Metabolism* 1980; **29**:1138–45.
64 Defronzo RA, Goldberg M, Agus Z. The effects of glucose and insulin on renal electrolyte transport. *J Clin Invest* 1976; **58**:83–90.
65 Defronzo RA, Cooke C, Andres R, Faloona GR, David PJ. The effect of insulin in renal handling of sodium, potassium, calcium and phosphate in man. *J Clin Invest* 1975; **55**:845–55.
66 De Chatel R, Weidmann P, Flammer J. Sodium, renin, aldosterone, catecholamines and blood pressure in diabetes mellitus. *Kidney Int* 1977; **12**:412–21.
67 Beretta-Piccoli C, Davies DL, Boddy K. Relation of arterial pressure with body sodium, body potassium and plasma potassium in essential hypertension. *Clin Sci* 1982; **63**:257–70.

68 Stout RW, Bierman EL, Ross R. The effect of insulin on the proliferation of cultured primate arterial smooth muscle cells. *Cir Res* 1975; **36**:319–27.

69 Chait A, Bierman EL, Albers JJ. Regulatory role of insulin in the degradation of low density lipoprotein by cultured human skin fibroblasts. *Biochem Biophys Acta* 1978; **529**:292–99.

70 Chait A, Bierman EL, Albers JJ. Low density lipoprotein receptor activity in fibroblasts cultured from diabetic donors. *Diabetes* 1979; **28**:914–18.

71 Colwell JA, Winocour PD, Lopes-Virella M, Halushka PV. New concepts about the pathogenesis of atherosclerosis in diabetes mellitus. *Am J Med* 1983; **75**:67–80.

72 Steinbrecher UP, Witztum JL. Glucosylation of low-density lipoproteins to an extent comparable to that seen in diabetes slows their catabolism. *Diabetes* 1984; **33**:130–4.

73 Kaji H, Kurasaki M, Ito K et al. Increased lipoperoxide value and glutathione peroxidase activity in blood plasma of type 2 (non-insulin-dependent) diabetic women. *Klin Wochenschr* 1985; **63**:765–68.

74 Bengtsson C, Blohme G, Lapidus L, Lindquist O, Lundgren H, Nystrom E, Petersen K, Sigurdsson JA. Do antihypertensive drugs precipitate diabetes? *BMJ* 1984; **289**:1495–97.

75 Murphy MC, Lewis PJ, Kohner E et al. Glucose intolerance in hypertensive patients treated with diuretics: a fourteen year follow-up. *Lancet* 1982; **ii**:1293–95.

76 Strathers AD, Murphy MB, Dollery CT. Glucose tolerance during antihypertensive therapy in patients with diabetes mellitus. *Hypertension* 1985; **7[Suppl 2]**:95–101.

77 Northcote RJ. Beta-blockers, lipids, and coronary atherosclerosis: fact or fiction? *BMJ* 1988; **296**:731–32.

78 Pollare T, Lithell H, Selinus I, Berne C. Sensitivity to insulin treatment with atenolol and metoprolol: a randomised double blind study of effects on carbohydrate and lipoprotein metabolism in hypertensive patients. *BMJ* 1989; **298**:1152–57.

79 Collins R, Peto R, MacMahon S et al. Blood pressure, stroke and coronary heart disease. Part 2, short term reductions in blood pressure: overview of randomised drug trials in their epidemiological context. *Lancet* 1990; **i**:827–38.

80 Bain SC, Dodson PM. The chronic cardiovascular risk factor syndrome (syndrome X): mechanisms and implications for atherogenesis. *Postgrad Med J* 1991; **67**:922–27.

81 Martin MFR, McKenna F, Bird HA, Surrall KE, Dixon JS, Wright V. Captopril: a new treatment for rheumatoid arthritis? *Lancet* 1984; **1**:1325.

82 Chopra M, Scott N, McMurray J, McLay J, Bridges A, Smith WE, Belch JJF. Captopril: a free radical scavenger. *Br J Clin Pharmacol* 1989; **27**:396–99.

83 Sinclair AJ, Barnett AH, Lunec J. Free radicals and antioxidant systems in health and disease. *Br J Hosp Med* 1990; **43**:334–44.

84 Jennings PE, Jones AF, Florkowski CM, Lunec J, Barnett AH. Increased diene conjugates in diabetic subjects with microangiopathy. *Diabetic Med* 1987; **4**:452–56.

85 Sinclair AJ, Gray L, Girling AJ, Le Guen C, Barnett AH. Diabetes is associated with abnormalities of ascorbic acid metabolism which are only partially corrected by ascorbate supplementation. *Diabetic Med* 1989; **6**:1A.

86 Jennings PE, Barnett AH. New approaches to the pathogenesis and treatment of diabetic microangiopathy. *Diabetic Med* 1988; **5**:111–17.

87 Wilson SK. Role of oxygen-derived free radicals in acute angiotensin II-induced hypertensive vascular disease in the rat. *Circ Res* 1990; **66**:722–34.

88 Halliwell B. Free radicals, reactive oxygen species and human disease: a critical evaluation with special reference to atherosclerosis. *Br J Exp Path* 1989; **70**:737–57.

89 Pi X, Chen X. Captopril and ramiprilat protect against free radical injury in isolated working rat hearts. *J Mol Cell Cardiol* 1989; **21**:1261–71.

90 Westlin W, Mullane K. Does captopril attenuate reperfusion-induced myocardial dysfunction by scavenging free radicals? *Circulation* 1988; **77[Suppl I]**:30–39.

91 Le Guen C, Bain SC, Barnett AH, Lunec J. Captopril inhibits fluorescence development in non-enzymatically glycosylated proteins. *Agents and Actions* 1992; **36**:264–70.

92 Bain SC, Le Guen CA, Lunec J, Barnett AH. Comparison of the free radical activity of captopril versus enalapril: a three month in vivo study in hypertensive diabetic patients. *J Hum Hypert* 1991; **5**:511–15.
93 Migdalof BH, Antonaccio MJ, McKinstry DN et al. Captopril: Pharmacology, metabolism and disposition. *Drug Metabol Rev* 1984; **15**:841–69.
94 Creasy WA, Funke PT, McKinstry DN, Sugarman AA. Pharmokinetics of captopril in elderly healthy male volunteers. *J Clin Pharmacol* 1986; **25**:264–68.
95 Richer C, Gilroux B, Plouin PF et al. Captopril: Pharmacokinetics, antihypertensive and biological effects in hypertensive patients. *Br J Clin Pharmacol* 1984; **17**:243–50.
96 Guidicelli JF, Chaignon M, Richer C et al. Influence of chronic renal failure on captopril pharmacokinetics and clinical and biological effects in hypertensive patients. *Br J Clin Pharmacol* 1984; **18**:749–58.
97 McKinstry DN, Willard DA, Cohen A et al. Blood levels of captopril in hypertensive patients during chronic therapy. In: Abstracts 2nd World Conference on Clinical Pharmacology and Therapeutics. Washington 1983; July/August:99.
98 Schneider R, Iscovitz H, Ilan Z, Bernstein K, Gros M, Iaina A. Oxygen free radical scavenger system intermediates in essential hypertensive patients before and immediately after sublingual captopril administration. *Isr J Med Sci* 1990; **26**:491–95.
99 Uysal M, Buler H, Sener D, Oz H. Lipid peroxidation in patients with essential hypertension. *Int J Clin Pharmacol Res* 1986; **24**:474–76.
100 Prabha et al. Free radical generation, lipid peroxidation and essential fatty acids in uncontrolled hypertension. *Prostag Leukot Essent Fatty Acids* 1990; **41**:27–33.
101 Przyklenk K, Kloner RA. Angiotensin converting enzyme inhibitors enhance contractile function of 'stunned' myocardium by different mechanisms of action. *Circulation* 1990; **82[Suppl III]**:III–264.

11

ACE inhibitors, reperfusion injury and cardiac arrhythmias

John J V McMurray and Simon G Ray

The concept of reperfusion injury

The term reperfusion injury is an experimental one that arises from the observation that coronary reperfusion, following a period of myocardial ischaemia, leads to further cell injury or death, distinct from that induced by the preceding ischaemia.[1–3]

Types of reperfusion injury

Three main types of experimental reperfusion injury have been proposed:[1–3]

i) *Reperfusion related arrhythmias.* In many animal models restoration of coronary flow to ischaemic myocardium leads to ventricular arrhythmias including premature beats, idioventricular rhythm, and ventricular fibrillation. These arrhythmias generally occur within seconds of reperfusion. The myocardium may be most vulnerable to reperfusion arrhythmias after only a few minutes of ischaemia and this susceptibility may decrease when ischaemia is prolonged (an important issue when clinical parallels are considered — see below).

ii) *Myocardial stunning.* The term myocardial stunning refers to the reversible mechanical dysfunction that is observed after the sequence of ischaemia followed by reperfusion. This is a phenomenon where viable myocytes, which have had their blood supply restored to normal continue to exhibit reduced contractile function. This impairment in contractility may persist for hours or even days. Full function, however, eventually returns because the injury is reversible. Return of function can be accelerated by inotropic stimulation (Figs 11.1 and 11.2).

iii) *Lethal reperfusion injury.* This is the most controversial form of reperfusion injury and indeed, many authorities question its existence. Potentially, however, it is the most important form of reperfusion injury, causing irreversible myocyte injury (that is, cell death). Clearly, preventive therapy at the time of reperfusion might lead to greater myocardial salvage, with better ventricular function and improved survival.

What causes reperfusion injury?

There are probably multiple mechanisms involved in reperfusion injury and the relative importance of these may be different for each of the adverse manifestations of reflow.[1–3]

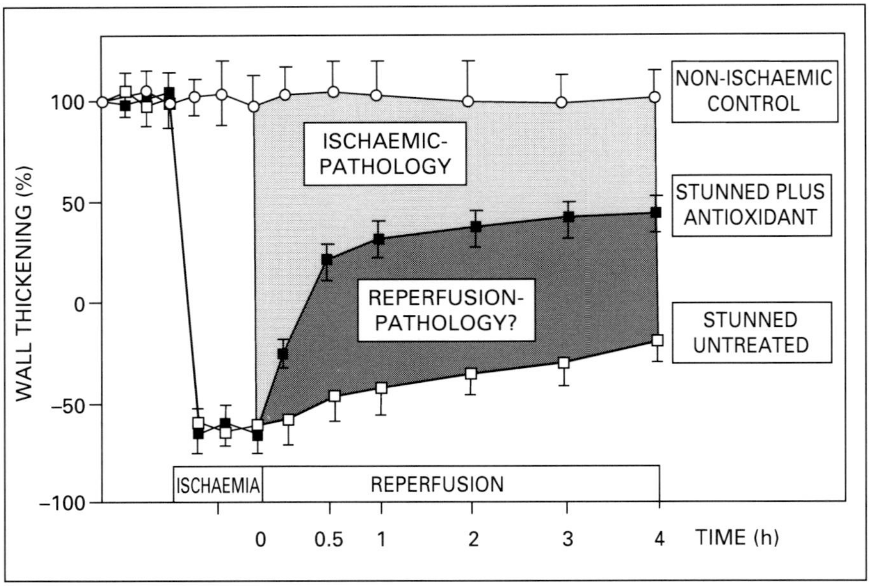

Figure 11.1
Myocardial stunning is represented by the area of dark shading (reperfusion pathology?). This component of post-ischaemic contractile dysfunction can be prevented or limited by use of an antioxidant intervention given before reperfusion.

Stunning is the best characterized, experimentally (and clinically — see below), of these injuries and there is persuasive evidence that free radicals (FR) are involved in its pathogenesis.[4] FR are short-lived cytotoxic chemicals produced by a number of normal and abnormal cellular metabolic and enzymatic pathways. FR are also generated by activated neutrophils. The sequence of ischaemia (by priming tissues) and reperfusion (by readmitting oxygen) appears to favour excessive oxygen-derived FR production, while simultaneously depressing cellular anti-FR 'scavenging' or 'antioxidant' defences (Fig. 11.3). The net effect is unopposed FR attack on biomolecules, for example in cell membranes, which ultimately leads to cellular dysfunction. Disturbance of calcium homeostasis is almost certainly an essential component of FR-mediated injury to myocytes. Other cells (for example endothelial cells) and tissues (for example the collagen interstitium) may also be damaged and contribute to reperfusion injury. FR generation occurs within the first few

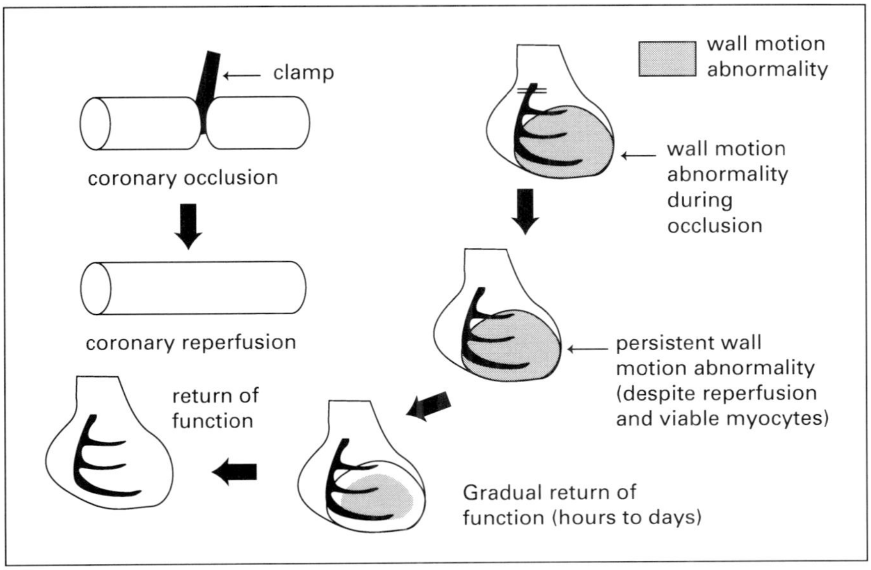

Figure 11.2
Schematic illustration of stunning where wall motion remains persistently abnormal for hours–days despite restoration of coronary blood flow to viable myocytes (a relief of myocardial ischaemia).

minutes of reperfusion, probably initially within the extracellular fluid or vessel wall.

There is increasing evidence that the highly reactive hydroxyl radical is a particular culprit in FR-mediated ischaemia-reperfusion injury.[5,6] FR scavengers, especially hydroxyl radical scavengers, administered just before coronary reperfusion, decrease FR formation and consistently and substantially reduce the degree of myocardial contractile impairment seen in experimental models. Similar protection can be obtained by prior administration of drugs, for example desferrioxamine, or chelating transition metals, for example iron, which may catalyse the formation of the hydroxyl radical.[4]

Clinical relevance of reperfusion injury

Clinical interest in experimental observations has grown with the widespread use of cardiopulmonary bypass for cardiac surgery and, particularly, with the advent of coronary

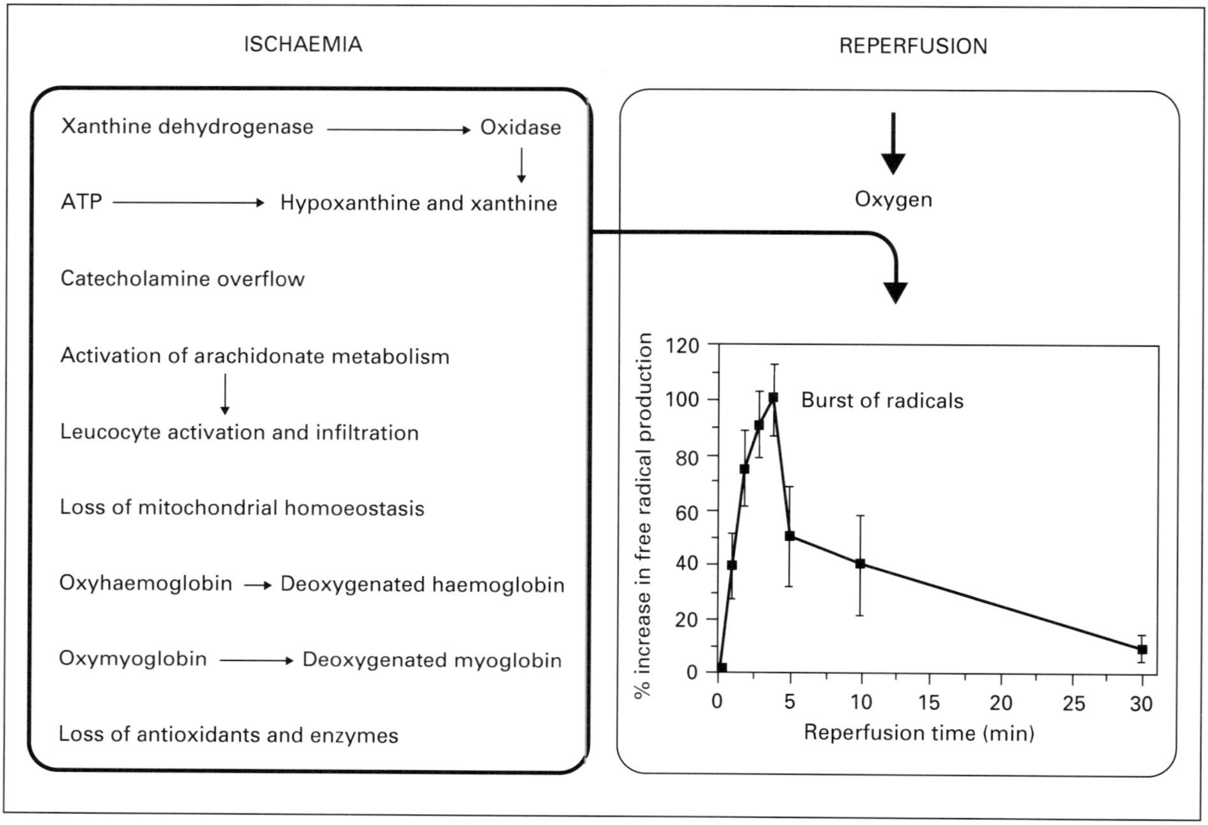

Figure 11.3
Ischaemia primes the myocardium for free-radical-induced injury, particularly during reperfusion (ATP = adenosine triphosphate). Reproduced by permission.

thrombolysis for the routine treatment of acute myocardial infarction.[7,8] Both situations lead to the sequence of myocardial ischaemia followed by reperfusion, although many differences exist between the experimental models and the clinical situation.[8] For example, most of these models bear little resemblance to successful coronary thrombolysis in the patient with myocardial infarction. In the usual model of stunning a snare is tightened to occlude a normal coronary artery. A brief period (usually 15 minutes) of ischaemia follows which is insufficient to cause myocyte necrosis. Rapid reperfusion is then induced by loosening the coronary snare. The animals are usually anaesthetized and often treated with antiarrhythmic drugs. In the clinical situation, by contrast, multivessel coronary disease with variable collateralization is frequently present. Myocyte necrosis has generally occurred as the patient

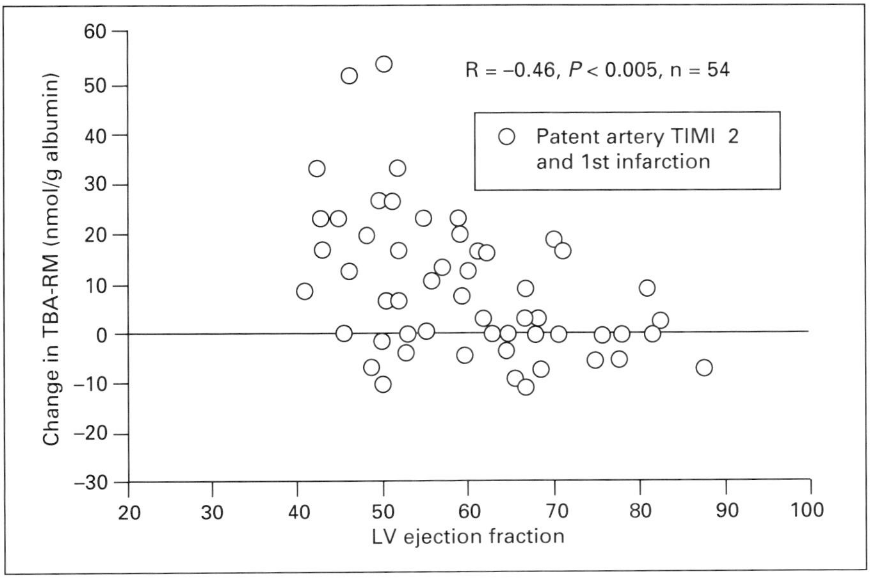

Figure 11.4
Relation between the increase in TBA-RM associated with thrombolysis and left ventricular (LV) ejection fraction for subjects with patent infarct related artery and first myocardial infarct. From Davies SW et al. Br Heart J *1993;* ***69:****114–20.*[18]

usually presents several hours after the onset of chest pain. Reperfusion, if it occurs, may be intermittent and gradual; subsequent coronary run-off may be poor due to significant residual stenosis. Nevertheless, there is some evidence to suggest that clinical counterparts of experimental reperfusion injury do exist in humans.

Reperfusion related arrhythmias

Early ventricular fibrillation may be more common in patients with myocardial infarction given thrombolytic therapy, especially if this is soon after the onset of chest pain. However, the most specific reperfusion arrhythmia in patients seems to be the relatively benign idioventricular rhythm.[9] Overall, reperfusion arrhythmias are not a major problem in the treatment of myocardial infarction in the thrombolytic era.

Myocardial stunning

Reversible left ventricular dysfunction has been described in a number of clinical settings associated with ischaemia-reperfusion.[10–17] These include angina pectoris, unstable angina pectoris, percutaneous transluminal coronary angioplasty, cardiac surgery and myocardial infarction. Biochemical evidence of FR generation and damage to biomolecules has also been detected in these settings. In one important study the greater this biochemical disturbance the lower the left ventricular ejection fraction (a measure of left ventricular contractile function) was immediately following thrombolysis for acute myocardial infarction (Fig. 11.4).[18] This is the first clinical evidence that FR generation may occur in patients and may be associated with myocardial impairment (either stunning or even lethal reperfusion injury — further work needs to be done to determine which).

Of course, the question about stunning is whether it really matters? If the dysfunction is reversible and can be corrected by inotropic stimulation, is it important? These questions are difficult to answer. A minority of patients (around 15–20%) suffer clinical heart failure at the time of their myocardial infarction and 2–4% develop cardiogenic shock. It is conceivable that in at least a proportion of these, prevention of stunning might have prevented cardiac decompensation. The success of inotropic therapy in these circumstances may also reflect, at least in some cases, the ability to recruit viable but stunned myocardium. Prevention of stunning, rather than its correction with inotropes or circulatory support with other drugs such as diuretics, is undoubtedly preferable. Inotropes, diuretics, etc. have many undesirable effects such as increasing heart rate and myocardial oxygen consumption (thereby increasing the risks of arrhythmias and further ischaemia) and causing neuroendocrine activation. So stunning probably does matter, but crucially so probably only in a small proportion of patients. The same line of reasoning applies to patients undergoing cardiac surgery. Stunning is unlikely to be important in unstable angina or angioplasty as clinically significant ventricular dysfunction is seldom seen.

Lethal reperfusion injury

There is little, if any, evidence of a clinical counterpart of this experimental possibility. Very early mortality is actually increased in patients receiving thrombolytic therapy for myocardial infarction, though subsequent reductions in mortality lead to a net substantial improvement in survival.[19] One possible explanation for this observation is that early lethal reperfusion injury occurs in some patients following successful thrombolysis. However, convincing proof of lethal reperfusion injury will require demonstration of improved late myocardial function (that is, after stunning has recovered) and/or survival with adjunctive therapy given at the time of thrombolysis.[8] Given the low mortality seen in trials of thrombolysis, such a clinical study would be a formidable challenge involving many thousands (or tens of thousands) of patients.[8]

Why should ACE inhibitors have a role in ischaemia-reperfusion injury? (Fig. 11.5)

Evidence for renin-angiotensin system (RAS) activation in myocardial ischaemia

There is some experimental evidence that RAS activation occurs during myocardial ischaemia.[20] Myocardial angiotensin II production increases and, if ischaemia causes myocardial dysfunction, systemic RAS stimulation may occur. There is preliminary clinical evidence of myocardial RAS activation during ischaemia.[21]

Evidence for RAS activation due to myocardial reperfusion

Ray et al. have shown that systemic plasma renin concentration rises following the administration of the thrombolytic agent streptokinase to patients with myocardial infarction.[22] This increase, however, was similar in patients who had evidence of coronary reperfusion and those who did not. The rise in plasma renin may therefore simply have reflected the hypotensive effect of streptokinase.

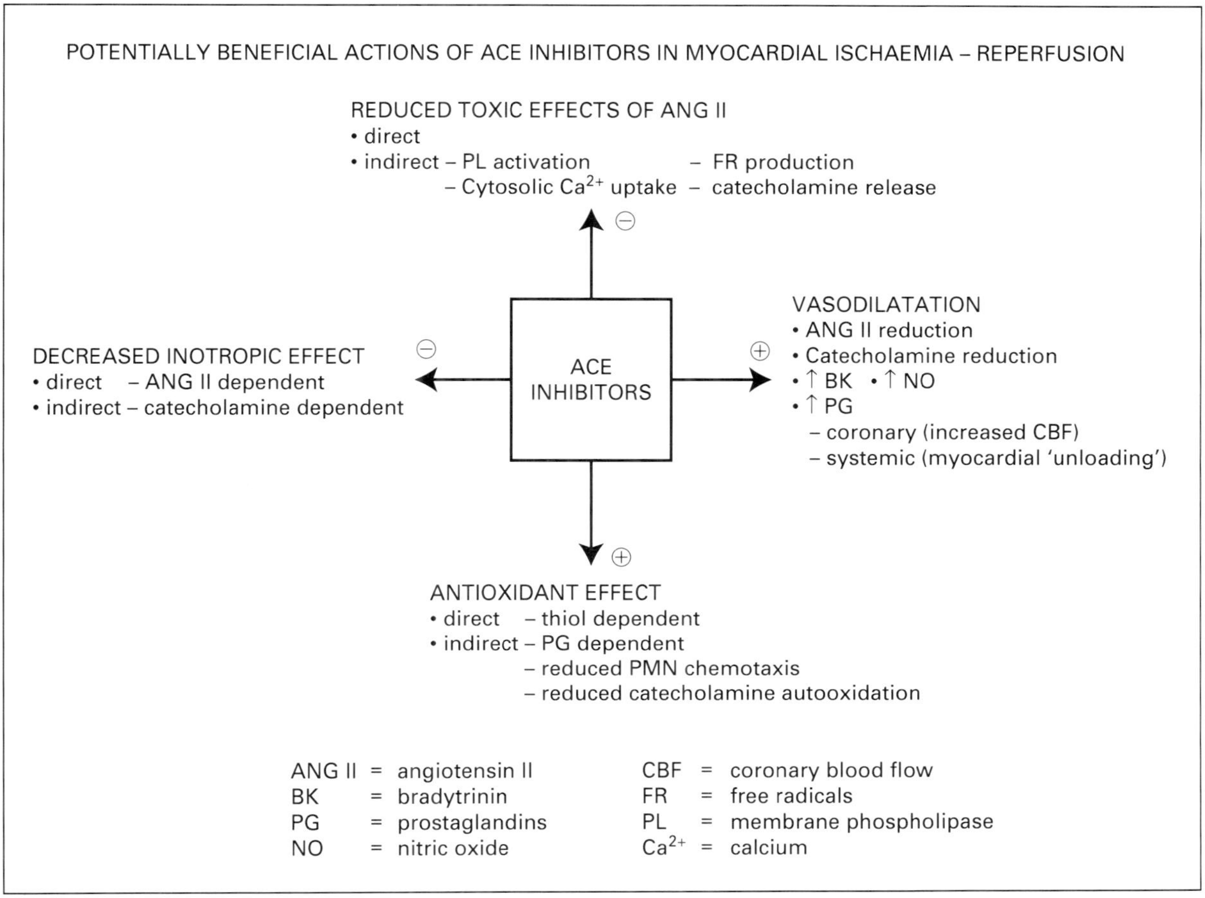

Figure 11.5
Potentially beneficial actions of ACE inhibitors in myocardial ischaemia – reperfusion

Potential effects of RAS activation during ischaemia — reperfusion

The RAS exerts a number of direct and indirect effects that could be detrimental in the setting of ischaemia/infarction. Systemic vasoconstriction (leading to increased afterload), coronary vasoconstriction, and enhanced inotropic stimulation (both directly and, indirectly, through facilitation of the sympathetic nervous system) all have the potential to increase myocardial ischaemia and extend the area of infarction. As the degree of stunning is related to the severity of preceding ischaemia (and the extent of collateral flow), it is possible that both systemic and myocardial RAS activation could aggravate reperfusion injury.

The RAS could also contribute to reperfusion injury in other ways, though these are more speculative. Rabkin has made the intriguing observation that angiotensin II formation during hypoxia in cultured myocardial cells leads to impaired recovery of contractile function during reoxygenation. This would suggest a direct toxic effect of angiotensin II on myocytes.[23] Angiotensin II may also contribute to free radical production by activating membrane bound phospholipases, which would initiate cascade reactions involving lipid peroxidation. Similarly, angiotensin II may increase catecholamine release which might also, directly (through auto-oxidation) and indirectly (via increased metabolic and mitochondral activity), lead to increased FR production.[8] The action of angiotensin II on membrane phospholipases may activate the phosphatidyl-inositol pathway leading to increased cytosolic calcium which could directly contribute to myocardial stunning.

Other mechanisms through which ACE inhibitors might limit ischaemia reperfusion injury

Bradykinin metabolism

Angiotensin converting enzyme is identical to kinase II, responsible for the breakdown of the vasodilator bradykinin to inactive metabolites. ACE inhibitors may therefore prevent bradykinin degradation. Bradykinin administered at the time of ischaemia—reperfusion is reported to reduce myocardial injury and reperfusion arrhythmias, and the beneficial effects of at least some ACE inhibitors in this setting seem to depend on bradykinin (see below).[24,25]

Prostaglandin metabolism

ACE inhibitors may increase production of vasoactive prostanoids at least partly through enhanced bradykinin concentrations. Pretreatment with ACE inhibitors increases myocardial prostaglandin production during ischaemia-reperfusion.[26] Inhibition of prostaglandin synthesis may reduce the beneficial effect of some ACE inhibitors in this situation (see below).[27] Administration of the stable synthetic epoprostenol analogue, Iloprost, reduced ischaemia-reperfusion injury in a rat model.[28]

Free radical scavenging

Sulphydryl or thiol containing ACE inhibitors, like other sulphydryl compounds, are efficient FR scavengers in vitro.[29,30] In vivo, other ACE inhibitors may also scavenge FR or prevent FR-mediated damage (Fig. 11.6). This may be because:

- though poor scavengers of other FR and related toxic oxygen species, non-sulphydryl (SH) ACE inhibitors may effectively scavenge the highly reactive hydroxyl radical;[31]
- they may chelate transition metals by virtue of their ability to bind zinc at the active moiety of ACE.[30,31] Transition metals are instrumental in the generation of the hydroxyl radical.
- by increasing prostanoid production non-sulphydryl ACE inhibitors may indirectly lead to protection against FR injury.[27,28]

The development of specific renin inhibitors and angiotensin II receptor antagonists may in the future allow clarification of the importance of these different postulated mechanisms.

Experimental evidence that ACE inhibitors reduce reperfusion injury

Reperfusion related arrhythmias

ACE inhibitors markedly reduce the incidence and duration of reperfusion related ventricular

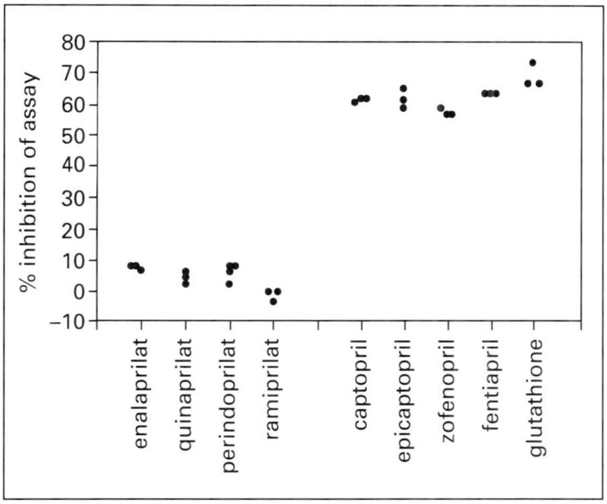

Figure 11.6
Effect of the eight angiotensin-converting enzyme inhibitors studied in the riboflavin-dianisidine photo-oxidation assay. The –SH-containing compounds (right) significantly inhibited the assay, i.e., were preferential scavengers of FR other than O_2^-. The non –SH compounds (left) were inactive in the assay. Derived from Chopra M et al. J Cardiovasc Pharmacol *1992;* ***19(3)****:333.*[30]

arrhythmias in a dose dependent fashion.[26] Intracoronary saralasin (an angiotensin II receptor partial antagonist) also reduces reperfusion arrhythmias, raising the possibility that myocardial RAS activation may directly contribute to this phenomenon.[32] There is some preliminary clinical evidence that ACE inhibition at the time of thrombolysis may reduce the incidence of ventricular arrhythmias (see below).

Myocardial stunning

Whole animal studies

A number of investigators have examined the effect of ACE inhibitors on experimental myocardial stunning. These reports are summarized in Table 11.1.

Westlin and Mullane first showed, in 1988, that ACE inhibitors might reduce myocardial stunning (Fig. 11.7).[33] Captopril 5 mg/kg (administered by systemic intravenous bolus injection), given 15 minutes before coronary occlusion in dogs, led to a considerably greater (to 65% of preocclusion value) recovery of contractility in the ischaemic segment following reperfusion than did placebo (10% of the preocclusion value). More importantly, captopril 5 mg/kg IV, given two minutes before reperfusion, also led to a greater recovery of contractility. SQ 14534, a stereoisomer of captopril with one hundredth of its ACE inhibiting activity, at a dose of 5 mg/kg IV, was equally effective in restoring mechanical function when given according to the same protocol. At first sight these observations seem to suggest that captopril is behaving in a manner analogous to other sulphydryl-containing compounds in this model (that is, as a FR scavenger), and that its beneficial effect is independent of ACE inhibition. This is supported by Przyklenk and Kloner's observations that zofenopril (SQ 26703), another sulphydryl-containing ACE inhibitor, given by intravenous bolus injection (5 mg/kg) one minute before reperfusion, also results in improved recovery of contractility in dogs (Fig. 11.8).[27,34,35]

Westlin and Mullane have observed that enalaprilat (1.6 mg/kg) resulted in a considerably worse recovery of function than occurred with placebo (–30% of preocclusion value), apparently confirming that sulphydryl-containing ACE inhibitors are uniquely beneficial because of their thiol moiety and not because of their ACE inhibiting activity. However, further work by Przyklenk and Kloner has directly contradicted the findings of Westlin and Mullane and rejected any conclusion

Author/Reference	Animal model	Drug	Index	Change in myocardial function
Westlin et al.[33]	Dog: 15 min CO + 3 h reflow	Captopril: before CO	Segment shortening	↑
Westlin et al.[33]	Dog: 15 min CO +	Captopril: at reflow	Segment shortening	↑
Westlin et al.[33]	Dog: 15 min CO + 3 h reflow	Enalaprilat: before CO	Segment shortening	↓
Przyklenk et al.[34]	Dog: 15 min CO +	Zofenopril: at reflow	Segment shortening	↑
Przyklenk et al.[27]	Dog: 15 min CO + 3 h reflow	Zofenopril: at reflow	Segment shortening	↑
Przyklenk et al.[27]	Dog: 15 min CO + 3 h reflow	Enalaprilat: at reflow	Segment shortening	↑
Bittar et al.[46]	Dog: 15 min CO + 3 h reflow	Captopril: before CO	Segment shortening	↑

CO = coronary occlusion
↓ = contractile dysfunction was exacerbated with enalaprilat treatment.
↑ = improved myocardial function

Table 11.1
ACE inhibitors in animal models of myocardial stunning

favouring a unique benefit of sulphydryl-containing drugs.[27]

In 1991 Przyklenk and Kloner showed that enalaprilat 1.5 mg/kg, given by intravenous bolus one minute before reperfusion, resulted in the return of segment shortening to 83% of baseline value in dogs. The control achieved –5%, zofenopril (5 mg/kg) achieved 54%, and SQ 14534 (5 mg/kg) achieved 33% (Fig. 11.8).[27] These authors went on to carry out further experiments to elucidate the mechanism of action of these compounds. They found that the beneficial effect of enalaprilat was reversed by the cyclo-oxygenase inhibitor indomethacin, although indomethacin did not reduce the effect of zofenopril.[27] It thus appears that sulphydryl-containing and non-sulphydryl-containing ACE inhibitors may have similar beneficial effects of myocardial stunning, but that these are mediated by different

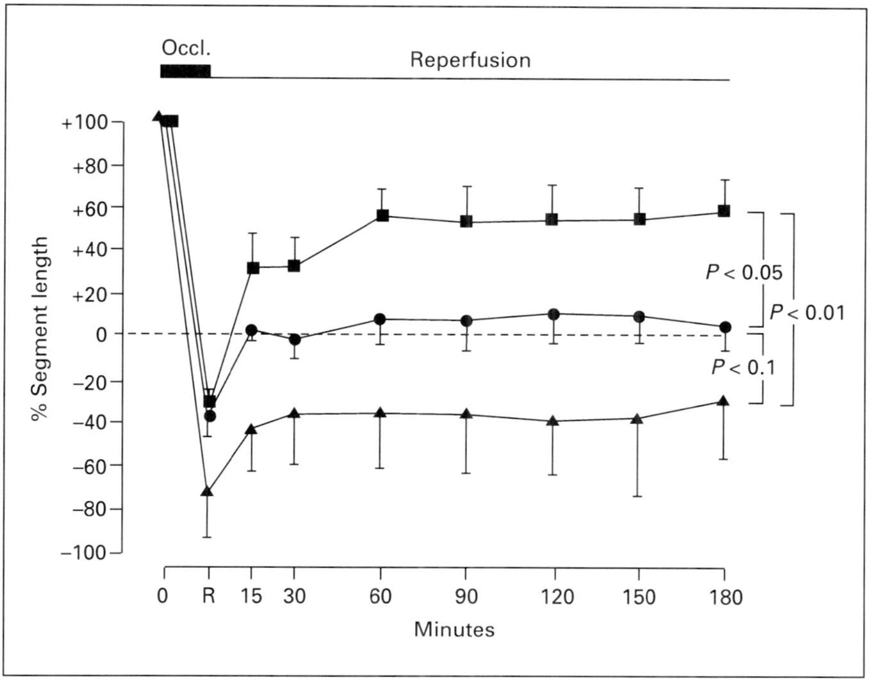

Figure 11.7
ACE inhibitors in models of brief occlusion/reperfusion not resulting in myocardial necrosis (models of 'stunned' myocardium). Captopril (5 mg/kg), given 15 min before ischaemia, significantly enhanced the recovery of % segment length (an index of regional contractile function) following a brief 15-min episode of coronary occlusion in the canine model. However, pretreatment with enalaprilat (mean dose 1.6 mg/kg) exacerbated contractile dysfunction both during occlusion and following reperfusion. Squares = captopril-treated group; triangles = enalaprilat-treated group; circles = controls. From Westlin W, Mullane K. Circulation *1988;* ***77[Suppl 1]****:130–139.*[33]

mechanisms. The reason for the discrepant effects of enalaprilat in Westlin and Mullane's original report and Przyklenk and Kloner's more recent paper remains unclear. However, support for the conclusions of Przyklenk and Kloner can be found from studies in isolated hearts.

Studies in isolated hearts

Pi and Chen used an isolated working heart model from a rat to examine the relevance of the sulphydryl group.[28] FR were generated by electrolysis of the perfusate, leading to biochemical damage (increased myocardial thiobarbituric acid reactive substances/

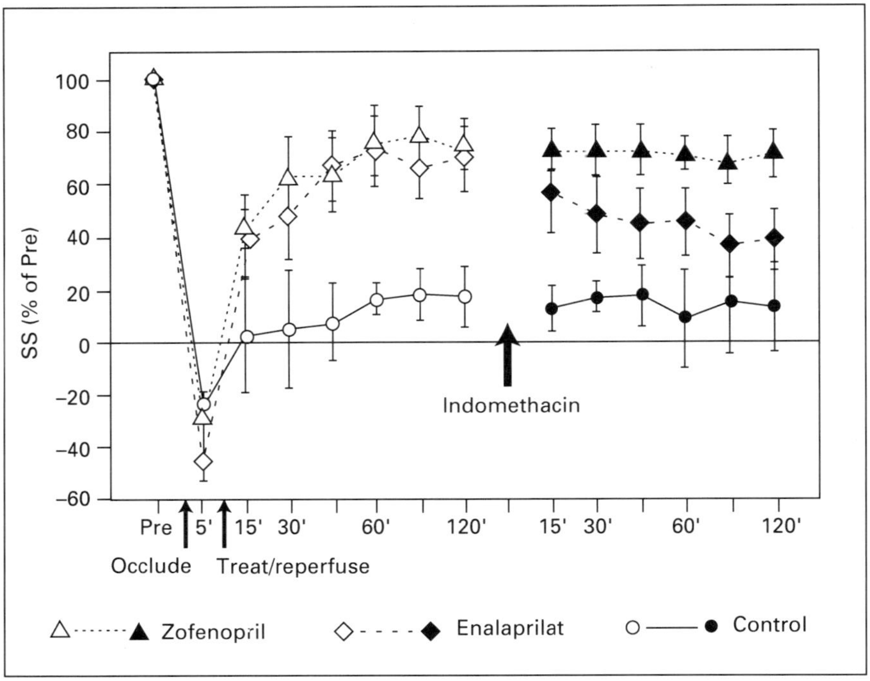

Figure 11.8
In contrast to Fig. 1.7, both zofenopril (5 mg/kg) and enalaprilat (1.5 mg/kg), given at the time of reperfusion, significantly attenuated post-ischaemic dysfunction following 15 min of brief transient ischaemia in the canine model. However, the mechanisms of action of the two agents appeared to differ: infusion of endomethacin essentially reversed the improvement in contractile function achieved with enalaprilat, but had no effect in dogs that had been treated with zofenopril. That is, enalaprilat — but not zofenopril — improved contractile function of stunned myocardium by a prostaglandin-mediated mechanism. SS = segment shortening, expressed as a % of preocclusion (Pre) values. From Pryzklenk, K, Kloner A. Am Heart J *1991;* ***121:****1319–30.*

decreased myocardial superoxide disumitase), decreased coronary flow, and mechanical dysfunction; the culprit FR in this model appeared to be the hydroxyl radical. Captopril gave dose dependent protection from all three types of injury; 90 µmol/l was the minimum effective concentration. Captopril was more effective than dimethylsuphoxide, a known hydroxyl radical scavenger. Another sulphydryl-containing agent, S8, offered the same degree of protection as captopril. Importantly, however, Pi and Chen also showed that hearts removed from rats given ramipril (a non-sulphydryl ACE inhibitor)

1 mg/kg intraperitoneally one hour before death, and then perfused with ramipril at 12.5 μmol/l, also had less biochemical and functional injury. The protective effect of ramipril was, however, abolished by coinfusion of indomethacin; by contrast that of captopril was only partially attenuated. Iloprost, 3 μmol/l, a stable synthetic analogue of epoprostenol and an in vitro FR scavenger, also showed marked protective effects in this model. If the effects of bradykinin are also mediated through prostaglandins, then the reports that bradykinin antagonists abolish the protective effects of some ACE inhibitors also concur with observations of Pi and Chen.

Menasché et al. found that pretreatment with captopril resulted in improved mechanical function in excised rat hearts subjected to ischaemia-reperfusion, compared to pretreatment with enalapril.[36] Enalapril pretreatment, however, led to greater recovery than in the placebo-treated hearts. Interestingly, both ACE inhibitors increased coronary blood flow equally, compared to placebo, suggesting that the superior effect of captopril was not due to enhanced myocardial perfusion.

Bagchi et al., using the isolated rat heart model, have shown that captopril is a FR scavenger and also affects arachidonic acid metabolism/prostaglandin production (its effects on mechanical function were not reported).[37]

Ferrai et al. compared the effects of zofenopril and captopril in isolated rabbit hearts.[38] Zofenopril was superior to captopril in terms of creatine kinase release, myocardial noradrenaline release, and mechanical recovery following ischaemia-reperfusion. Interestingly, in this model, the authors did not find that either ACE inhibitor reduced oxidative stress.

These studies, taken as a whole, suggest that ACE inhibitors may ameliorate experimental stunning. Surprisingly, different ACE inhibitors appear to work by different mechanisms, namely, sulphydryl dependent and prostaglandin dependent. Captopril (according to the work of Bagchi et al.) may work through both mechanisms, though may not be as potent as zofenopril. The beneficial effects of ACE inhibitors do not appear to depend on systemic or coronary haemodynamic effects. An interesting clinical question that arises from these conclusions is what might be the effect of aspirin (like indomethacin, a cyclooxygenase inhibitor)? Aspirin is routinely given with thrombolysis in the clinical setting, principally for an antiplatelet action.

Lethal reperfusion injury

Przyklenk and Kloner have recently reviewed the evidence that ACE inhibitors might reduce the extent of myocardial necrosis occurring after permanent coronary occlusion or after prolonged 'ischaemia' (that is, infarction) followed by reperfusion.[39]

While several studies suggested a 'cardiprotective' effect of ACE inhibitors, several found no benefit and in two infarct size was increased. One possible explanation for the latter finding, clinically alarming, was that the extent of collateralization was a determinant of the action of the ACE inhibitor.[40] In poorly collateralized infarcts the amount of necrosis appears to be less. However, where there is good collateralization, ACE inhibitor therapy may induce coronary steal, through vasodilatation, and thus reduce flow to the infarcting area. Hence, these agents may be a double edged sword in the very unstable situation of severe ischaemia and ongoing infarction.

This raises the whole question of the role of vasodilators in the acute infarction setting. Reduction in myocardial work is theoretically attractive, and with, for example, intravenous β-blockers, is probably beneficial. If ACE inhibitors are also beneficial, one must ask whether this is so because they are ACE inhibitors or because they are vasodilators. In one study, the non-specific vasodilator hydralazine appeared to be superior to enalapril in promoting functional recovery, when given after reperfusion.[41] However, as suggested above, there is a potential 'downside' for vasodilators, especially in the clinical setting. Infarct size may be increased when coronary perfusion is jeopardised by excessive blood pressure reduction, and this has been shown with non-ACE-inhibitor vasodilators. The study of de Lorgeril et al. even suggests that intracoronary redistribution of flow, related to steal, could have an adverse effect.[40]

Further complex studies taking account of collateral flow and coronary perfusion pressure will be necessary definitively to address the question of benefit from ACE inhibitors in prolonged ischaemia/infarction-reperfusion. Comparative studies with non-ACE-inhibitor vasodilators will also be necessary to dissect out the contribution of vasodilation (for example load reduction) per se, and any specific benefit of the ACE inhibitor itself. Further studies comparing ACE inhibitors to renin inhibitors and angiotensin II receptor-antagonists could also be required to elucidate the mechanism of any special benefit of ACE inhibitors.

A clinical perspective

Problems with the dose of drugs used in experimental studies

Captopril and zofenopril have been used successfully at doses of 5 mg/kg IV, massive by clinical standards.[27,33,35] A beneficial effect of lower doses has not been reported. Other thiol compounds are experimentally protective, however, at concentrations of 1–8 μmol/l, which could be reached by doses of ACE inhibitors similar to those already used therapeutically.[8]

Problems with the drugs

While ACE inhibitors have many attractive haemodynamic, neuroendocrine, and, possibly, FR scavenging actions, their unpredictable and potentially large hypotensive action probably makes them unsuitable for intravenous administration in the setting of thrombolysis for acute myocardial infarction. This is especially the case if very high concentrations (that is, large doses) are required for a therapeutic effect from the point of view of ischaemia-reperfusion injury. Indeed, simple thiols appear to be equally beneficial in this respect. As these are already used in clinical practice, do not have marked haemodynamic effects and can be used in large doses (for example, N-acetyl cysteine is used at a dose of 150 mg/kg for paracetamol poisoning), they would seem preferable for this purpose.[8]

Relevant clinical experience

Despite the above reservations, there are two reports of clinical administration of ACE inhibitors at the time of thrombolysis. Kingma et al. gave captopril 2.5 mg IV bolus followed by 10 mg IV over five minutes, followed by a further 10 mg 30 minutes later in conjunction with intravenous streptokinase. Marked hypotension proved to be a serious clinical problem in this study, and no further trials of this type have been carried out.[42] The authors also suggested that there might be an adverse hypotensive interaction between streptokinase (as opposed to other thrombolytics) and ACE inhibitors. The observations of Nabel et al.[43] support this conclusion. These authors randomized 38 patients receiving the thrombolytic agent tissue plasminogen activator (tPA) to either intravenous captopril (2 mg over one minute, plus a further 8 mg, blood pressure permitting) or placebo. Only one patient became notably hypotensive, though captopril was not given until one and a half to two and a half hours after the starting of tPA. In both these studies it is important to note the low doses of intravenous captopril given (10–12.5 mg initially) compared to those in the experimental studies (5 mg/kg or approximately 350 mg in a 'typical' 70 kg patient).

CONSENSUS II — a further concern

In the Cooperative New Scandinavian Enalapril Survival Study (CONSENSUS II), intravenous enalapril or placebo was given to nearly 600 patients in the very early stages of acute myocardial infarction after thrombolysis.[44] Hypotension was a frequent problem and led to a protocol adjustment. Even then hypotension remained a problem and was associated with increased mortality. Overall, mortality was slightly raised in the enalapril group. This is in striking contrast to subsequent studies giving oral ACE inhibitors either in the first 24 hours post-myocardial infarction or within the first one to two weeks of recovery.[45]

Conclusions

Experimentally, large doses of intravenous ACE inhibitors appear consistently to reduce the form of reperfusion injury known as stunning. Stunning probably does occur in patients, and in a minority of these (for example post-myocardial infarction) may be clinically important. However, the clinical situation seems to differ from the experimental one, and the doses of ACE inhibitor known to be beneficial in animals are likely to cause severe and harmful hypotension. While ACE inhibitors may have other actions shortly after myocardial infarction, the clinical evidence available suggests that the risks of concomitant administration with thrombolysis probably outweigh any benefit. The role of ACE inhibitors in other situations where myocardial reperfusion injury might be important, such as during cardiac surgery, remains to be determined.

References

1 Hearse J, Bolli R. Reperfusion induced injury: manifestations, mechanisms, and clinical relevance. *Cardiovas Res* 1992; **26**:101–8.

2 Yellon DM, Downey JM. Current research views on myocardial reperfusion and reperfusion injury. *Cardioscience* 1990; **1**:89–98.

3 Opie LH. Reperfusion injury and its pharmacological modification. *Circulation* 1989; **80**:1049–62.

4 Bolli R. Mechanism of myocardial 'stunning'. *Circulation* 1990; **82**:723–38.

5 Sekili S, McCay PB, Xiao-Ying L et al. Direct evidence that the hydroxyl radical plays a pathogenetic role in myocardial 'stunning' in the conscious dog and demonstration that stunning can be markedly attenuated without subsequent adverse effects. *Circ Res* 1993; **73**:705–23.

6 Sun JZ, Kaur H, Halliwell B et al. Use of aromatic hydroxylation of phenylalanine to measure production of hydroxyl radicals after myocardial ischemia in vivo: direct evidence for a pathogenetic role of the hydroxyl radical in myocardial stunning. *Circ Res* 1993; **73**:534–49.

7 Fox KAA. Reperfusion injury: laboratory phenomenon or clinical reality? *Cardiovas Res* 1992; **26**:656–59.

8 McMurray J, Chopra M. Influence of ACE inhibitors on free radicals and reperfusion injury: pharmacological curiosity or therapeutic hope? *Br J Clin Pharmacol* 1991; **31**:373–79.

9 Editorial. Reperfusion or persisting occlusion? *Lancet* 1989; **i**:114–15.

10 Scott BD, Kerber RE. Clinical and experimental aspects of myocardial stunning. *Prog Cardiovas Dis* 1992; **35**:61–76.

11 Kloner RA, Allen J, Cox TA et al. Stunned left ventricular myocardium after exercise treadmill testing in coronary artery disease. *Am J Cardiol* 1991; **68**:329–34.

12 Bolli R. Myocardial 'stunning' in man. *Circulation* 1992; **86**:1671–91.

13 Roberts MJD, Young IS, Trouton TG. Transient release of lipid peroxides after coronary artery balloon angioplasty. *Lancet* 1990; **336**:143–45.

14 Ferrari R, Alfieri O, Curello S et al. Occurrence of oxidative stress during reperfusion of the human heart. *Circulation* 1990; **81**:201–11.

15 Janssen M, Koster JF, Bos E et al. Malondialdehyde and glutathione production in isolated perfused human and rat hearts. *Circ Res* 1993; **73**:681–88.

16 Young IS, Purvist JA, Lightbody JH et al. Lipid peroxidation and antioxidant status following thrombolytic therapy for acute myocardial infarction. *Eur Heart J* 1993; **14**:1027–33.

17 Pfisterer M, Zuber M, Wenzel R et al. Prolonged myocardial stunning after thrombolysis: can left ventricular function be assessed definitely at hospital discharge? *Eur Heart J* 1991; **12**:214–17.

18 Davies SW, Ranjadayalan K, Wickens DG et al. Free radical activity and left ventricular function after thrombolysis for acute infarction. *Br Heart J* 1993; **69**:114–20.

19 Mauri F, De Baise AM, Franzosi MG et al. G.I.S.S.I.: Analisi delle cause di morte introspedaliera. *G Ital Cardiol* 1987; **17**:37–44.

20 Er HG, Alexander RW, Kloner RA. Interaction between coronary occlusion and the renin-angiotensin system in the dog. *Bas Res Cardiol* 1983; **78**:518–33.

21 Remme WJ, Look MP, Bootsma M et al. Neurohumoral activation during acute myocardial ischaemia: effects of ACE inhibition. *Eur Heart J* 1990; **11[suppl B]**:162–71.

22 Ray SG, Morton JJ, Dargie HJ. Neuroendocrine activation and markers of early reperfusion in the acute phase of myocardial infarction. *Eur Heart J* 1993; **14**:1615–21.

23 Rabkin SW. Cilazapril and captopril accelerate recovery from hypoxia in myocardial cell aggregates in culture. *J Cardiovas Pharmacol* 1992; **19**:394–401.

24 Linz W, Martorana PA, Wiemer G et al. The cardioprotective effects of the ACE inhibitor ramipril resemble those of bradykinin in ischaemic hearts. In: MacGregor GA, Sever PS, eds. Current advances in ACE inhibition. Edinburgh: Churchill Livingstone, 1989; 283–50.

25 Hartman J, Wall TM, Hullinger G. Reduction of myocardial infarct size in rabbits by ramiprilat: reversal by the bradykinin antagonist HOE 140. *J Cardiovas Pharmacol* 1993; **21**:996–1003.

26 Arad M, Shotan A, Horowitz L. Effects of captopril on metabolic and hemodynamic alterations in global ischemia and reperfusion in the isolated working rat heart. *J Cardiovas Pharmacol* 1992; **19**:319–23.

27 Przyklenk K, Kloner A. Angiotensin converting enzyme inhibitors improve contractile function of stunned myocardium by different mechanisms of action. *Am Heart J* 1991; **121**:1319–30.

28 Pi XJ, Chen X. Captopril and ramiprilat protect against free radical injury in isolated working hearts. *J Mol Cell Cariol* 1989; **21**:1261–71.

29 Smith WE, Belch JJF. Captopril: a free radical scavenger. *Br J Clin Pharmacol* 1989; **27**:396–99.

30 Chopra M, Beswick H, Clapperton M et al. Antioxidant effects of angiotensin-converting enzyme (ACE) inhibitors: free radical and oxidant scavenging are sulfhydryl dependent, but lipid peroxidation is inhibited by both sulfhydryl- and nonsulfhydryl-containing ACE inhibitors. *J Cardiovas Pharmacol* 1992; **19**:330–40.

31 Mira ML, Silva MM, Queiroz MJ. Angiotensin converting enzyme inhibitors as oxygen free radical scavengers. *Free Radic Res Commun* 1993; **19**:173–81.

32 Ribuot C, Yamaguchi N, Godin D et al. Intracoronary administration of saralasin: effects on cardiac arrhythmias inducted by ischaemia and reperfusion in the anaesthetised dog. *Cardiovasc Res* 1992; **26**:968–72.

33 Westline W, Mullane K. Does captopril attenuate reperfusion-induced myocardial dysfunction by scavenging free radicals? *Circulation* 1988; **77[Suppl 1]**:I30–I39.

34 Przyklenk K, Kloner RA. Relationships between structure and effects of ACE inhibitors: comparative effects in myocardial ischaemic/reperfusion injury. *Br J Clin Pharmacol* 1989; **28**:1675–755.

35 Przyklenk K, Whittaker P, Kloner RA. Zofenopril, a newly developed sulphydryl-containing converting enzyme inhibitor, enhances contractile function of 'stunned' myocardium. In: MacGregor GA, Sever P, eds. Current advances in ACE inhibition. Edinburgh: Churchill Livingstone, 1989; 279–820.

36 Menasché P, Grousset C, Peynet J. Pretreatment with captopril improves myocardial recovery, after cardioplegic arrest. *J Cardiovasc Pharmacol* 1992; **19**:402–7.

37 Bagchi D, Iyengar J, Stockwell P et al. Enhanced prostaglandin production in the ischemic reperfused myocardium by captopril linked with its free radical scavenging action. *Prostaglandins, Leuko Essent Fatty Acids* 1989; **38**:145–50.

38 Ferrari R, Cargnoni A, Curella S et al. Protection of the ischemic myocardium by the converting enzyme inhibitor zofenopril: insight into its mechanism of action. *J Cardiovasc Pharmacol* 1992; **20**:694–704.

39 Przyklenk K, Kloner RA. "Cardioprotection" by ACE-inhibitors in acute myocardial ischemia and infarction? In: Gobecker H, Heusch G, Strauer BE, eds. *Basic research in cardiology*. New York: Springer-Verlag, 1993; **88**[Suppl 1]:139–540.

40 de Lorgeril M, Ovize M, Delaye J et al. Importance of the flow perfusion deficit in the response to captopril in experimental myocardial infarction. *J Cardiovasc Pharmacol* 1992; **19**:324–29.

41 Przyklenk K, Kloner RA. Acute effects of hydralazine and enalapril on contractile function of post-ischaemic 'stunned' myocardium. *Am J Cardiol* 1987; **60**:934–36.

42 Kingma JH, van Gilst WH, de Graeff PA et al. Captopril during thrombolysis in acute myocardial infarction: feasibility, tolerance and beneficial neurohumoral effects. In: MacGregor GA, Sever PS, eds. *Current advances in ACE inhibition*. Edinburgh: Churchill Livingstone, 1989: 291–95.

43 Nabel EG, Topol EJ, Galeana A et al. A randomized placebo-controlled trial of combined early intravenous captopril and recombinant tissue-type plasminogen activator therapy in acute myocardial infarction. *J Am Coll Cardiol* 1991; **17**:467–73.

44 Swedberg K, Held P, Kjekshus J et al. Effects of the early administration of enalapril on mortality in patients with acute myocardial infarction — results of the cooperative new Scandinavian enalapril survival study II (CONSENSUS II). *N Engl J Med* 1992; **327**:678–84.

45 Pfeffer MA, Braunwald E, Moye LA, et al. for the SAVE Investigators. Effect of captopril on mortality and morbidity in patients with left ventricular dysfunction after myocardial infarction. Results of the survival and ventricular enlargement trial. *N Engl J Med* 1992; **327**:669–77.

46 Bittar N, Zheng HN, Nwosu EA. Captopril is more effective than dipyridamole or labetolol in attenuating myocardial stunning in the dog. Proceedings of the Second International Symposium on ACE Inhibition. London, UK, 1991. Abstract: 179.

12

ACE inhibitors and vascular injury following therapeutic intervention

D deBlois, SM Schwartz and AW Clowes

Introduction

Angiotensin converting enzyme (ACE) is a carboxydipeptidase that cleaves the inactive decapeptide angiotensin I to release the vasoactive octapeptide angiotensin II (ANG II). Since the introduction of captopril in 1980, a number of related compounds that inhibit ACE have been widely used to reduce high blood pressure in humans. They have also been employed experimentally in animals and humans to control injury-induced intimal hyperplasia and restenosis after angioplasty. These studies were undertaken based on the assumption that not only is ANG II a vasoconstrictor, but it may also regulate smooth muscle cell (SMC) hypertrophy and replication.[1]

The purpose of this chapter is to review the known effects of ACE inhibitors in vascular injury. Possible mechanisms for the effect of ACE inhibitors on intimal lesion development will be discussed. Finally we will discuss how the complexity of the intimal lesion might affect the ability of ACE inhibitors to reduce intimal development.

The injured carotid artery in the rat: a model of fibromuscular intimal hyperplasia

The cellular and molecular mechanisms involved in the rapid progression of the atherosclerotic plaque after angioplasty in humans are poorly understood although the response of SMC to angioplastic injury is considered an important etiologic factor for restenosis.[2] This response of SMC has been studied in some detail in animals. The rat carotid artery injured with a balloon catheter is one of the most thoroughly characterized animal models of vascular injury and the contribution of SMC to the development of intimal lesions. What follows is a description of significant events leading to lesion formation after injury to the normal rat carotid artery.

Intimal thickening does not occur spontaneously in rat arteries. A fibromuscular intima develops, however, after mechanical dilatation of the common carotid artery with an

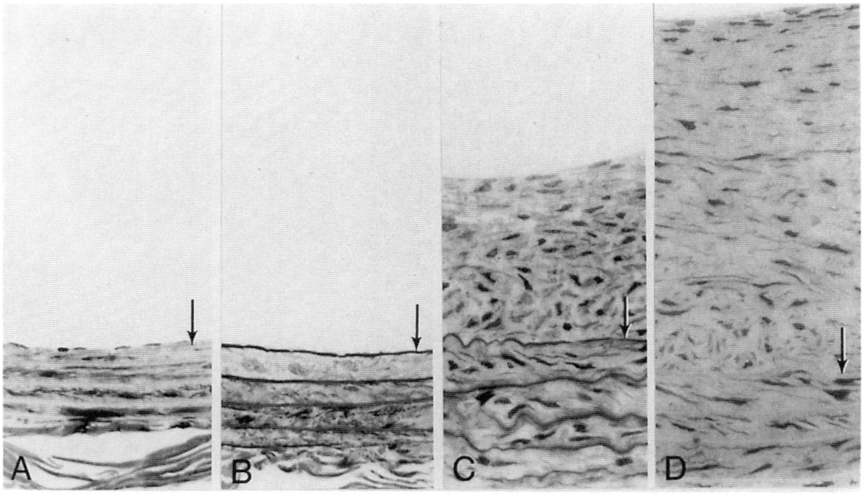

Figure 12.1
Histologic cross sections of the region lacking endothelium of injured rat common carotid arteries. A: Normal vessel. Note single layer of endothelium in intima. B: Denuded vessel at 2 days. Note loss of endothelium. C: Denuded vessel at 2 weeks. Intima is now markedly thickened due to smooth muscle proliferation. D: Denuded vessel at 12 weeks. Further intimal thickening has occurred. Intimal elastic lamina indicated by arrow. Lumen is at the top. ×360. Reprinted with permission from Clowes et al.[3]

embolectomy balloon catheter (Fig. 12.1). This procedure causes complete destruction of the endothelium over the length of the vessel as well as medial SMC cell death (15%–25%).[3,4] During the first 24 hours after denudation, platelets adhere and form a monolayer on the exposed subendothelial surface.[3] Although coagulation enzymes from the plasma are probably activated by exposure to the subendothelium, accumulation of insoluble fibrin is not observed by scanning or transmission electron microscopy of injured rat arteries.[5] In rabbit arteries, an increase in thrombin activity associated with the vessel wall is observed following balloon injury.[6,7] Ultimately, the SMC response to balloon injury culminates in the formation of a new layer of muscular tissue, the neointima, inside the internal elastic lamina (Figs 12.1C and 12.1D). Luminal narrowing occurs after injury to the rat carotid artery.[8] It is more pronounced after two weeks (75%) than after 12 weeks (35%). Early stenosis is largely due to smooth muscle contraction of the vessel while late stenosis is due only to structural changes in the vascular wall such as intimal thickening.

The SMC response to injury follows a pattern that can conveniently be broken down into three phases or waves (Figs 12.2 and 12.3). The first wave is the burst of medial

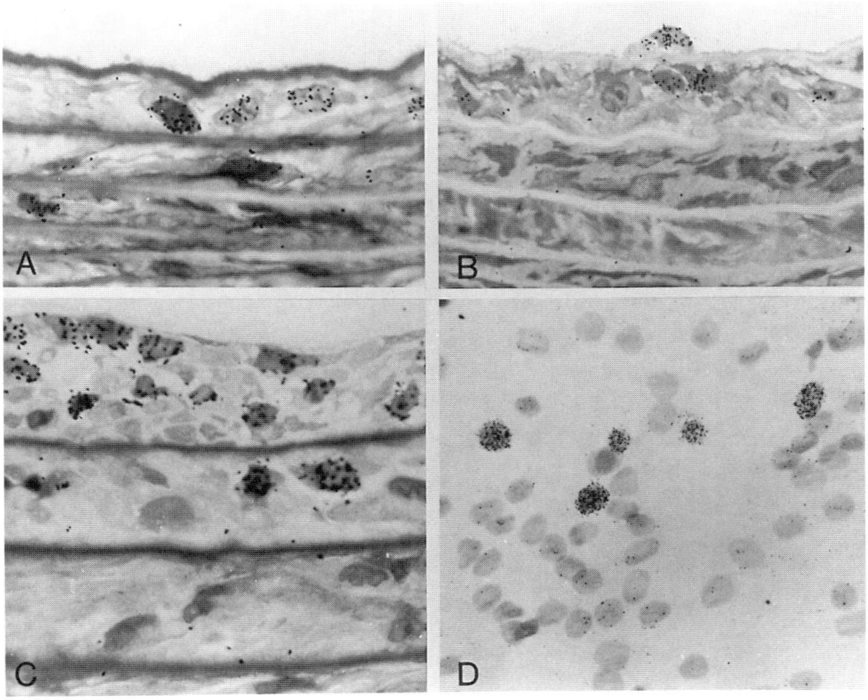

Figure 12.2
Light microscopic autoradiograms demonstrating proliferating SMC in rat common carotid arteries early after injury. Rats received 3 intra-peritoneal injections of tritiated thymidine at 17 hours, 9 hours and 1 hour before death. A: Histologic cross-section at 2 days. Note the absence of intimal cells and presence of silver grains over medial SMC. ×850. B: Histologic cross-section at 4 days. Note SMC in media and intima. ×850. C: Histologic cross-section at 7 days. Note extensive labeling of intimal SMC. ×850. D: Hautchen en face *preparation of left carotid at 12 weeks demonstrating proliferating luminal SMC. ×340. Intimal elastic lamina indicated by arrow. Reprinted with permission from Clowes et al.*[3]

SMC proliferation that peaks within a few days after injury and subsides within the next 10 days following repopulation of the media to the pre-injury level. This proliferative response is caused by the synchronous entry into the cell cycle of 15% to 25% of the SMC remaining in the media.[3,9] Basic fibroblast growth factor (bFGF) plays a major role in the initiation of DNA synthesis in the injured media. The proliferative response is stimulated by bFGF infusion[10–12] and is greatly reduced (>80%) by an antibody to bFGF.[13] Lindner et al. proposed that bFGF is released from damaged SMC and endothelial cells. Exogenous heparin, which binds bFGF tightly, inhibits medial SMC proliferation in part by displacing bFGF from the injured arterial wall.[14] Replication in the media two days after balloon injury is similar whether or not rats are rendered thrombocytopenic with an antibody against circulating platelets.[15] This result suggests that platelet-derived products, such as platelet-derived growth factor (PDGF), do not play a major role in initiation of SMC proliferation in the media following injury.

The migration of SMC across the internal elastic lamina to form the intima constitutes the second wave. SMC are readily observed on the luminal side of the internal elastic lamina four days after injury (Fig. 12.2B). The duration of the second wave in the rat injured carotid is not known. However, SMC migration is probably negligible once medial SMC have stopped replicating because cellular depletion of the media does not occur. Three lines of evidence strongly suggest that a potent stimulant of the second wave is PDGF. First, elimination of circulating platelets, a treatment that does not affect the first wave, prevents the appearance of SMC on the luminal side of the internal elastic lamina at day four after balloon injury.[15] Depletion of circulating platelets also significantly reduces the size of the intimal lesion at day seven after injury. Second, daily treatment with a polyclonal antibody to PDGF does not affect SMC proliferation, but significantly inhibits (40%) intimal thickening in the carotid artery of nude rats nine days after injury.[16] Finally, a seven-day administration of PDGF-BB, which recognizes all isoforms of PDGF receptors, results in a stimulation (up to 20-fold) of intimal lesion development, while having a modest or no effect at all on the rate of SMC proliferation in the media (two- to five-fold) and the intima, respectively.[17]

The third wave refers to the sustained proliferation of intimal SMC which contributes to the rapid growth of the lesion during the first four weeks after injury (Fig. 12.2C). Chronic proliferation in the intima is particularly striking at the luminal surface of the lesion where the number of SMC incorporating tritiated thymidine is consistently higher than in the deeper layers of the intima as late as 12 weeks after injury (Fig. 12.3). The pharmacology of the third wave is still poorly defined. Treatment with an antibody to bFGF four and five days after injury does not affect intimal SMC proliferation.[18] Large doses of TGF given two weeks after injury modestly increases the percentage of proliferating cells in the intima.[19] In vivo infusion of ANG II in the third and fourth week after injury markedly increases DNA synthesis of intimal and medial SMC as well as cross-sectional area of intimal and medial lesions.[20]

Finally, active deposition of extracellular matrix by intimal SMC further contributes to the growth of the fibromuscular lesion until three months following injury, after which time the intima appears stable.[21] (Figs 12.1C and 12.1D).[21] Matrix components synthesized in the injured vessel include collagen type I, elastin, osteopontin, and proteoglycans.[22,23,24] Increased extracellular matrix accumulation in the lesion may be important in the stimulating effect of ANG II on lesion size.[20] In fact, SMC synthesis of collagen, osteopontin and proteoglycans is stimulated by ANG II in vitro.[23,25,26]

Coronary restenosis as a response to injury

Percutaneous transluminal coronary angioplasty (PTCA) is a commonly used treatment for coronary insufficiency syndrome. Although PTCA is associated with a high primary success rate (90–95%), angiographic restenosis

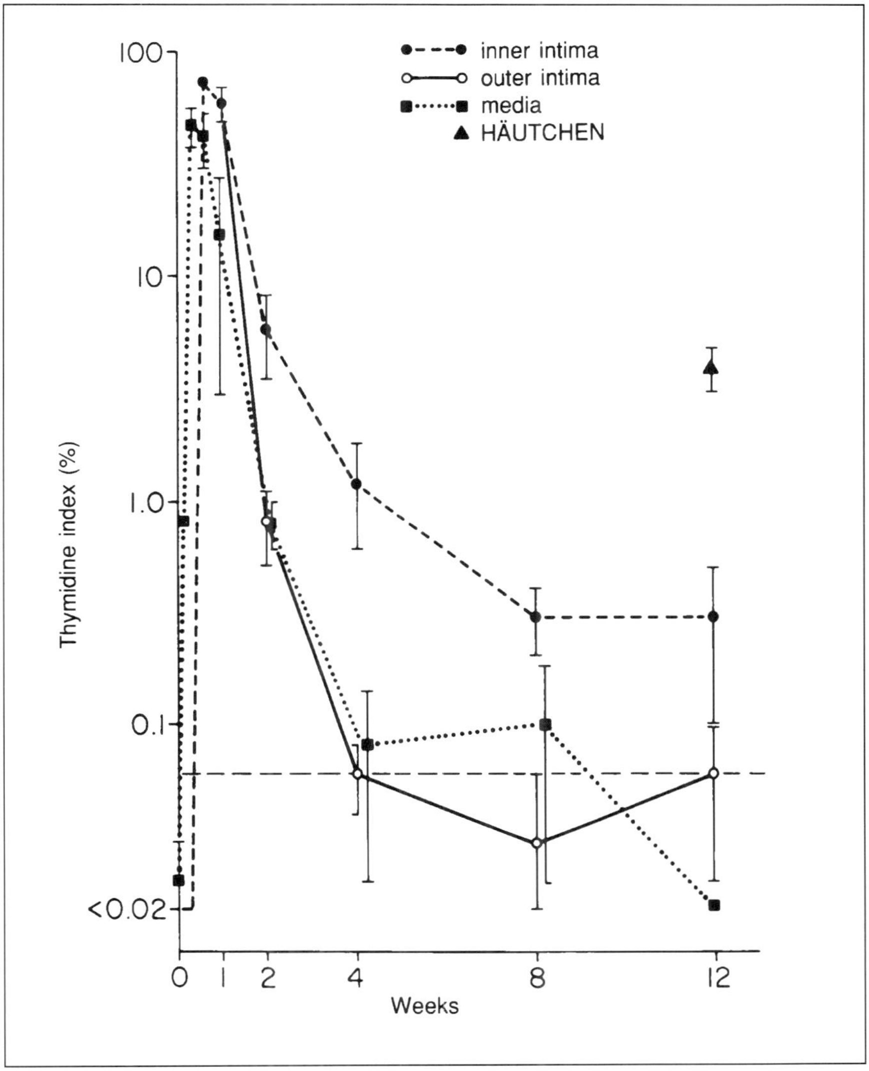

Figure 12.3

Graphic representation of smooth muscle proliferation between 0 and 12 weeks after injury to rat common carotid artery. Proliferation is expressed as thymidine index %, or percentage of radiolabeled cell nuclei over total number of cell nuclei. Rats received three intraperitoneal injections of tritiated thymidine at 17 hours, 9 hours and 1 hour before death. Each point is the mean ± SEM of measurements on three animals. ● = inner intima (or entire intima at 4 and 7 days); ○ = outer intima; ■ = first layer of media (or entire media at 1 day and 4, 8, and 12 weeks); ▲ = Häutchen en face *preparation of luminal smooth muscle cells; --- = mean thymidine index for all uninjured vessels (0.06%). In the time interval 4 to 12 weeks only the thymidine index of the luminal SMC in the Häutchen preparation was significantly different from control (*$P < 0.05$*, t-test). Reprinted with permission from Clowes et al.*[3]

typically develops in 30–40% of the patients within six months after the intervention.[27,28] There is as yet no effective therapy to reduce the high incidence of restenosis after PTCA in humans.[29,30] Accelerated structural remodeling of the arterial wall in restenotic patients is thought to result, among other causes, from the stretching and tearing injury to the vessel during PTCA.[2] Although several definitions of restenosis have been proposed, all of them are based on lumen diameter measurements obtained by angiography. They are descriptive and do not distinguish between the different possible causes of lumen narrowing. Thus, restenosis is a poorly defined term used to describe lumen narrowing at any time after vascular dilatation. Early restenosis which occurs within a few days after surgery may be caused by acute local thrombosis, elastic recoil of the vessel wall or vasospasm at the site of dilatation.

The cellular and molecular mechanisms leading to late (several weeks) restenosis are poorly understood. The importance of cell proliferation in the restenotic response of reconstructed arteries is still a matter of controversy. Quantitative studies of cell proliferation using the proliferating cell nuclear antigen (PCNA) in primary versus restenotic lesions obtained by atherectomy have yielded conflicting results, with the indices of cell proliferation in the restenotic specimens ranging from less than 1% to more than 15%.[31,32] O'Brien et al.[32] reported no difference in the frequency and degree of proliferating cells between primary and restenotic specimens. They also reported no difference when they compared the number of PCNA-positive cells in restenotic specimens obtained a few days versus a few months after the primary coronary intervention.[32] This observation suggests that important differences may exist in the proliferative response to injury between diseased human arteries and normal animal arteries devoid of a pre-existing lesion. Studies with atherectomy specimens raise the possibility that late restenosis after PTCA depends on aggravating factors that are not dependent on DNA synthesis, such as SMC migration, extracellular matrix accumulation or chronic vasospasm. It is interesting to note in that regard that, in the rat model of balloon injury, reducing the first wave of SMC proliferation by 80% with a neutralizing antibody against bFGF does not inhibit intimal thickening eight days after injury.[13] Thus, drugs that inhibit SMC migration, extracellular matrix accumulation, chronic vasospasm or a combination of these aggravating factors should be considered in the design of an effective therapy to prevent restenosis after PTCA. Because ANG II stimulates several of the cellular responses mentioned above, the potential role of the renin-angiotensin system in vascular wound healing has been under intense investigation in recent years.

Evidence linking the RAS and vascular wound healing in rat arteries

The ability of ACE inhibitors to prevent intimal lesion formation in rat injured arteries (carotid, aorta) is well established (Table 12.1). The list of ACE inhibitors tested to date in the rat include captopril, cilazapril, enalapril and ramipril. All of these inhibitors were able to reduce lesion size after injury, thus suggesting that their effect on vascular remodeling is a class effect associated with inhibition of ACE.

Inhibition by cilazapril of intimal thickening in the rat carotid is dose-dependent.[33] However, higher doses are needed to achieve significant suppression of intimal thickening (3–30 mg/kg/day) than to obtain a maximal (20%)

Species	Artery	Drug	Dose × 2 weeks	Inhibition of intimal thickening	Ref
Rat	Carotid	Cilazapril	3–30 mg/kg/d PO	20–78%	33
		Captopril	100 mg/kg/d PO	67%	
Rat	Carotid	Cilazapril	10 mg/kg/day PO	77%	35
		Captopril	100 mg/kg/d PO	62%	
Rat	Carotid	Cilazapril	10 mg/kg/d PO	70%	36
Rat	Aorta	Ramipril	10 mg/kg/d PO	40%	86
Rat	Carotid	Cilazapril*	1–25 mg/kg/d PO	47–81%	87
Rat	Aorta	Captopril	100 mg/kg/d PO	59%	41
Rat	Carotid	Benazeprilat	3 mg/kg/d IP	35%	37
Rat	Carotid	Perindopril	3 mg/kg/d PO	75%	49
Rat	Carotid	Ramipril†	5 mg/kg/d IP	50–60%	73
		Enalapril†	10 mg/kg/d IP	50%	
Guinea Pig	Carotid	Cilazapril	30 mg/kg/d PO	42%	54
Rabbit	Iliac	Cilazapril	3 mg/kg/d PO	NS	54
Pig	Coronary	Trandolapril	0.4 mg/kg/d PO	NS‡	57
Pig	Carotid	Cilazapril	20 mg/kg bid	NS	56
Baboon	Femoral	Cilazapril	20 mg/kg/d	NS	58
	Carotid§	Cilazapril	20 mg/kg/d	NS	
Human	Coronary	Cilazapril	5 mg bid	NS‖	59

*Inhibitory effects of heparin and cilazapril were additive when the drugs were used together
†Inhibitory effect of ramipril and enalapril were partially reversed by the kinin receptor antagonist Hoe 140 (73% and 67% reversal, respectively)
‡The fractional increase in mean neointimal thickness per unit of injury was significantly less for the trandolapril-treated animals
§Carotid endarterectomy was the experimental injury
‖Angiographic restenosis 6 months after PTCA was the end-point of the study

Table 12.1
ACE inhibitors and intimal thickening after arterial injury with a balloon catheter

reduction of blood pressure in the same strain of normotensive rats (3 mg/kg/day).[33] This difference in dosage requirements may reflect the higher doses necessary to achieve inhibition of vascular tissue ACE[34] or, as discussed below, to effectively block degradation of bradykinin at the site of injury. The vasodilators minoxidil and hydralazine were given at doses that reduced blood pressure to a similar extent as cilazapril but failed to affect intimal lesion formation.[35,36] Taken together, these data suggest that the suppression by ACE inhibitors of intimal thickening after injury is not a consequence of blood pressure reduction.

Suppression of intimal thickening by ACE inhibitors depends on when and how long the drug is administered. Lesion size two weeks after injury was significantly reduced when daily administration of cilazapril was started six days before injury (77% inhibition of intima to media area ratio), but also when daily treatment was initiated as late as one hour before injury (58% inhibition) or even two days after injury (58% inhibition). The effectiveness of the earlier administration of the drug may reflect the time needed to achieve complete inhibition of ACE. Intimal thickening at two weeks was not affected by a single administration of cilazapril one hour before injury or when the drug was given six days before until two days after injury. Finally, prolongation of the treatment with cilazapril for eight weeks after injury did not result in better inhibition as compared with a two-week treatment followed by six weeks of placebo. Taken together, these data suggest that the vascular response to injury is particularly sensitive to ACE inhibitors at the time of intense SMC migration and proliferation shortly after trauma, rather than at later times when SMC contribution to intimal lesion growth is mainly due to extracellular matrix production.

ACE inhibitors may affect SMC migration in response to injury. Prescott et al.[37] reported that the ACE inhibitor benazeprilat reduced intimal lesion size by 35% 12 days after balloon catheterization of the rat carotid. Treatment with benazeprilat did not affect medial SMC proliferation at day two after injury, but reduced by 68% the number of SMC appearing on the luminal side of the internal elastic lamina at day four, suggesting that SMC migration was suppressed. Interestingly, ANG II has been shown to stimulate migration of cultured SMC in an in vitro wound assay system.[38]

Several additional lines of evidence strongly suggest that ANG II plays an important role in intimal lesion development (Table 12.2). First, in vivo infusion of a hypertensive dose of ANG II during the third and fourth week after injury induces SMC proliferation in the normal and injured arterial wall.[20] The mitogenic effect of hypertension induced by ANG II appears to depend on the proliferative status of vascular SMC, as it was more marked in neointimal SMC than in SMC from the underlying media. The cumulative labeling fraction of the injured carotid artery was increased from 5% to 20% in the intima, whereas it was increased from 0.2% to 2.5% in the underlying media. The reasons for the increased propensity of SMC to proliferate in the neointima are not known, but this property may reflect the altered phenotype of intimal SMC which include increased levels of AT_1 receptor expression in vivo[39] or the modulating influence of local factors such as matrix components or growth factors. Infusion of ANG II during the first two weeks of lesion development increases intimal lesion size and seems to reverse at least partially the suppressive effect of cilazapril.[40]

Second, several structurally unrelated antagonists of the ANG II receptors were shown to be effective in suppressing neointimal thickening in rat injured arteries. These include peptide analogs of ANG II that recognize all receptors for ANG II. Pan et al.[41] reported that saralasin (360 mg/kg/day, IV) inhibited by 45% of the intimal thickening response in the balloon-injured rat aorta at 14 days, compared with 59% inhibition by captopril (100 mg/kg/day PO). Laporte and Escher[42] infused the irreversible, long-acting ANG II antagonist [Sar^1, $Phe(Br_5)^8$]-ANG II (10 nmol/kg/h, IA) and observed an 80% inhibition of intima development at two weeks in the injured common carotid.

Two subtypes of receptor for ANG II, termed AT_1 and AT_2, have been classified on the basis of their binding interactions with

Species	Artery	Drug	Dose × 2 weeks	Effect on intimal thickening	Ref
Rat	Carotid Aorta	ANG II	435 mg/kg/min SC*	59% increase	20
Rat	Carotid	[Sar,Phe(Br_5)[8]] ANG II	10 nmol/kg/h IA	86% decrease	42
Rat	Aorta	Saralasin[†]	360 g/kg/h IV	45% decrease	41
Rat	Carotid	Losartan[‡]	10 mg/kg/d IP	49% decrease	37
Rat	Carotid	Losartan	5 mg/kg/day	NS[§]	45
		Losartan	15 mg/kg/day	48%	
Rat	Carotid	Losartan	5 mg/kg/day PO	NS	49
		Losartan	30 mg/kg/day PO	47% decrease	
		CGP 42112A[‖]	1 mg/kg/d SC	NS	
		CGP 42112A	1 mg/kg/d perivascular	78% decrease	
Rat	Carotid	Losartan	10 mg/kg/d IP	30% decrease	73
Rabbit	Carotid	Losartan	10 mg/kg/d PO	58% decrease	55

*ANG II was infused from week 2 to week 4 after injury
[†]Antagonist for all ANG II receptors
[‡]AT_1 receptor antagonist
[§]NS: No significant effect
[‖]AT_2 receptor ligand (antagonist or agonist activity not known)

Table 12.2
Ligands for ANG II receptors and intimal thickening after arterial injury with a balloon catheter

selective non-peptide ligand molecules. These ligands were also tested against the response to injury in rat arteries. The AT_1 receptor, which has been associated with all known biological effects of ANG II, binds the antagonist losartan (previous nomenclature DUP 753).[43,44] In the rat carotid, intimal thickening two weeks after injury was reduced by half with IP losartan (10 mg/kg/day)[37] or with IV losartan (15 mg/kg/day).[45] Prescott et al.[37] reported that losartan reduced by 94% SMC migration to the intima at day four. These data support the hypothesis that ACE inhibitors suppress intimal thickening by reducing levels of ANG II in the injured wall. Losartan, however, also suppressed DNA synthesis in medial SMC two days after injury.[37] This is in contrast with the lack of effect of benazaprilat on medial SMC proliferation. This may reflect differences in the pharmacokinetics of the two drugs. Alternatively, losartan may possess additional pharmacological effects beyond ANG II receptor blockade, as suggested by Ohlstein et al.[46]

AT_2 receptors bind the non-peptide ligand CGP 42122A, but not losartan.[47] AT_2 receptors appear to be developmentally regulated, as they are predominant (81% of ANG II receptors) in the aorta of two-week-old rats, but not three-month-old rats (20% of ANG II receptors).[39,48] No change in basal levels of AT_2 receptor binding activity was observed two weeks following injury.[39] The only evidence linking AT_2 receptors and the response to vascular injury comes from a study by Janiak et al.[49] in which CGP 42112A (1 mg/kg/day) inhibited by 73% intimal lesion development when continuously infused perivascularly for two weeks, but not when given systemically to rats. Because no biological activity of ANG II has yet been associated with AT_2 receptors in SMC, it is not known whether CGP 42112A acted as a blocker or as an agonist on AT_2 receptors in the arterial wall. Moreover, beside the obvious possibility of toxic effects associated with the local infusion of a concentrated drug solution, it is not clear that occupation of a greater number of vascular AT_2 receptors by CGP 42112A is the sole explanation for the greater effectiveness of the perivascular infusion over the systemic infusion. The local concentration of CGP 42112A during perivascular infusion was not determined. This raises questions about the specificity of the perivascular treatment, because high concentrations of CGP 42112A (>5 μmol) will displace more than 50% of the binding of ^{125}I-[Sar^1, Ile^8]ANG II to AT_1 receptors in the rat adrenal glomerulosa zone.[47] Thus, the molecular basis for the possible effect of selective AT_2 receptors ligands on vascular lesion development is still lacking and requires further study.

Finally, studies showing upregulation of certain components of the renin-angiotensin system in the vasculature after injury have emerged. Rakugi et al.[50] claim that they were able to use quantitative in situ hybridization to demonstrate an increase in angiotensinogen gene expression in both the media and the intima during the first two weeks after balloon injury to the rat carotid. Since angiotensin I generation by renin is submaximal at normal plasma concentrations of angiotensinogen, higher levels of angiotensinogen in the vasculature are likely to result in a higher rate of angiotensin I generation and an increase in local levels of ANG II.[51] AT_1 receptors for ANG II are also increased at two weeks in the injured rat carotid.[39] Increased levels of AT_1 receptors in the intima may contribute to the enhanced proliferative effect of ANG II observed in this layer of the arterial wall.[20] Whether vascular ACE is upregulated following injury is still a matter of controversy. Rakugi et al.[52] reported that two weeks after injury the carotid expressed increased levels of ACE activity as well as immunoreactivity using a polyclonal human and monoclonal rat ACE antibody. Viswanathan,[39] however, used quantitative autoradiography and reported that binding to ACE in the aortic neointima 15 days after injury was not different from binding in the underlying media or the media of sham operated animals. Taken together, these observations suggest that ACE activity per unit of wall mass is not increased in the vascular wall two weeks after injury. Whether ACE activity is acutely increased earlier after injury remains to be determined. Preliminary data from Berk et al.[53] suggest that ACE activity in cultured SMC is stimulated by bFGF, a growth factor that has been associated with the first wave of SMC proliferation after vascular injury.

Effect of ACE inhibitors on vascular wound healing in other animal models and humans

In guinea-pig carotid arteries, treatment with a high dose of cilazapril (30 mg/kg/day) resulted in a 42% inhibition of intimal hyperplasia after injury.[54] On the other hand, intimal thickening in balloon-injured rabbit iliac arteries was not reduced by cilazapril given at a dose that is marginally active in blocking the response in the rat model (3 mg/kg/day PO).[54] Because higher doses of cilazapril are not well tolerated by rabbits, it is not clear whether ACE inhibitors can alter intimal thickening after injury in rabbit arteries. It is important to note, however, that in vivo blockade of AT_1 receptors with losartan (10 mg/kg/day PO) reduced by 58% intimal thickening in rabbit denuded carotids.[55] This observation suggests that ANG II may play a role in intimal lesion development in rabbit arteries.

In larger species, ACE inhibitors have a modest effect or no effect at all on intimal lesion development while effectively blocking circulating ACE activity. After mild or severe balloon injury to pig carotid artery, ACE inhibition with cilazapril (20 mg/kg bid) did not affect intimal thickening three weeks after injury.[56] Huber et al.[57] studied the effect of the ACE inhibitor trandolapril in pig coronary arteries injured with oversized metallic coils. Quantitative comparisons were conducted using the following vessel injury score: (0) no injury noted, (1) internal elastic lamina disrupted but media not lacerated, (2) internal elastic lamina disrupted and media lacerated, and (3) external elastic lamina disrupted.[57] A reduction in the fractional increase in mean intimal thickness per unit of injury was observed in animals treated with a low dose of trandolapril (0.3 mg/kg/day PO) for four weeks. This effect, however, did not translate into a significant reduction in percentage of luminal area stenosis. Captopril (6–7 mg/kg/day) failed to prevent luminal narrowing in the same model, but this may reflect the low dose tested by Huber et al.[57] Finally, in the baboon, cilazapril (20 mg/kg/day) did not significantly reduce intimal hyperplasia three weeks after various types of vascular injury including balloon catheterization of the superficial femoral artery and endarterectomy of the carotid artery.[58]

Thus, despite their proven effectiveness in suppressing the response to vascular injury in rat arteries, ACE inhibitors gave disappointing results in several other animal species, including humans. Low doses of cilazapril (1–5 mg bid) in humans did not reduce the rate of angiographic restenosis after PTCA.[59] The failure of cilazapril to favorably alter the overall clinical outcome of angioplasty in humans may have various explanations. The dose of cilazapril may have been much too low to inhibit ACE in the vasculature, which is more resistant to inhibitors.[60,61] In fact, preliminary data suggest that inhibition by cilazapril of intimal thickening in the rat carotid artery follows more closely inhibition of vascular ACE than plasma ACE.[34] The suppression of intimal thickening in the rat was observed with daily doses of cilazapril which, on a gravimetric basis, are 65 times greater than the dose used in the MERCATOR Study. Studies in the baboon and the pig, however, revealed that high doses of ACE inhibitor similar to those used in the rat studies had little, if any, effect against the response to arterial injury. Because of this, it is not clear whether higher doses of cilazapril would prevent human restenosis.

The vascular response to injury and the ability of ACE inhibitors to alter its key processes may be vessel- or species-specific. It is well known, for example, that the degree of sensitivity to specific vasoactive mediators varies between arteries of different size and location. The response to injury of coronary arteries may be specific in aspects yet to be determined. The availability of atherectomy specimens from both primary and restenotic lesions should prove extremely useful in that regard. It is also necessary to consider differences between human and other animal responses to vascular injury. In particular, thrombus formation and leukocyte infiltration are minimal in the rat, but prominent in larger animals and humans. Libby et al.[62] suggested that in humans the leukocyte content of a particular lesion could determine in part propensity to develop restenosis after angioplasty. In pig coronary arteries, injury caused by distention with an oversized stent initially causes thrombus formation, followed within a few days by significant macrophage and lymphocyte infiltration, and finally SMC migration and proliferation in the intima.[63] Leukocyte infiltration is also common in rabbit lesions.[64,65] By contrast, only 1.2% of cells in the rat carotid intima express the leukocyte common antigen at two weeks after injury. No cells expressing this antigen have been observed at one week or at three weeks and later after injury, suggesting that leukocyte infiltration in the rat intima occurs transiently and at a low level.[66] Finally, the response to injury may be quite different in diseased and healthy vessels. Clearly, there is a need for a better characterization of the vascular response to injury in vessels with pre-existing lesions.

Potential effects of ACE inhibition in injured vessels

The proposed mechanism for the suppressive effect of ACE inhibitors in vascular lesion formation is the inhibition of ANG II generation.[35,59] This hypothesis relies on two assumptions. The first assumption is that ACE is the major enzyme able to generate ANG II from its precursors. While this assumption may be correct with nondiseased vessels, the complex cellular and molecular composition of intimal lesions and, in particular, of the atherosclerotic plaque, may provide alternative pathways for ANG II generation even in the absence of ACE activity. Enzymes different from ACE and insensitive to its inhibitors may substitute for ACE in processing angiotensin I. These include serine proteases such as human heart chymase,[67] leukocyte-derived tonin,[68] and carboxypeptidases from human platelets[69] or from the tunica media.[70]

The second assumption is that the blockade of ANG II generation is the only relevant effect of ACE inhibition. ACE, however, has a poor substrate specificity and its inhibition does affect other humoral pathways. Of particular interest is the fact that ACE is a key enzyme in the inactivation of kinins, a family of blood-derived vasoactive peptides. Kinins are produced in association with the activated intrinsic coagulation cascade,[71] that is, in a situation that most likely occurs at the site of arterial injury. In the rat model, the effective doses of ACE inhibitor that reduce intimal thickening are significantly higher than those necessary to obtain an antihypertensive effect.[36] The affinity of ACE (K_m) is twenty times higher for bradykinin than for angiotensin I.[72] Since ACE binds more tightly to bradykinin ($K_m < 3$ μmol) than to angiotensin I ($K_m = 70$ μmol), it is possible that lower doses of ACE inhibitors may be more selective for the angiotensin pathway and that higher doses are needed to significantly inhibit metabolism of kinins and enhance their accumulation at the surface of the injured vessel wall.

Results obtained with the injured rat carotid model suggest that kinins released at sites of vascular injury may be involved in the reduction of intimal hyperplasia by large doses of ACE inhibitors.[73–75] The effects of kinins on vascular cells include stimulation of prostacyclin and nitric oxide (NO) production in endothelial cells, resulting in a decrease in vascular resistance.[76] Vascular effects of kinins also include the direct activation of the inositol phosphate pathway and intracellular calcium movements in vascular SMC, which may lead to vasoconstriction in the absence of a functional endothelium. Kinins act through two types of receptors, termed B_1 and B_2. Outside the renal vasculature, vasoactive effects of kinins are mainly mediated by receptors of the B_2 subtype. The B_2 receptors are selectively antagonized by the bradykinin analog Hoe 140.[77] Because of its resistance to degradation by several peptidases including ACE, Hoe 140 was used in the rat carotid model to investigate the following hypothesis: the protective effect of ACE inhibitors in the response to arterial injury is partly dependent on increased stimulation of receptors for kinins. Farhy et al.[75] showed that suppression by ramipril and by enalapril in this model was reduced by 75% and 62%, respectively, in animals receiving a continuous SC infusion of Hoe 140 (70 μg/kg/day). Preliminary data from deBlois et al.[74] also suggested that Hoe140 reduced the suppressive effect of ramipril on intimal thickening in the rat model. It is of interest that Hoe 140 did not prevent the reduction in blood pressure induced by the ACE inhibitors.[74,75] These data provide additional evidence indicating that the reduction by ACE inhibitors of systemic blood pressure and of intimal hyperplasia are not causally related. More importantly, these data suggest a double mechanism of action for ACE inhibitors in vascular wound healing; that is, by reducing levels of the growth promoter ANG II and increasing levels of kinins, which inhibit intimal growth by a mechanism as yet undefined.

Stimulation of prostaglandin production is a well-known response of vascular cells to kinins.[76] However, Powell et al.[33] reported that very high doses of aspirin (300 mg/kg/day) failed to alter intimal growth in the presence or absence of ACE inhibitors. This suggests that prostaglandins are not involved in the suppressive effect of ACE inhibitors on intimal growth. Farhy et al.[75] reported that L-NMMA, a blocker of nitric oxide generation, could also partially reverse the suppressive effect of ramipril on intimal growth. Thus, they suggested that kinin-induced NO may act as a growth inhibitor in the intimal lesion. It is unclear, however, how kinins could generate NO in a vessel stripped of its endothelium. Vascular SMC express an isoform of NO synthase that, by contrast with the NO synthase from the endothelium, is constitutively active and is not further activated by kinins.[78] Joly et al.[79] reported the induction of such a constitutive NO synthase activity in the rat carotid artery within hours after de-endothelialization with a balloon catheter. They speculated that vascular cytokines such as interleukin-1 (IL-1) may be responsible for this phenomenon. In fact, they demonstrated that the constitutive production of NO could be further enhanced by exposure to IL-1 ex vivo. Whether kinins can induce the constitutive NO synthase isoform has not been reported.

The suggestion that ACE inhibitors may increase the activity of kinins is not new.[80] Since the development of the kinin antagonist Hoe 140, new evidence has been accumulating in support of this hypothesis. Chronic blockade of B_2 receptors for kinins attenuates the antihypertensive effect of ramipril in 2-kidney–1-clip hypertensive rats.[81] It also abolishes the antihypertrophic effect of ramipril, independently of its hypotensive effects, in a model of left ventricular hypertrophy in rats with aortic stenosis.[82] Treatment of rats with Hoe 140 reduces the aggravation by captopril of paw edema induced by urate crystals.[83]

Understanding the role of kinins in vascular injury and structural remodeling of vessels may provide insights in the mechanism of action of ACE inhibitors in these pathologies. The activated coagulation cascade has generally been considered a source of factors accelerating intimal lesion progression. The potential role of kinins as suppressors of intimal thickening requires further investigation. Inhibition of intimal thickening by kinin agonists of B_2 receptors needs to be demonstrated and characterized in terms of the SMC response to injury, for example migration, proliferation and matrix secretion. The role of B_1 receptors for kinins is also of potential interest, particularly because vascular responses mediated by this receptor subtype are upregulated in certain blood vessels within a few hours following exposure to the proinflammatory cytokine IL-1 in vivo or in vitro.[84,85]

Future prospects

It is interesting that rat arteries develop a much simpler lesion than that observed in rabbits, pigs, primates or humans. As noted earlier, inflammation and invasion of the lesion by leukocytes, for instance, are much more prominent in the latter species than in the rat. These responses may provide alternative pathways for ANG II generation and effective kinin degradation even in the presence of ACE inhibitors. The finding that kinins may act as inhibitors of intimal thickening suggests the interesting possibility that ACE inhibition combined with local kinin administration could yield better results than ACE inhibitors alone. Agonist experiments with kinins are needed to test this hypothesis.

It is unclear whether increased doses of ACE inhibitors would be more effective in treating human restenosis. It is possible, however, that a therapy combining an ACE inhibitor and a blocker of AT_1 receptors for ANG II would be more effective against restenosis than either therapy alone. There are several reasons for this. First, ACE inhibitors may act by blocking generation of a stimulant (ANG II) as well as by blocking degradation of inhibitors (kinins) of intimal thickening. Second, because of alternative pathways for the metabolism of angiotensin I, the use of ACE inhibitors alone may not prevent completely the generation of ANG II in the lesion. Third, the production of renin by the kidney is under a negative feedback control loop through AT_1 receptors for ANG II. Because of this, the use of an AT_1 blocker alone is likely to result in a dramatic increase in levels of circulating ANG II. This, in turn, is likely to result in an enhanced stimulation of AT_2 receptors and with unknown consequences.

Understanding the cellular and molecular mechanisms of the vascular response to injury may provide a basis for the development of therapies designed to reduce the high incidence of restenosis after PTCA in humans. ACE inhibitors seem to be particularly effective against the development of a myofibrotic lesion after vascular injury in rat arteries. On the other hand, it is possible that the failure of ACE inhibitors to alter lesion formation in several other animal models of restenosis reflects the higher complexity of the lesions in those species. The molecular and cellular basis for these differences may lie in alternative metabolic pathways for the formation of ANG II and for the degradation of kinins. Large thrombi and leukocytes infiltrating the lesions are a potential source of such alternative metabolic pathways, as well as potential sources of growth factors for lesion progression independent of the renin-angiotensin or the kinin systems.

References

1 Jackson CL, Schwartz SM. Pharmacology of smooth muscle cell replication. *Hypertension* 1992; **20**:713--736.

2 Forrester JS, Fishbein M, Helfant R, Fagin J. A paradigm for restenosis based on cell biology. Clues for the development of new preventive therapies. *J Am Coll Cardiol* 1991; **17**:758–69.

3 Clowes AW, Reidy MA, Clowes MM. Kinetics of cellular proliferation after arterial injury. I. Smooth muscle growth in the absence of endothelium. *Lab Invest* 1983; **49**:327–33.

4 Clowes AW, Clowes MM, Fingerle J, Reidy MA. Kinetics of cellular proliferation after arterial injury. V. Role of acute distension in the induction of smooth muscle proliferation. *Lab Invest* 1989; **60**:360–4.

5 Clowes AW, Reidy MA. Prevention of stenosis after vascular reconstruction: Pharmacologic control of intimal hyperplasia — A review. *J Vasc Surg* 1991; **13**:885–91.

6 Hatton MW, Moar SL, Richardson M. Deendothelialization in vivo initiates a thrombogenic reaction at the rabbit aorta surface. Correlation of uptake of fibrinogen and antithrombin III with thrombin generation by the exposed subendothelium. *Am J Pathol* 1989; **135**:499–508.

7 Pasche B, Swedenborg J, Hedin U, Olsson P, Ljungqvist A. Thrombin-inhibitory capacity of the injured vessel wall. *Thrombosis Res* 1991; **62**:531–44.

8 Clowes AW, Reidy MA, Clowes MM. Mechanisms of stenosis after arterial injury. *Lab Invest* 1983; **49**:208–15.

9 Majesky MW, Schwartz SM, Clowes MM, Clowes AW. Heparin regulates smooth muscle S phase entry in the injured rat carotid artery. *Circ Res* 1987; **61**:296–300.

10 Lindner V, Lappi DA, Baird A, Majack RA, Reidy MA. Role of basic fibroblast growth factor in vascular lesion formation. *Circ Res* 1991; **68**:106–13.

11 Cuevas P, Gonzalez AM, Carceller F, Baird A. Vascular response to basic fibroblast growth factor when infused onto the normal adventitia or into the injured media of the rat carotid artery. *Circ Res* 1991; **69**:360–9.

12 Edelman ER, Nugent MA, Smith LT, Karnovsky MJ. Basic fibroblast growth factor enhances the coupling of intimal hyperplasia and proliferation of vasa vasorum in injured rat arteries. *J Clin Invest* 1992; **89**:465–73.

13 Lindner V, Reidy MA. Proliferation of smooth muscle cells after vascular injury is inhibited by an antibody against basic fibroblast growth factor. *Proc Natl Acad Sci USA* 1991; **88**:3739–43.

14 Lindner V, Olson NE, Clowes AW, Reidy MA. Inhibition of smooth muscle cell proliferation in injured rat arteries. Interaction of heparin with basic fibroblast growth factor. *J Clin Invest* 1992; **90**:2044–49.

15 Fingerle J, Johnson R, Clowes AW, Majesky MW, Reidy MA. Role of platelets in smooth muscle cell proliferation and migration after vascular injury in rat carotid artery. *Proc Natl Acad Sci USA* 1989; **86**:8412–16.

16 Ferns GAA, Raines EW, Sprugel KH, Motani AS, Reidy MA, Ross R. Inhibition of neointimal smooth muscle accumulation after angioplasty by an antibody to PDGF. *Science* 1991; **253**:1129–32.

17 Jawien A, Bowen-Pope DF, Lindner V, Schwartz SM, Clowes AW. Platelet-derived growth factor promotes smooth muscle migration and intimal thickening in a rat model of balloon angioplasty. *J Clin Invest* 1992; **89**:507–11.

18 Olson NE, Chao S, Lindner V, Reidy MA. Intimal smooth muscle proliferation after balloon catheter injury. The role of basic fibroblast growth factor. *Am J Pathol* 1992; **140**:1017–23.

19 Majesky MW, Lindner V, Twardzik DR, Schwartz SM, Reidy MA. Production of transforming growth factor β_1 during repair of arterial injury. *J Clin Invest* 1991; **88**:904–10.

20 Daemen MJAP, Lombardi DM, Bosman FT, Schwartz SM. Angiotensin II induces smooth muscle cell proliferation in the normal and injured rat arterial wall. *Circ Res* 1991; **68**:450–6.

21 Clowes AW, Clowes MM, Reidy MA. Kinetics of cellular proliferation after arterial injury. III. Endothelial and smooth muscle growth in chronically denuded vessels. *Lab Invest* 1986; **54**:295–303.

22 Majesky MW, Giachelli CM, Reidy MA, Schwartz SM. Rat carotid neointimal smooth muscle cells reexpress a developmentally regulated mRNA phenotype during repair of arterial injury. *Circ Res* 1992; **71**:759–68.

23 Giachelli CM, Bae N, Almeida M, Denhardt DT, Alpers CE, Schwartz SM. Osteopontin is elevated during neointima formation in rat arteries and is a novel component of human atherosclerotic plaques. *J Clin Invest* 1993; **92**:1686–96.

24 Clowes AW, Clowes MM, Gown AM, Wight TN. Localization of proteoheparan sulfate in rat aorta. *Histochemistry* 1984; **80**:379–84.

25 Scott-Burden T, Hahn AW, Resink TJ, Buhler FR. Modulation of extracellular matrix by angiotensin II: Stimulated glycoconjugate synthesis and growth in vascular smooth muscle cells. *J Cardiovasc Pharmacol* 1990; **16[Suppl 4]**:S36–41.

26 Kato H, Suzuki H, Tajima S, Ogata Y, Tominaga T, Sato A, Saruta T. Angiotensin II increases collagen synthesis. *J Hypertens* 1991; **9**:17–22.

27 Serruys PW, Luitjen HE, Beatt KJ, Geuskens R, deFeyter PJ, van den Brand M, Reiber JH, ten Katen HJ, van Es GA, Hugenholtz PG. Incidence of restenosis after successful coronary angioplasty: A time-related phenomenon. A quantitative angiographic study in 342 consecutive patients at 1, 2, 3 and 4 months. *Circulation* 1988; **77**:361–71.

28 Nobuyoshi M, Kimura H, Nosaka H, Mioka S, Ueno K, Hamasaki N, Horiuchi H, Oshishi H. Restenosis after successful percutaneous transluminal coronary angioplasty: Serial angiographic follow-up of 299 patients. *J Am Coll Cardiol* 1988; **12**:616–23.

29 Dehmer GJ, Popma JJ, van den Berg EK, Eichhorn EJ, Prewitt JB, Campbell WB, Jennings L, Willerson JT, Schmitz JM. Reduction in the rate of early restenosis after coronary angioplasty by a diet supplemented with n-3 fatty acids. *N Engl J Med* 1988; **319**:733–40.

30 Hermans WRM, Rensing BJ, Strauss BH, Serruys PW. Prevention of restenosis after percutaneous transluminal coronary angioplasty: The search for the "magic bullet". *Am Heart J* 1991; **122**:171–87.

31 Pickering JG, Weir L, Jekanowski J, Kearney MA, Isner JM. Proliferative activity in peripheral and coronary atherosclerotic plaque among patients undergoing percutaneous revascularization. *J Clin Invest* 1993; **91**: 1469–80.

32 O'Brien ER, Alpers CE, Stewart DK, Ferguson M, Tran N, Gordon D, Benditt EP, Hinohara T, Simpson JB, Schwartz SM. Proliferation in primary and restenotic coronary atherectomy tissue: Implications for antiproliferative therapy. *Circ Res* 1993; **73**:223–31.

33 Powell JS, Muller RKM, Rouge M, Kuhn H, Hefti F, Baumgartner HR. The proliferative response to vascular injury is suppressed by angiotensin-converting enzyme inhibition. *J Cardiovasc Pharmacol* 1990; **16[Suppl 4]**:S42–S49.

34 Rakugi H, Dazu VJ, Pratt RE. Importance of tissue angiotensin converting enzyme (ACE) in neointimal hyperplasia. *Circulation* 1992; **86[Suppl I]**:I–169.

35 Powell JS, Clozel J-P, Muller RKM, Kuhn H, Hefti F, Hosang M, Baumgartner HR. Inhibitors of angiotensin-converting enzyme prevent myointimal proliferation after vascular injury. *Science* 1989; **245**:186–88.

36 Powell JS, Muller RKM, Baumgartner HR. Suppression of the vascular response to injury: The role of angiotensin-converting enzyme inhibitors. *J Am Coll Cardiol* 1991; **17**:137B–142B.

37 Prescott MF, Webb RL, Reidy MA. Angiotensin-converting enzyme inhibitor versus angiotensin II, AT1 receptor antagonist: Effects on smooth muscle cell migration and proliferation after balloon catheter injury. *Am J Pathol* 1991; **139**:1291–1302.

38 Bell L, Madri JA. Influence of the angiotensin system on endothelial and smooth muscle cell migration. *Am J Pathol* 1990; **137**:7–12.

39 Viswanathan M, Stromberg C, Seltzer A, Saavedra JM. Balloon angioplasty enhances the expression of angiotensin II AT1 receptors in neointima of rat aorta. *J Clin Invest* 1992; **90**:1707–12.

40 Osterrieder W, Muller RK, Powell JS, Clozel JP, Hefti F, Baumgartner HR. Role of angiotensin II in injury-induced neointima formation in rats. *Hypertension* 1991; **18[Suppl 4]**:II60–4.

41 Pan X-M, Nelken N, Colyvas N, Rapp JH. Inhibition of injury induced intimal hyperplasia by saralasin in rats. *J Vasc Surg* 1992; **15**:693–98.

42 Laporte S, Escher E. Neointima formation after vascular injury is angiotensin II mediated. *Biochem Biophys Res Commun* 1991; **187**:1510–16.

43 Wong PC, Hart SD, Zaspel AM, Chiu AT, Ardecky RJ, Smith RD, Timmermans PBMWM. Functional studies of nonpeptide angiotensin II receptor subtype-specific ligands: DuP 753 (AII-1) and PD123177 (AII-2). *J Pharmacol Exp Therapeut* 1990; **255**:584–91.

44 Timmermans PBMWM, Wong PC, Chiu AT, Herblin WF. Nonpeptide angiotensin II receptor antagonists. *TIPS* 1991; **12**:55–62.

45 Kauffman RF, Bean JS, Zimmerman KM, Brown RF, Steinberg MI. Losartan, a nonpeptide angiotensin II (ANG II) receptor antagonist, inhibits neointima formation following balloon injury to rat carotid arteries. *Life Sci* 1991; **49**:PL223–PL228.

46 Ohlstein EH, Gellai M, Brooks DP, Vickery L, Jugus J, Sulpizio A, Ruffolo RR, Weinstock J, Edwards RM. The antihypertensive effect of the angiotensin II receptor antagonist DuP 753 may not be due solely to angiotensin II receptor antagonism. *J Pharmacol Exp Therapeut* 1992; **262**:595–601.

47 Balla T, Baukal AJ, Eng S, Catt KJ. Angiotensin II receptor subtypes and biological responses in the adrenal cortex and medulla. *Molec Pharmacol* 1991; **40**:401–6.

48 Viswanathan M, Tsutsumi K, Correa FMA, Saavedra JM. Changes in expression of angiotensin receptor subtypes in the rat aorta during development. *Biochem Biophys Res Commun* 1991; **179**:1361–67.

49 Janiak P, Pillon A, Prost J-F, Vilaine J-P. Role of angiotensin subtype 2 receptor in neointima formation after vascular injury. *Hypertension* 1992; **20**:737–45.

50 Rakugi H, Jacob HJ, Krieger JE, Ingelfinger JR, Pratt RE. Vascular injury induces angiotensinogen gene expression in the media and neointima. *Circulation* 1993; **87**:283–90.

51 Poulsen K, Jacobsen J. Enzymic reactions of the renin-angiotensin system. In: Robertson JI, Nicholls MG, eds. The Renin-Angiotensin System, vol. 1. New York: Gower Medical Publishing, 1993: 5.1–5.12.

52 Rakugi H, Krieger J, Wang DS, Dzau VJ, Pratt RE. Induction of angiotensin converting enzyme in neointima after balloon injury. *Arterioscler Thrombos* 1991; **11**:1396a.

53 Fishel RS, Redden RA, Bernstein KE, Berk BC. Angiotensin-converting enzyme and vascular injury: Synergistic upregulation by stress steroids and basic fibroblast growth factor. *Circulation* 1992; **86[Suppl I]**:I–85.

54 Clozel J-P, Hess P, Michael C, Schietinger K, Baumgartner HR. Inhibition of converting enzyme and neointima formation after vascular injury in rabbits and guinea pigs. *Hypertension* 1991; **18[Suppl II]**:II55–II59.

55 Azuma H, Niimi Y, Hamasaki H. Prevention of intimal thickening after endothelial removal by a nonpeptide angiotensin II receptor antagonist, losartan. *Br J Pharmacol* 1992; **106(3)**:665–71.

56 Lam JYT, Lacoste L, Bourassa MG. Cilazapril and early atherosclerotic changes after balloon injury of porcine carotid arteries. *Circulation* 1992; **85**:1542–47.

57 Huber KC, Schwartz RS, Edwards WD, Camrud AR, Bailey KR, Jorgenson MA, Holmes DR. Effects of angiotensin converting enzyme inhibition on neointimal proliferation in a porcine coronary injury model. *Am Heart J* 1993; **125**:695–701.

58 Hanson SR, Powell JS, Dodson T, Lumsden A, Kelly AB, Anderson JS, Clowes AW, Harker LA. Effects of angiotensin-converting enzyme inhibition with cilazapril on intimal hyperplasia in injured arteries and vascular grafts in the baboon. *Hypertension* 1991; **18[Suppl II]**:II70–II76.

59 MERCATOR Study Group. Does the new angiotensin-converting enzyme inhibitor cilazapril prevent restenosis after percutaneous transluminal coronary angioplasty? *Circulation* 1992; **86**:100–110.

60 Dzau VJ. Significance of the vascular renin-angiotensin pathway. *Hypertension* 1986; **8**:553–59.

61 Cushman DW, Wang FL, Fung WC, Harvey CM, DeForrest JM. Differentiation of angiotensin-converting enzyme (ACE) inhibitors by their selective inhibition of ACE in physiologically important target organs. *Am J Hypertens* 1989; **2**:294–306.

62 Libby P, Schwartz D, Brogi E, Tanaka H, Clinton SK. A cascade model for restenosis: A special case of atherosclerosis progression. *Circulation* 1992; **86[Suppl III]**:III47–III52.

63 Schwartz RS, Edwards WD, Huber KC, Antoniades LC, Bailey KR, Camrud AR, Jorgenson MA, Holmes DR. Coronary restenosis: Prospects for solution and new perspectives from a porcine model. *Mayo Clin Proc* 1993; **68**:54–62.

64 Weidinger FF, McLenachan JM, Cybulsky MI, Fallon JT, Hollenberg NK, Cooke JP, Ganz P. Hypercholesterolemia enhances macrophage recruitment and dysfunction of regenerated endothelium after balloon injury of the rabbit iliac artery. *Circulation* 1991; **84**:755–67.

65 Tanaka H, Swanson SJ, Lee C, Clinton SK, Simon-Morrissey E, Ganz P, Cybulsky MI, Libby P. Sustained activation of vascular cells and leukocytes in rabbit aorta after balloon injury. *Circulation* 1992; **86[Suppl I]**:I–86.

66 Jonasson L, Holm J, Hansson GK. Smooth muscle cells express Ia antigens during arterial response to injury. *Lab Invest* 1988; **58**:310–15.

67 Urata H, Kinoshita A, Misono KS, Bumpus FM, Husain A. Identification of a highly specific chymase as the major angiotensin II-forming enzyme in the human heart. *J Biol Chem* 1990; **265**:22348–57.

68 Reilly CF, Tewsksbury DA, Schechter NM, Travis J. Rapid conversion of angiotensin I to angiotensin II by neutrophil and mast cell proteinases. *J Biol Chem* 1982; **257**:8619–22.

69 Snyder RA, Watt KW, Wintroub BU. A human platelet angiotensin I-processing system. Identification of components and inhibition of angiotensin-converting enzyme byproduct. *J Biol Chem* 1985; **260**:7857–60.

70 Johnson H, Drummer OH. Hydrolysis of angiotensin I by peptidases in homogenates of rat lung and aorta. *Biochem Pharmacol* 1988; **37**:1131–36.

71 Proud D, Kaplan AP. Kinin formation: Mechanisms and role in inflammatory disorders. *Ann Rev Immunol* 1988; **6**:49–83.

72 Bunning P, Holmquist B, Riordan JF. Substrate specificity and kinetic characteristics of angiotensin-converting enzyme. *Biochem* 1983; **22**:103–10.

73 Farhy RD, Ho KL, Carretero OA, Scicli AG. Kinins mediate the antiproliferative effect of ramipril in rat carotid artery. *Biochem Biophys Res Commun* 1992; **182**:283–88.

74 deBlois D, Lombardi DM, Garvin MA, Schwartz SM. Inhibition by ramipril of intimal hyperplasia in the denuded rat carotid is reversed by HOE 140, a kinin B_2 receptor antagonist. *Circulation* 1992; **86[Suppl 4]**:I–226.

75 Farhy RD, Carretero OA, Ho K-L, Scicli AG. Role of kinins and nitric oxide in the effects of angiotensin converting enzyme inhibitors on neointima formation. *Circ Res* 1993; **72**:1202–10.

76 Regoli D, Barabe J. Pharmacology of bradykinin and related kinins. *Pharmacol Rev* 1980; **32**:1–46.

77 Wirth K, Hock FJ, Albus U, Linz W, Alpermann HG, Anagnostopoulos H, Henke S, Breipohl G, Konig W, Knolle J, Scholkens BA. Hoe 140, a new potent and long-acting bradykinin-antagonist: In vivo studies. *Br J Pharmacol* 1991; **102**:774–77.

78 Moncada S, Palmer MJ, Higgs EA. Nitric oxide: Physiology, pathophysiology, and pharmacology. *Pharmacol Rev* 1991; **43**:109–42.

79 Joly GA, Schini VB, Vanhoutte PM. Balloon injury and interleukin-1 induce nitric oxide synthase activity in rat carotid arteries. *Circ Res* 1992; **71**:331–38.

80 Erdos EG. Angiotensin I-converting enzyme and the changes in our concepts through the years. Lewis K. Dahl memorial lecture. *Hypertension* 1990; **16**:363–70.

81 Bao G, Gohlke P, Qadri F, Unger T. Chronic kinin receptor blockade attenuates the antihypertensive effect of ramipril. *Hypertension* 1992; **20**:74–79.

82 Linz W, Scholkens BA. A specific B_2-bradykinin receptor antagonist HOE 140 abolishes the antihypertrophic effect of ramipril. *Br J Pharmacol* 1992; **105**:771–72.

83 Damas J, Remacle-Volon G. Influence of a long-acting bradykinin antagonist, Hoe 140, on some acute inflammatory reactions in the rat. *Eur J Pharmacol* 1992; **211**:81–86.

84 Marceau F, Lussier A, Regoli D, Giroud JP. Pharmacology of kinins: Their relevance to tissue injury and inflammation. *Gen Pharmacol* 1983; **14**:209–29.

85 deBlois D, Bouthillier J, Marceau F. Pulse exposure to protein synthesis inhibitors upregulates vascular tissue responses to des-Arg^9-bradykinin: Possible role of interleukin-1. *Br J Pharmacol* 1991; **103**:1057–66.

86 Capron L, Heudes D, Chajara A, Bruneval P. Effect of ramipril, an inhibitor of angiotensin converting enzyme, on the response of rat thoracic aorta to injury with a balloon catheter. *J Cardiovasc Pharmacol* 1991; **18**:207–11.

87 Clowes AW, Clowes MM, Vergel SC, Muller RKM, Powell JS, Hefti F, Baumgartner HR. Heparin and cilazapril together inhibit injury-induced intimal hyperplasia. *Hypertension* 1991; **18[Suppl II]**:II65–II69.

Index